Remaking the North American Food System

Our Sustainable Future

SERIES EDITORS

Charles A. Francis
University of Nebraska–Lincoln

Cornelia Flora
Iowa State University

Paul A. Olson
University of Nebraska–Lincoln

Remaking the North American Food System

Strategies for Sustainability

EDITED BY C. CLARE HINRICHS
AND THOMAS A. LYSON

University of Nebraska Press • Lincoln and London

 ¶ Manufactured in the United States of America ¶ An expanded version of chapter 15, "Be Careful What You Wish For: Democratic Challenges and Political Opportunities for the Michigan Organic Community," appeared in *Culture & Agriculture* 27 (2): 131–43. ¶ ♾ ¶ Library of Congress Cataloging-in-Publication Data ¶ Remaking the North American food system : strategies for sustainability / edited by C. Clare Hinrichs and Thomas A. Lyson. ¶ p. cm. — (Our sustainable future) ¶ Includes bibliographical references and index. ¶ ISBN 978-0-8032-2438-4 (cloth : alk. paper) ¶ 1. Food industry and trade—North America. 2. Produce trade—North America. 3. Food consumption—North America. 4. Sustainable development—North America. I. Hinrichs, C. Clare. II. Lyson, Thomas A. III. Series. ¶ HD9005.R46 2007 ¶ 338.1'9097—dc22 ¶ 2007022094 ¶ Set in Quadraat & Quadraat Sans by Bob Reitz. ¶ Designed by R. W. Boeche.

This book is dedicated to
the memory of Tom Lyson
(January 20, 1948–December 28, 2006)
who encouraged and enlivened
this conversation
and so many others.

Contents

Acknowledgments xi

Introduction
Practice and Place in Remaking the Food System 1
C. Clare Hinrichs

Part I
What's Wrong with the Food System?
Orienting Frameworks for Change

1. Civic Agriculture and the
North American Food System 19
Thomas A. Lyson

2. Warrior, Builder, and Weaver Work
Strategies for Changing the Food System 33
G. W. Stevenson, Kathryn Ruhf,
Sharon Lezberg, and Kate Clancy

Part II
Institutions and Practices to Remake the Food System

3. Farmers' Markets as Keystones in
Rebuilding Local and Regional Food Systems. 65
Gilbert Gillespie, Duncan L. Hilchey,
C. Clare Hinrichs, and Gail Feenstra

4. Practical Research Methods to
Enhance Farmers' Markets 84
Larry Lev, Garry Stephenson,
and Linda Brewer

5. Community Supported Agriculture as an Agent of Change
Is It Working? 99
Marcia Ruth Ostrom

6. Food Policy Councils
Past, Present, and Future 121
Kate Clancy, Janet Hammer, and Debra Lippoldt

7. The "Red Label" Poultry System in France
Lessons for Renewing an Agriculture-of-the-Middle in the United States 144
G. W. Stevenson and Holly Born

8. Eating Right Here
The Role of Dietary Guidance in Remaking Community-Based Food Systems 163
Jennifer Wilkins

9. Community-Initiated Dialogue
Strengthening the Community through the Local Food System 183
Joan S. Thomson, Audrey N. Maretzki, and Alison H. Harmon

Part III
The Importance of Place and Region in Remaking the Food System

10. Retail Concentration, Food Deserts, and Food-Disadvantaged Communities in Rural America 201
Troy C. Blanchard and Todd L. Matthews

11. Localization in a Global Context
Invigorating Local Communities in Michigan through the Food System 216
Michael W. Hamm

12. Assessing the Significance of Direct Farmer-Consumer Linkages as a Change Strategy in Washington State
Civic or Opportunistic? 235
Marcia Ruth Ostrom and
Raymond A. Jussaume, Jr.

13. Emerging Farmers' Markets and the Globalization of Food Retailing
A Perspective from Puerto Rico 260
Viviana Carro-Figueroa and Amy Guptill

14. The Lamb That Roared
Origin-Labeled Products as Place-Making Strategy in Charlevoix, Quebec 277
Elizabeth Barham

15. Be Careful What You Wish For
Democratic Challenges and Political Opportunities for the Michigan Organic Community 298
Laura B. DeLind and Jim Bingen

16. The Social Foundation of Sustainable Agriculture in Southeastern Vermont 315
Matthew Hoffman

17. Community Food Projects and Food System Sustainability 332
Audrey N. Maretzki and
Elizabeth Tuckermanty

Conclusion
A Full Plate
Challenges and Opportunities in Remaking the Food System 345
C. Clare Hinrichs and Elizabeth Barham

List of Contributors 357

Index 363

Acknowledgments

This book emerges in large part from the collective work of two related United States Department of Agriculture Multi-State Research Projects: NE-185, Local Food Systems in a Globalizing Environment: Commodities, Communities, and Consumers (1997–2002), and NE-1012, Sustaining Local Food Systems in a Globalizing Environment: Forces, Responses, and Impacts (2002–7). Through these two sequential projects, a shifting assortment of anthropologists, community nutritionists, agricultural economists, political scientists, sociologists, extensionists, and others from across the United States have been meeting at least once a year to coordinate and share their research and outreach activities on the challenges, processes, and impacts of creating more sustainable local food systems. The organizational framework provided by the Multi-State Research Project has been instrumental in allowing the individual ideas, the regional research efforts, and the diverse practical projects reported in this book to come into rich and productive conversation.

Most of the chapters in this volume were first publicly presented in special paper sessions organized for the 2003 joint annual meetings of the Agriculture, Food, and Human Values Society and the Association for the Study of Food and Society, held in Austin, Texas. Many have undergone extensive revision in conversation with the editors and other contributors to this book. The Pennsylvania State University College of Agricultural Sciences provided resources that supported the final editing and assembly of this book. Special thanks to Debra Putt and Dara Bloom for careful and consistent attention to the details.

As editors we also thank our families, friends, and colleagues for their continued patience and support while we worked on this project. We hope it has been worth the wait.

Introduction

Practice and Place in Remaking the Food System

C. Clare Hinrichs

We live in a time when food attracts growing scrutiny. Long taken for granted, food now gives many people pause. They ask where it comes from, how it is grown and prepared, and what implications it has for our health and the environment. A dairy cow found to have mad cow disease unleashes troubling questions about an international system of industrialized meat production, processing, and distribution. Lawsuits brought by obese teens against fast food companies that offer super-sized fare and parental campaigns to take the "junk" out of school lunches and vending machines highlight questionable commercial influences on food choices. Rural regions awash in a sea of commodity agriculture but without groceries or markets selling fresh or nutritious food suggest the sad ironies of our current agricultural "abundance." And the visual perfection but disappointing taste of a Delicious apple prompts yearning for the irregular shapes and in-season novelty of regional and old varieties—those that may not pack well, travel far, or keep but that bloom with distinctive flavor.

Having both material presence and symbolic charge, food now figures prominently in struggles for power, negotiations about policy, possibilities for partnership, and new and renewed expressions of pleasure and identity. Consequently, food provides a unique analytical and experiential nexus, drawing together and crystallizing many urgent, complicated problems facing society. No longer taken for granted or viewed in isolation, food can and should be connected to community vitality, cultural survival, economic development, social justice, environmental quality, ecological integrity, and human health.

This book explores the widening circles of connection emanating from food by examining the diverse efforts now underway to remake the North American food system. Such circles link food to agricultural and nonagricultural uses of the land on the one hand and to human bodies and spirits, individual lives, and community experiences on the other (Friedman 1999). The broader systemic nature of agriculture, food, and nutrition is a compelling, but not widely considered, view (Feenstra 1997; Sobal 1999). Farms operate within production systems that include families, nearby communities, and the surrounding environment. Furthermore, farms are linked to upstream suppliers of tools, equipment, seed, and knowledge and to downstream brokers, buyers, processors, distributors, and retailers, who in turn link the products of those farms to their end consumers. The influences, pressures, questions, and responses move up and down the chain and circulate through the system. Components of the food system can be analyzed in isolation, but they do not exist in isolation from other levels and stages (Jackson and Jackson 2002).

It is a formidable and perhaps impossible task to describe the food system in its entirety. It is possible, however, and also necessary to describe and analyze facets of the food system from a perspective that considers links and relationships and how the parts combine in particular configurations. This book takes such a systemic view in surveying the landscape of efforts now occurring in North America to craft food system alternatives that are designed to improve social, economic, environmental, and health outcomes. Although the book focuses primarily on U.S. examples, it offers a broader North American perspective attuned to globalizing tendencies at the level of markets and governance, as well as looking at how these factors intersect with experience in specific localities.

Remaking the North American Food System emerges from a collaborative research initiative begun in the late 1990s by a group of social and nutritional scientists and practitioners based primarily, but not exclusively, at U.S. land-grant universities. Working in their respective regions, members of the group sought to document and analyze the emergence of so-called local food systems within a wider context of globalization. They paid particular attention to the diversity of local strategies, practices, initiatives, and outcomes. From their differing academic and practitioner orientations, the contributors to this book share reservations about the homogenizing and industrializing course of the dominant food system with its growing evidence of social and economic vulnerabilities and harmful environmental effects. At the same time, some are discomfited by occasionally glib enthusiasm and uncritical endorsements of seeming alternatives to that dominant system. Although contributors are committed to a transition to a more sustainable agriculture and more just and equitable patterns of development, they suggest that viable alternatives can best take root and flourish through a process of careful description, empirical evaluation, monitoring, and critical appraisal. A culture of continual shared learning is necessary if we hope to remake the food system in substantial or sustainable ways.

This book then has several aims. It puts forward concepts and frameworks that can help us better understand the similarities and differences between diverse activities now taking place that arguably contribute to remaking the food system. It charts in detail some of the opportunities and barriers facing new community-based food system institutions. Such information can guide new research and inform practice. It shifts the discourse on food system alternatives from simple boosterism to constructive assessment by critically evaluating the outcomes of certain initiatives. Finally, and perhaps most crucially, the book speaks to multiple audiences, including students, practitioners, and policy makers, out of the conviction that the challenges facing the food and agricultural system are more than academic and cannot be resolved in isolated conversations.

Remaking the North American Food System is divided into three sections. In the first, "What's Wrong with the Food System? Orienting Frameworks for Change," two chapters lay out fundamental patterns in the conventional food system along with key concepts and frameworks for understanding how and why individuals and communities are challenging that system. These chapters set up the argument for remaking the food system and offer some guidance as to how we might think about that process and its goals.

The second section, "Institutions and Practice to Remake the Food System," critically examines specific organizational forms and practices, some of which are seen as hallmarks of more local, community-based food systems. The chapters in this section draw on both empirical research and program assessments in order to chart the outcomes of such initiatives and to explore strategies for their enhancement. This section also explores how policies at the local and national levels can facilitate a turn toward more sustainable local and regional food systems.

The third section, "The Importance of Place and Region in Remaking the Food System," looks further at important institutions and practices that work to change the food system in more sustainable directions while asking how these elements are brought to bear in particular places and regions. This section's chapters pay attention to how the distinctive socioeconomic and geographic configurations of places and regions shape both the opportunities and obstacles people encounter as they seek to remake the food system. They underscore that context filters the flow of possibilities and remind us that general solutions must be thoughtfully tailored in response to the particularities of place and region.

Remaking the Food System

The broad idea of remaking the food system organizes this book. With a nod perhaps to *homo faber*, the contributors generally see human beings as crafters, inventors, shapers, and experimenters. Although people are constrained by the history that precedes them and the geography in which they find themselves, they are not entirely bound by either. People harbor independent and changing desires and motivations. Some will act on these alone, while some will join together with others. As they identify shared interests and concerns, more people are engaging more forthrightly with the food system. Many are responding to disenchanting and even disturbing encounters with the food system by attempting to change some aspect of what they have experienced. The social location and resource endowments of different individuals and groups certainly afford different skills and opportunities for such work and, indeed, different understandings of what exactly the work should be. Overall, remaking first involves deliberate, sometimes dogged efforts simply to grasp what currently exists, and it requires second a refashioning of some of the institutions and practices of agriculture and food in more desirable ways. The process is dialectical in that changing the food system generally proceeds from the starting point of openings or vulnerabilities associated with the dominant conventional food system (Hendrickson and Heffernan 2002), and in that it occurs in continual dialogue with that conventional food system.

Some critiques of the dominant, industrial food system have questioned the transformative potential of grassroots efforts to launch local and sustainable food system alternatives (Magdoff, Foster, and Buttel 2000). In this view such alternatives are populist attempts to ameliorate the shortcomings of the dominant system but fail to address root causes and logics of that system. Many alternative food system initiatives center more on local consumer education and farmer entrepreneurship than on social justice issues or needed and challenging policy reforms (Allen et al. 2003). However, others argue that a pragmatic politics consisting of incremental steps may be the best and perhaps the only realistic route to "food democracy" (Hassanein 2003). In this view alternative food system initiatives—of whatever scale and scope—reflect what is currently possible in an overwhelming situation and should not be dismissed when they lack coherence or consistency or fail to correspond perfectly to movement ideals. In any case, definitive judgments about the transformative nature of efforts to change the food system are problematic in the short term, given uncertainties and disagreements about how to assess the achievement of sustainability.

Remaking the food system then suggests neither a revolutionary break nor a radical transformation but rather deliberate, sometimes unglamorous multipronged efforts in areas where openings exist to do things differently. Supporting a farmers' market may never shut down the local big box supermarket, but it does divert consumer dollars to local food producers, consequently helping them stay in business and providing some consumers with fresher, local foods. Such activities quietly and modestly remake parts of the food system. Whether pursued by individuals, by groups, or by communities, such remaking is not a linear or foreordained process that possesses some clear, known endpoint. It is instead movement in what is hoped to be a more promising direction. Remaking shifts us from a paralyzing focus on what is worrying, wrong, destructive, and oppressive about our current food system to a wide-angle view that takes in the broader landscape, whose troubling contours, we begin to notice, are punctuated by encouraging signs of change. Seen together, these initially isolated and spontaneous efforts to remake the food system begin to link and form a platform from which people might continue to work, step by step, toward a more sustainable food system.

The burgeoning literature on alternative and local food systems tends to frame change following one of two main approaches—one emphasizing civic renewal and redemocratization and the other stressing resistance and social mobilization. Although by no means incompatible, these two approaches have different legacies, inflections, and implications. A civic approach, highlighted in the chapter by Thomas Lyson, finds its heirs in mid-twentieth-century research on farming communities and on small business by social scientists Walter Goldschmidt and C. Wright Mills. It also draws on more recent scholarship in the 1980s and 1990s on economic restructuring and social capital. With its strong rural roots, Lyson's notion of civic agriculture links alternative food initiatives to local economic development, capacity building, and community problem solving. Although civic agriculture has a populist flavor, it also stresses the importance of ensuring a rooted, stable small business class that will act out of enlightened self-interest. As it reconsiders the links from land to economy and from food to health, civic agriculture provides a democratic counterweight to the excesses of an industrialized, corporately controlled food system. Accordingly, it informs an approach to change and suggests development of particular types of initiatives. Civic agriculture represents an encouraging alternative model, but as Lyson notes, it has thus far operated *beside* rather than in place of the conventional food system.

A social mobilization approach, as highlighted in the chapter by G. W. Stevenson and colleagues, is framed more in terms of social resistance to industrializing and globalizing trends in the food system and represents a potentially more combative stance than what is offered in Lyson's chapter. Based on their close observation of activities taking place across North America, Stevenson and colleagues develop a comprehensive typology of strategies for changing the food system. This includes warrior work that explicitly challenges the harms and excesses of the industrialized, corporately controlled food system; builder work that designs and constructs more promising ways of producing, marketing, and experiencing food; and weaver work that creates and nurtures linkages across activities, sectors, and groups. This social mobilization framework has different points of reference than a civic agriculture framework. It draws more directly on the social movements literature, much of which is more urban than rural in focus, and it often implies a translocal rather than locality-specific orientation.

Both chapters in part 1 offer frameworks that help orient our thinking about remaking the food system. Our idea of remaking the food system rides on a current of cautious optimism, fueled by evidence gathered in this book from investigations in different parts of North America. Nor is the evidence limited to North America. Indeed, diverse forms of sustainable farming and food systems are now emerging worldwide, many of them led by people on the margins of society and showing preliminary evidence of positive agroecological and socioeconomic impacts (Pretty 2002). While negative trends in the food system remain undeniable, there is currently a clear and urgent basis for hope when one considers the growing density and diversity of initiatives undertaken by ordinary people to reconstruct the relations between land, agriculture, and food (Lappé and Lappé 2002). This book documents such initiatives across North America, particularly in the United States, while emphasizing the importance of practice and place.

Practice

Related to the notion of remaking, the idea of practice underlies many of the contributions to this book. Practice is relevant, as both a philosophical and a political stance. "Thoughtful practice," for example, is a concept rooted in the work of John Dewey that can help us describe and understand emerging strategies and activities by agricultural producers and food consumers (Heldke 1992). Western philosophical and educational tradition has long distinguished "knowing" activities, such as science and art, from other "doing" activities involving practical, manual labor. The former, presumed to deal with fixed, eternal truths, have always been more valued than the latter, which are seen as transient and repetitive activities, subject to flux, growth, and decay.

Treating foodmaking—the growing, harvesting, preparing, and eating of foods—as a thoughtful practice can reconcile this problematic, binary formulation (Heldke 1992). Foodmaking is firmly rooted in the mundane, material world; anchored to natural cycles; and linked to senses and bodies. But at the same time, it requires skills of careful observation and the application of human knowledge, judgment, artistry, and care. In this sense many activities connected to food have the potential to bridge and unite head work and hand work. Philosophically, "foodmaking activities are valuable because of, not in spite of, the fact that they ground us in the concrete, embodied present" (Heldke 1992, 211). The contributors to this volume acknowledge in various ways the crucial role of thoughtful practice by producers, consumers, and other stakeholders in the food system.

However, the idea of thoughtful practice is not simply a philosophical justification for more attention to human activity surrounding food or agriculture. Beyond the self-reflection that thoughtful practice might engender for the person kneading bread or planting a row of peas, practice can also be understood as an open-ended, generative, social activity. Indeed, in new food system initiatives this is particularly evident. The formation of formal and informal networks around specific local quality products or the creation of a new market venue requires social exchange—practical give-and-take. As a number of commentators have pointed out, processes of innovation in business and in civil society are highly dependent on the capacity for social learning (see Pretty 2002; Sirianni and Friedland 2001). In order to launch new initiatives that might remake the food system, existing social relationships may be mobilized in new ways. Entirely new relationships may need to be launched as well when, for example, farmers partner with other players in the food system or link to nonagricultural sectors like education, business, health, or tourism. Practice here involves a process of social engagement and exchange.

Additional features of practice evident in this book are experimentation and adaptation. In keeping with a pragmatic politics (Hassanein 2003), those seeking to remake the food system examine what exists, assess opportunities or openings, and try to implement new ideas and approaches, oftentimes in small, incremental ways. This pragmatic approach seems to hold strategic promise (Lappé and Lappé 2002). Yet issues of social inclusion in food system practice and the development of alternative food system institutions need to be addressed in more than nominal fashion. Involving diverse stakeholders and participants can ensure multiple viewpoints and solidify the democratic foundations of practice. Local food initiatives, for instance, sometimes begin quite logically by appealing to educated elites, who represent a market segment with a ready interest in better food and the discretionary income to pursue it. Expanding beyond this group to address the interests and needs of farm workers or low-income consumers pushes the boundaries of truly innovative practice. The complexity and uncertainty of the food system demand an emphasis on practice, based on continual learning and readjustment of effort by people working on working together.

Food system practices and institutions are the theme of the second section of the book. Drawing on case studies of farmers' markets in New York, Iowa, and California, Gilbert Gillespie and colleagues examine the community economic development role of farmers' markets, which have burgeoned in recent years. They present four processes by which farmers' markets have the potential to be the keystones of more localized food systems and reflect on conditions that help to realize that potential and those that may thwart it. Larry Lev and associates report on low-cost, collaborative learning–based approaches that have been developed and used in Oregon to enhance the organization and business performance of farmers' markets. Lev's chapter illustrates the potential of innovative models of participatory research and practice for strengthening food system institutions. Community supported agriculture (CSA) is a relatively new direct marketing arrangement that has grown in popularity in part due to its ability to link farmers and consumers more closely. Marcia Ostrom provides a detailed and sometimes sobering assessment of farmer and CSA member experiences, based on a ten-year study of twenty-four CSA farms in the upper Midwest. Local food policy councils are another institutional innovation associated with remaking the food system. Kate Clancy and colleagues offer a much needed critical appraisal of the goals and outcomes over time of North American government-sanctioned food policy councils and suggest ways that the practice might be supported.

Many of the institutions and practices outlined in this section of the book involve the development of different approaches to food marketing and education. G. W. Stevenson and Holly Born examine the *Label Rouge* poultry system in France as a possible model for a process-based labeling approach in the United States that also could dovetail with conventional food retailers and fit the production needs and interests of imperiled mid-sized farms. Under this model production process verification and labeling represent new forms of institutionalization and food practice. Dietary guidance is another practice of crucial importance for local food system development. Jennifer Wilkins examines the disjuncture between current nutritional recommendations and the possibilities and constraints of place-based food systems, and offers some suggestions for how such tensions might be resolved. The final chapter in this second section is by Joan Thomson and associates. They discuss recent process-based approaches to community dialogue for the purpose of identifying local food system concerns and planning and implementing alternatives. Their grounded account offers useful evidence from Pennsylvania on how civic agriculture, in a wider systemic view, might be approached.

Place

One irony of recent popular enthusiasm for local food system development is the presumption that it proceeds in a fairly predictable way once certain institutions or practices are in place (see Norberg-Hodge, Merrifield, and Gorelick 2000). Thus starting a farmers' market, creating a direct marketing farm alliance, or developing and implementing an ecolabel each entail certain steps that, if faithfully followed, will result in desired outcomes, typically including more economically viable family farms, better availability of fresh produce, and a healthier, less degraded natural environment. The irony consists in unwittingly replicating popular assumptions about the development of the large-scale conventional food system as an inexorable and uniform process of industrialization, standardization, and environmental decline (Kimbrell 2002). If globalization is seen as one master process with largely negative effects, then localization becomes its reverse, a process that will neatly and predictably turn the bad to good. Reality, however, appears far more complicated. While the broad contours of such assessment about a globalizing, conventional food system versus a localizing, alternative food system may be accurate, the precise workings on the ground are variable and complex.

In the mid-1990s social scientists studying food systems began to reevaluate their unilinear notions of how globalization of the food system proceeds (Buttel 2001). One result is greater attention to how globalization affects and is conditioned by local contexts (Ward and Almäs 1997). In the same way, although there are surely general principles and patterns that can be identified in how more localized food system alternatives emerge and the challenges that they may encounter, the particulars are bound to vary. In large part, they will vary for reasons associated with the significance of place. These place-based differences need more careful highlighting and thus are one important focus of this book.

Academic fields like rural sociology, urban sociology, geography, and planning have long stressed that closer consideration of space and place improves our understanding of social processes and their outcomes (Lobao and Saenz 2002). Researchers interested in social and environmental inequality note that similar types of people who live in regions with different natural resource endowments, economic compositions, or political infrastructures may end up experiencing very different opportunities or burdens. Such contextual factors have important implications for how people in different regions experience poverty and the likelihood that interventions initiated by government, outside nongovernmental organizations or local residents will succeed and just how they will succeed (Lyson and Falk 1993). Similarly, the agricultural and land use histories, demographic patterns, and political cultures of different regions combine to create different contexts for the viability of farmers' markets, of CSA, and food policy councils (Selfa and Qazi 2005).

If we think of *space* in simple terms of distance and configuration, the notion of *place* incorporates more: the specificity of location, particular material forms, associated meanings, and values (Gieryn 2000). Place encompasses the history and culture of the human-built environment in its most generous sense, but as it is understood in terms of its complex and dynamic interplay with nature. In this respect, place, and more broadly region, provide crucial context, and context matters, as agroecologists have long recognized. Food system practice should take into account place and region; general principles and prescriptions must be thoughtfully and deliberately adapted to particular circumstances. These needs require careful observation and effort to adapt and integrate the approaches to practice described in part 2 of this book so they can be relevant and effective in real places. As shown throughout part 3, the unique features of places and regions can represent assets or barriers to practices intended to remake the food system.

Troy Blanchard and Todd Matthews provide a national-scale analysis that draws out regional variation in the patterns of food deserts—areas with limited or no access to affordable and nutritious foods. They present a spatially sensitive picture of how food retail concentration corresponds to inequalities in food access, a condition with particularly troubling ramifications for many rural regions of the United States. Michael Hamm examines the environmental and nutritional changes that would ensue if residents of Michigan were to increase the proportion of their food acquired from local sources within that state. Grounded in the current empirical realities of Michigan's agricultural production and food consumption patterns, Hamm's analysis underscores the vital importance of context for regions considering or attempting transition to more reliance on local or regional food sources. Drawing on extensive survey data from the state of Washington, Marcia Ostrom and Raymond Jussaume assess both producer and consumer interest in more direct agricultural marketing linkages and situate the potential for development of more localized food systems within an account of the geographic and historical particularities of that northwest state. The distinctive workings of alternative food system institutions and their interplay with global economic change are taken up by Viviana Carro-Figueroa and Amy Guptill, who link the recent development of farmers' markets in Puerto Rico to the simultaneous expansion of large-scale food retailing chains on the island.

Place comes specifically into play in Elizabeth Barham's chapter detailing an initiative by lamb producers in Charlevoix, Quebec, to develop an origin-based label for their product. Barham's account stresses the specific resources and endowments of a region and underscores how local actors can use these to their advantage to claim their role in larger global markets and actively engage such seemingly remote institutions as the World Trade Organization. Laura DeLind and Jim Bingen discuss the developing organizations and strategic choices of organic farmers and activists in Michigan, noting in particular both the risks and the opportunities presented by increasing receptivity to organic agriculture in land-grant universities. Matthew Hoffman draws on detailed ethnographic data to take a closer look at the actual practices and concerns of sustainable farmers in Vermont, an account that highlights both general patterns and specificities related to the unique social and agricultural circumstances of that state. Finally, Audrey Maretzki and Elizabeth Tuckermanty offer an indepth analysis of a sample of community food projects from across the United States that were funded by the U.S. Department of Agriculture. All widely viewed as successes, the projects nonetheless suggest insights about the relevance of place for how community food projects are organized and how they achieve their goals. Taken together, the chapters in part 3 underscore the inherent diversity, as well as the underlying coherence, in efforts to remake the food system.

Informed by both the academy and the field, this book presents fresh approaches for thinking about, understanding, and working on food and agricultural issues. In its chapters the reader will find practice-informed research and research-informed practice that illuminate ways of remaking the food system. The contributors to this volume may refer variously to creating local food systems, to community-based food systems, or to alternative agrifood systems as they discuss new institutions and practices and situate these developments in particular places and regions. We have retained such terminological distinctions because they correspond to the diverse disciplines and differing project experiences of the many contributors to this book. What matters is the continuing vibrancy of this conversation about food, agriculture, land, and people—intertwined subjects that rightly give us pause, that compel us to examine how practice and place intersect with pleasure and power in order to create new possibility in the food system.

Acknowledgement

Support for this work came from Iowa State University Agricultural Experiment Station in conjunction with USDA/CSREES Multi-State Research Project NE-1012 and from Pennsylvania State University College of Agricultural Sciences. The author particularly wishes to acknowledge the critical support of Elizabeth Barham and Tom Lyson.

References

Allen, P., M. FitzSimmons, M. Goodman, and K. Warner. 2003. "Shifting plates in the agrifood landscape: The tectonics of alternative agrifood initiatives in California." *Journal of Rural Studies* 19:61–75.

Buttel, F. H. 2001. "Some reflections on late 20th century agrarian political economy." *Sociologia Ruralis* 41:165–81.

Feenstra, G. W. 1997. "Local food systems and sustainable communities." *American Journal of Alternative Agriculture* 12:28–36.

Friedmann, H. 1999. "Circles of Growing and Eating: The Political Ecology of Food and Agriculture." In *Food in Global History*, edited by R. Crew, 33–57. Boulder CO: Westview.

Gieryn, T. F. 2000. "A space for place in sociology." *Annual Review of Sociology* 26:463–96.

Hassanein, N. 2003. "Practicing food democracy: A pragmatic politics of transformation." *Journal of Rural Studies* 19:77–86.

Heldke, L. M. 1992. "Foodmaking as a Thoughtful Practice." In *Cooking, Eating, Thinking: Transformative Philosophies of Food*, edited by D. W. Curtin and L. M. Heldke, 203–29. Bloomington: Indiana University Press.

Hendrickson, M. K., and W. D. Heffernan. 2002. "Opening spaces through relocalization: Locating potential resistance in the weaknesses of the global food system." *Sociologia Ruralis* 42:347–69.

Jackson, D. L., and L. L. Jackson, eds. 2002. *The Farm as Natural Habitat: Reconnecting Food Systems with Ecosystems*. Washington DC: Island.

Kimbrell, A., ed. 2002. *Fatal Harvest: The Tragedy of Industrial Agriculture*. Washington DC: Island.

Lappé, F. M., and A. Lappé. 2002. *Hope's Edge: The Next Diet for a Small Planet*. New York: Tarcher/Penguin.

Lobao, L., and R. Saenz. 2002. "Spatial inequality and diversity as an emerging research area." *Rural Sociology* 67:497–511.

Lyson, T. A.., and W. W. Falk, eds. 1993. *Forgotten Places: Uneven Development in Rural America*. Lawrence: University Press of Kansas.

Magdoff, F., J. B. Foster, and F. H. Buttel, eds. 2000. *Hungry for Profit: The Agribusiness Threat to Farmers, Food and the Environment*. New York: Monthly Review.

Norberg-Hodge, H., T. Merrifield, and S. Gorelick. 2000. *Bringing the Food Economy Home: The Social, Ecological, and Economic Effects of Local Food.* Bloomfield CT: Kumarian.

Pretty, J. 2002. *Agri-Culture: Reconnecting People, Land and Nature.* London: Earthscan.

Selfa, T., and J. Qazi. 2005. "Place, taste, or face-to-face? Understanding producer-consumer networks in 'local' food systems in Washington State." *Agriculture and Human Values* 22 (4): 451–64.

Sirianni, C., and L. Friedland. 2001. *Civic Innovation in America: Community Empowerment, Public Policy, and the Movement for Civic Renewal.* Berkeley: University of California Press.

Sobal, J. 1999. "Food System Globalization, Eating Transformations and Nutrition Transitions." In *Food in Global History*, edited by R. Crew, 171–93. Boulder CO: Westview.

Ward, N., and R. Almäs. 1997. "Explaining change in the international agro-food system." *Review of International Political Economy* 4:611–29.

Part I

What's Wrong with the Food System?

Orienting Frameworks for Change

1. Civic Agriculture and the North American Food System

Thomas A. Lyson

As agriculture in North America enters the twenty-first century, the economically independent, self-reliant farmer of the last century is rapidly disappearing from the countryside. Farmers, once the centerpiece of the rural economy, often have been reduced to producers of basic commodities for large agribusiness corporations. The real value in agriculture no longer rests in the commodities produced by farmers but instead is captured by the corporately controlled and integrated sectors of the agrifood system that bracket producers with high-priced inputs on one side and tightly managed production contracts and marketing schemes on the other.

Although various shortcomings of a corporately controlled and managed food system in North America and elsewhere have been revealed by sociologists and environmentalists, only recently has an alternative agricultural paradigm emerged to challenge the wisdom of conventional production agriculture. The new paradigm, labeled civic agriculture, is associated with a relocalizing of production. From the civic perspective, agriculture and food endeavors are seen as engines of local economic development and are integrally related to the social and cultural fabric of the community. Fundamentally, civic agriculture represents a broad-based movement to democratize the agriculture and food system.

In this chapter, I outline the contours of the conventional or commodity model of agricultural production. This model is the organizational and technological one that undergirds most of production agriculture today. After introducing the concept of civic agriculture, I illustrate central differences between the conventional and civic forms of agricultural development. Some examples of civic agriculture endeavors are provided as well as suggestions for ways to strengthen and sustain a more locally based agriculture and food system.

The Conventional Model of Production Agriculture

In a set of publications, I have drawn the distinction between conventional/commodity agriculture on the one hand and civic agriculture on the other (Lyson 2000, 2002, 2004; Lyson and Guptill 2004). Conventional/commodity agriculture represents a set of practices, procedures, and techniques that are designed to produce as much food and fiber as possible for the least cost. The underlying biological paradigm for conventional/commodity agriculture is experimental biology, while the underlying social science paradigm is neoclassical economics.

The logic of experimental biology dictates that increasing output is the primary goal of scientific agriculture (Lyson and Welsh 1993). Neoclassical economics posits that optimal efficiency and presumably maximum profitability in production agriculture can be achieved by balancing the four factors of production: (1) land, (2) labor, (3) capital, and (4) management or entrepreneurship. These four factors form the basis of the production function.

The conventional/commodity model of farm production is supported by agricultural colleges and universities, government agencies such as the U.S. Department of Agriculture and Agriculture Canada, and more recently large, multinational agribusiness firms. At agricultural colleges throughout North America, the teaching and research activities of plant scientists center on increasing yields by enhancing soil fertility, reducing pests, and developing new genetic varieties. For animal scientists at these institutions, the focus is on health, nutrition, and breeding. The scientific and technological advances wrought by agricultural scientists are then filtered through a farm management paradigm in agricultural economics that champions sets of so-called best management practices as the blueprints for economic success.

Agribusiness Corporations

A truly global food system began to emerge in the 1950s as nationally organized food corporations grew in size. Beginning in the 1980s, a wave of mergers among food processors, input suppliers, and marketers resulted in a tremendous consolidation of power in the food sector (Heffernan 1997, 1999; Lyson and Raymer 2000). The large multinational food corporations that were formed by these mergers have taken on the task of organizing and coordinating the production, processing, and distribution of food. Today, globally oriented food processors and distributors along with mass market retailers are becoming dominant fixtures in the North American food economy. The degree of concentration has reached the point where the ten largest U.S.-based multinational corporations control almost 60 percent of the food and beverages sold in the United States, while the four largest retail conglomerates in Canada account for about 25 percent of all sales.

As the large multinational food corporations grew in size, the food choices facing consumers decreased. A food system dominated by a small number of large corporations offers consumers little real choice in terms of foods (Lyson and Raymer 2000). The so-called innovation of these firms largely involves designing better marketing strategies for a narrow range of basic products that may be differentiated in only superficial ways. For example, between 10,000 and 15,000 new food products are introduced to the U.S. consumer each year. The fact that only a few hundred of these products gain market acceptance shows that large agribusiness firms are not responding to consumer demand but rather trolling for consumer dollars by offering repackaged, reformulated, and reengineered products in the hope that some of them will be profitable.

Contract Farming

A globally orchestrated food system requires large quantities of standardized and uniform products that are available year around. To insure that they have sufficient supplies of basic commodities, food processors have entered into formal contracts with individual farmers (see Lyson 2004). Although no systematic data are available on contract production in the United States, Welsh notes that "since 1960, contracts and vertically integrated operations have accounted for an ever-larger share of total U.S. agricultural production" (1996, 20). In the United States about 85 percent of the processed vegetables are grown under contract and 15 percent are produced on large corporate farms (Welsh 1996, 4).

Through contract farming, food processors gain control over their agricultural suppliers. The major disadvantage to the farmer is a loss of independence. Many contracts specify quantity, quality, price, and delivery date, and in some instances the processor is completely involved in the management of the farm, including input provision.

Contract farming and farm size are tightly linked. Economies of scale dictate that processors are more inclined to work with large farmers whenever possible. Hart (1992) has suggested that the processor's ability to award or refuse a contract has contributed to differences in profitability between large and small producers and accelerated the process of farm concentration. Mark Drabenstott, formally an economist with the Federal Reserve Bank of Kansas City, sees contract farming leading to a much more tightly choreographed food system in the future. "The key component in this choreography is a business alliance known as a supply chain. In a supply chain, farmers sign a contract with a major food company to deliver precisely grown farm products on a preset schedule" (Drabenstott 1999). In some cases contract farming is reconfiguring production at the local level because it is the processor and not the farmer who determines what commodity is produced and where.

For farmers in North America and elsewhere, the globalization of the food system means that a much smaller number of producers will articulate with a small number of processors in a highly integrated business alliance. Drabenstott (1999) estimates that "40 or fewer chains will control nearly all U.S. pork production in a matter of a few years, and that these chains will engage a mere fraction of the 100,000 hog farms now scattered across the nation." The consequences are clear: "supply chains will locate in relatively few rural communities. And with fewer farmers and fewer suppliers where they do locate, the economic impact will be different from the commodity agriculture of the past" (Drabenstott 1999).

Wherever the conventional model of agricultural development takes hold, the commodities upon which the food system is built become "cheap" (Lyson and Raymer 2000). The conventional model takes value out of the commodity and moves it to the input suppliers, the processors, the distributors, and the marketers. Consider that a box of Wheaties that sells for $3.00 in the United States contains only three cents of wheat. There is consequently very little profit for farmers who produce most basic commodities including corn, soybeans, wheat, and rice for the global marketplace.

Simply stated, the emphasis of agricultural colleges, departments or ministries of agriculture, and large food corporations on producing as much food as possible for the least cost has resulted in an abundant supply of a relatively narrow set of commodities (that is, those that are easy to produce and process) (Critser 2003). However, as Marty Strange notes, "there is little need for more food output from American agriculture" (1988, 221). Continuing down the current path of conventional/commodity agriculture is likely to lead to greater concentration of production in large-scale corporate hands, the further erosion of rural communities and culture, and continued resource depletion and environmental degradation. A turn away from the production of low cost, undifferentiated commodities and toward civic agriculture requires the reintegration of agriculture and food into local communities.

A Theory of Civic Agriculture

A theory of civic agriculture has been laid out by Lyson (2000, 2004), DeLind (2002), and others (Lyson and Guptill 2004). Civic agriculture is one component of a larger theory of civic community (see Lyson and Tolbert 2003; Robinson, Lyson, and Christy 2002; and Tolbert, Lyson, and Irwin 1998). Proponents of civic agriculture look for an explanation of agricultural development that is driven by social processes other than economics.

Historically, Alexis de Tocqueville (1836) provides an important starting point for contemporary inquiries into the civic community. Writing about America in the nineteenth century, Tocqueville argues that the norms and values of civic community are embedded in voluntary associations, churches, and small businesses. It is within these venues that local problem solving occurs as citizens come together to discuss and debate the important social issues of the day. Contemporary scholars such as Robert Putnam have empirically demonstrated the relationships among "dense networks of secondary associations" and economic and political development (Putnam 1993, 90). As Putnam notes, "[p]articipation in civic organizations inculcates skills of cooperation as well as a sense of shared responsibility for collective endeavors. Moreover, when individuals belong to 'cross-cutting' groups with diverse goals and members, their attitudes will tend to moderate as a result of group interaction and cross-pressure" (1993, 90).

The literature on industrial districts, especially in Europe, provides further evidence that agriculture and food economies organized around smaller-scale, locally oriented production and distribution systems are possible. However, to be successful, agricultural/production districts "require a broad set of infrastructural institutions and services to coordinate relationships among economic actors" and to compensate for the inefficiencies of a fragmented system of production (Zeitlin 1989, 370). The success and survival of locally based economic systems is directly tied to the democratic efforts of the community to which they belong (Sabel 1992).

Proponents of civic agriculture contend that sound agricultural development emerges from attention to social processes in communities rather than from economics' narrower focus on profit-maximization. For example, the civic agriculture approach is oriented toward establishing, maintaining, and strengthening local social and economic systems, while the conventional/commodity agriculture approach is directed toward economic globalization. The desired outcome for conventional agriculture development is a global (mass) market for food articulating with standardized, low-cost, mass production of basic commodities. Development guided by civic agriculture principles is predicated on food production and consumption maintaining at least some linkages to the local community (Lyson 2004.)

In the conventional/commodity agriculture model, the ideal form of production is the large farm. Large farms are able to capture economies of scale and hence produce goods more cheaply than smaller, and presumably less efficient, farms. Following the precepts of neoclassical economics, large producers then link with large wholesalers, large wholesalers forge relationships with large retailers, and large retailers serve the mass market. In the food industry, large multinational agribusiness corporations across the supply chain become the driving engines of development (Lyson and Raymer 2000).

The civic agriculture perspective, however, favors smaller, well-integrated firms/farms cooperating with each other in order to meet the food needs of consumers in local (and global) markets. The ideal form is the production district, similar to the industrial district notion mentioned earlier (Piore and Sabel 1984). Producers share information and combine forces to market their products. The state supports these economic ventures by ensuring that all firms have access to the same resources such as information, labor, and infrastructure.

Civic agriculture flourishes in a democratic environment. Community problem solving around agriculture and food issues requires that all citizens have a say in how, where, when, and by whom their food is produced, processed, and distributed. Food and agriculture issues are an integral part of community life and recognized as such. Indeed, citizen participation in agriculture- and food-related organizations and associations stands as a cornerstone of civic agriculture.

In contrast, the conventional system of agricultural production, premised on notions of the free market, does not necessarily benefit from democracy. Barber has noted, "[c]apitalism requires consumers with access to markets; such conditions may or may not be fostered by democracy" (1995, 15). The conventional/commodity agriculture paradigm is compatible with a wide range of political regimes. However, it may be challenged in places where widespread democratic participation prevails. The motors for civic agriculture development are civic engagement and social movements. Civic communities are places where problem-solving can occur because residents come together in various formal and informal associations in order to address common interests and concerns. Communities that have dense associational and organizational structures nurture civic engagement among their residents and are best able to meet their social and economic needs. Instead of self-satisfying, individual rational actors being the foundation of the community, groups of engaged individuals organized into social movements are core to the civic agriculture approach (Lyson 2004).

Civic Community, Civic Agriculture, and Socioeconomic Welfare

A small but growing body of empirical research has examined the relationships among characteristics of civic community, civic agriculture, and socioeconomic welfare. This line of research and inquiry dates back to two studies commissioned by the U.S. Senate during World War II. One study by C. Wright Mills and Melville Ulmer (1946) examined the relationship between the economically independent middle class, civic engagement, and community welfare in six manufacturing cities in the Midwest and Northeast. Mills and Ulmer were particularly interested in evaluating the "effects of big and small business on city life." They found that cities in which the economically independent middle class was strong had higher levels of civic engagement and also higher levels of social and economic welfare than cities in which large corporations crowded out the independent middle class. In the forward to their report, Senator James E. Murray, chairman of the special committee that commissioned the study, noted that "for the first time objective scientific data show that communities in which small businesses predominate have a higher level of civic welfare than comparable communities dominated by big business" (cited in Mills and Ulmer 1946, v).

The other study by Walter Goldschmidt (1978) focused on farming communities in California. Goldschmidt used a comparative community framework similar to that employed by Mills and Ulmer but limited his study to only two communities. He found that residents in the community dominated by large-scale, corporately controlled farming had a lower standard of living and quality of life than residents in the community where production was dispersed among a large number of smaller farms.

Based on extensive field work in both communities, Goldschmidt concluded that the scale of operations that developed in the community dominated by large agribusiness "inevitably had one clear and direct effect on the community: It skewed the occupation structure so that the majority of the population could only subsist by working as wage labor for others. . . . The occupation structure of the community, with a great majority of wage workers . . . [,] has had a series of direct effects upon social conditions in the community" (1978, 415–16). In other words, differences in social and economic welfare between the large-farm and the small-farm community were directly the result of worker exploitation.

In recent years the theoretical linkages among the economically independent middle class, civic engagement, and social and economic well-being have been empirically tested by Tolbert et al. (2002); Tolbert, Lyson, and Irwin (1998); Lyson, Torres, and Welsh (2001); and Robinson, Lyson, and Christy (2002). In all instances the findings from these researches support the core findings reported by Mills and Ulmer more than fifty years ago and later by Goldschmidt. Places dominated by a small handful of very large farms or firms manifest significantly lower levels of community welfare than places in which the economy is organized around smaller family farms or firms.

Evidence of a New Civic Agriculture in North America

Communities can buffer and shelter themselves from the homogenizing and destabilizing forces of the global food system only if they develop the infrastructure, maintain a farmland base, and provide the technical expertise so that farmers and processors can successfully compete against the highly industrialized, internationally organized corporate food system in the local marketplace. There are several important characteristics associated with civic agriculture in North America (see Lyson 2004). First, farm production is locally oriented. Farmers emphasize producing for local and regional marketplaces rather than for national or international mass markets. Second, farming and food production is integrated into communities. Farming is not merely the production of undifferentiated commodities to meet the needs of agribusiness corporations. Third, farmers compete on the basis of quality of their products, not on who can be the least-cost producer. Fourth, civic agriculture production practices are often more labor and land intensive, but less capital intensive and land extensive. As a result civic agriculture enterprises tend to be considerably smaller in scale than conventional commodity enterprises. Fifth, farmers often rely more on local, site-specific, and shared knowledge and less on a uniform set of best management practices prescribed by outside experts. Finally, civic agriculture producers are more likely to forge direct market links to consumers rather than indirect links through middlemen such as wholesalers, brokers, and processors.

Green and Hilchey (2002) and others (Lyson 2000) have begun to identify different types of civic agriculture enterprises and activities. Farmers' markets provide immediate, low-cost, direct contact between local farmers and consumers and are an effective first step for communities seeking to develop stronger local food systems. Community and school gardens produce fresh vegetables for underserved populations, teach food production skills, and can increase agricultural literacy. Farm-to-school programs nurture relationships between farmers, youth, and local communities. By enacting local ordinances or reducing red tape to facilitate local purchasing by public institutions, community officials can strengthen agriculture in their areas.

Community supported agriculture (CSA) projects are forging direct links between groups of member-consumers (often urban) and their CSA farms. Restaurant agriculture describes a system of production and marketing in which farmers target their products directly to restaurants. New grower-controlled marketing cooperatives are emerging to more effectively tap regional markets. Marketing and trading clubs are groups of farmers who share marketing information and may invest in commodity futures contracts. Agricultural districts organized around particular commodities (such as wine) have served to stabilize farms and farmland in many areas of the United States. Community kitchens provide the infrastructure and technical expertise necessary to launch new food-based enterprises. Specialty producers and on-farm processors of products for which there are not well-developed mass markets (deer, goat or sheep cheese, free-range chickens, organic dairy products, and so on) and small-scale, off-farm, local processors add value in local communities and provide markets for civic agriculture farmers. What all of these efforts have in common is their potential to nurture local economic development, maintain diversity, and quality in products as well as to provide forums in which producers and consumers can come together to solidify bonds of community.

Although reliable data on some of these types of civic agriculture enterprises are still difficult to find in North America, the U.S. Department of Agriculture has reported a dramatic increase in the number of farmers who are selling directly to the public. According to the U.S. Census of Agriculture, the number of farms selling directly to the public increased by 35 percent between 1992 and 2002, while the total value of products sold to the public increased by 101 percent during the same time period. While not all civic agriculture producers sell directly to the public, a large proportion does. These data from the Census suggest a continuing and strengthening trend for civic agriculture in the years ahead.

An Agenda for Civic Agriculture

Civic agriculture activities are now expanding across North America. As the number of farmers' markets, CSA's, on-farm processors, community kitchens, and other civic enterprises increase, the balance between local food self-sufficiency and global food dependence appears to be shifting back to the local. Of course, a totalizing civic agriculture characterized by complete local or regional self-sufficiency would be neither practical nor desirable. Some level of international and interregional trade can be beneficial to both importing and exporting communities.

It is important to recognize that control of the food system today rests in the hands of economically powerful and highly concentrated corporate interests (Heffernan 1999; Lyson and Raymer 2000). Furthermore, the current system of farm subsidies, agricultural finance practices, and global trade rules do nothing to advance civic agriculture. Nevertheless, communities, organizations, local governments, and even individuals have many tools with which to effect change and promote a more civic agriculture.

As a beginning, community officials need to understand and communicate the economic impact of agriculture in their communities. Local farms provide livelihoods for farm families but also for farm-related businesses. Because farmers purchase inputs from local businesses and provide the products for food-processing firms, they produce a large economic multiplier effect by recirculating money in local economies (Green and Hilchey 2002). In addition to the economic impacts of civic agriculture, local farms preserve open space, maintain rural character, and make communities more attractive to tourists and to nonfarm employers. As Green and Hilchey note, "[f]arms can also benefit the environment by protecting watersheds, enhancing wildlife habitat and fostering biodiversity. They provide fresh, wholesome foods with superior taste and nutrition. In short, they contribute to community quality of life" (2002, 89).

It is time to put agriculture and food on the political agendas of local communities. Locally organized agriculture and food enterprises must be fully integrated into a community's general planning and economic development efforts. This integration means that local agriculture and food businesses need the same access to economic development resources—such as grants, tax incentives, and loans—as nonfarm-related businesses. Additionally, communities should ensure that agricultural constituencies are represented on community boards, task forces, and governing bodies. Likewise, local agriculture and food systems activities should be addressed and integrated into any comprehensive planning processes.

When civic agriculture is effectively integrated with local planning and development efforts, community leadership becomes much more knowledgeable about agriculture and its needs. Local policies and programs become more supportive of the agricultural community. At the same time, the agricultural community itself, oriented to civic concerns, develops more effective leadership, and builds capacity for directing its own future.

At present civic agriculture seems to represent more of a consumer-driven alternative rather than an economic challenge to conventional/commodity agriculture. However, the currently dominant commodity system of production faces growing pressure to address and accommodate more of the environmental and community dimensions that are embodied in civic agriculture. Not only are consumers (Rifkin 1992) and environmentalists (Buttel 1995) calling for a more civic agriculture but so too are farmers (Kirshenmann 1995). In the near term, however, the agricultural landscape will likely continue to be characterized by two rather distinct systems of food production.

Acknowledgments

Support for this research was provided by Cornell University Agricultural Experiment Station in conjunction with USDA/CSREES Multi-State Research Projects NE-1012 and NC-1001.

References

Barber, B. R. 1995. *Jihad vs. McWorld*. New York: Times Books.

Buttel, F. H. 1995. "Twentieth Century Agricultural-Environmental Transitions: A Preliminary Analysis." In *Research in Rural Sociology and Development: Sustainable Ag-*

riculture and Rural Communities, edited by H. K. Schwarzweller and T. A. Lyson, 1–21. Vol. 6. Greenwich CT: JAI Press.

Critser, G. 2003. *Fat Land*. New York: Houghton Mifflin.

DeLind, L. B. 2002. "Place, work, and civic agriculture: Common fields for cultivation." *Agriculture and Human Values* 19:217–24.

Drabenstott, M. 1999. "New Futures for Rural America: the Role for Land-Grant Universities." William Henry Hatch Memorial Lecture presented at the annual meeting of the National Association of State Universities and Land-Grant Colleges. San Francisco.

Goldschmidt, W. R. 1978. *As You Sow*. Montclair NJ: Allanheld, Osmun.

Green, J., and D. Hilchey. 2002. *Growing Home: A Guide to Reconnecting Agriculture, Food, and Communities*. Ithaca NY: Community Food and Agriculture Program, Department of Development Sociology, Cornell University.

Hart, P. 1992. "Marketing Agricultural Produce." In *The Geography of Agriculture in Developed Market Economies*, edited by I. R. Bowler, 162–206. New York: Wiley.

Heffernan. W. D. 1997. "Domination of World Agriculture by Transnational Corporations." In *For All Generations*, edited by J. P. Madden and S. G. Chaplowe, 173–81. Glendale CA: OM Publishing.

———. 1999. *Consolidation in the Food and Agriculture System*. Report to the National Farmers Union. http://www.nfu.org/wp-content/uploads/2006/03/1999.pdf (last accessed April 9, 2006).

Kirschenmann, F. 1995. "Reinvigorating Rural Economies." In *Research in Rural Sociology and Development: Sustainable Agriculture and Rural Communities*, edited by H. K. Schwarzweller and T. A. Lyson, 215–25. Vol. 6. Greenwich CT: JAI Press.

Lyson, T. A. 2000. "Moving toward civic agriculture." *Choices* 15 (3): 42–45.

———. 2002. "Advanced agricultural biotechnologies and sustainable agriculture." *TRENDS in Biotechnology* 20:193–96.

———. 2004. *Civic Agriculture: Reconnecting Farm, Food, and Community*. Medford MA: Tufts University Press.

Lyson, T. A., and A. Guptill. 2004 "Commodity agriculture, civic agriculture, and the future of U.S. farming." *Rural Sociology* 69:370–85.

Lyson, T. A., and A. L. Raymer. 2000. "Stalking the wily multinational: Power and control in the U.S. food system." *Agriculture and Human Values* 17:199–208.

Lyson, T. A., and C. M. Tolbert. 2003. "The civic community and balanced economic development." *Research in Rural Sociology and Development* 9:103–20.

Lyson, T. A., R. Torres, and R. Welsh. 2001. "Scale of agricultural production, civic engagement, and community welfare." *Social Forces* 80:311–27.

Lyson, T. A., and R. Welsh. 1993. "Crop diversity, the production function and the debate between conventional and sustainable agriculture." *Rural Sociology* 58:424–39.

Mills, C. W., and M. Ulmer. 1946. *Small Business and Civic Welfare*. Report of the Smaller

War Plants Corporation to the Special Committee to Study Problems of American Small Business. 79th Congress, 2nd sess., February 13, S. Doc. 135.

Piore, M. J., and C. F. Sabel. 1984. *The Second Industrial Divide*. New York: Basic Books.

Putnam, R. M. 1993. *Making Democracy Work*. Princeton NJ: Princeton University Press.

Rifkin, J. 1992. *Beyond Beef: The Rise and Fall of the Cattle Culture*. New York: Dutton.

Robinson, K. L., T. A. Lyson, and R. D. Christy. 2002. "Civic community approaches to rural development in the South: Economic growth with prosperity." *Journal of Agricultural and Applied Economics* 34:327–38.

Sabel, C. F. 1992. "Studied Trust: Building New Forms of Cooperation in a Volatile Economy." In *Explorations in Economic Sociology*, edited by R. Swedberg, 104–44. New York: Russell Sage Foundation.

Strange, M. 1988. *Family Farming: A New Economic Vision*. Lincoln NE: University of Nebraska Press.

Tocqueville, A. de. 1836. *Democracy in America*. London: Saunders & Otley.

Tolbert, C., M. Irwin, T. A. Lyson, and A. Nucci 2002. "Civic community in small town USA." *Rural Sociology* 67:90–113.

Tolbert, C. M., T. A. Lyson, and M. Irwin. 1998. "Local capitalism, civic engagement, and socioeconomic well-being." *Social Forces* 77:401–28.

Welsh, R. 1996. *The Industrial Reorganization of U.S. Agriculture*. Policy Studies Report. No. 6. Greenbelt MD: Henry A. Wallace Institute for Alternative Agriculture.

Zeitlin, J. 1989. "Introduction." *Economy and Society* 18:367–73.

2. Warrior, Builder, and Weaver Work

Strategies for Changing the Food System

G. W. Stevenson, Kathryn Ruhf,
Sharon Lezberg, and Kate Clancy

Until recently, analyses of the modern agrifood system have focused more on the dynamics of the prevailing system, particularly the corporate sector, rather than on activities aimed at building alternative agrifood paradigms and initiatives. Researchers have examined the industrialization of agriculture (Welsh 1996), increasing corporate concentration and integration in food transportation, processing and retailing (Heffernan, Hendrickson, and Gronski 1999; Hendrickson et al. 2001; Mehegan 1999), and elite globalization[1] as it redefines international food-related economic development (McMichael 1996a; Perlas 2000; Shiva 2000). These analyses share a critical stance toward the corporate trajectory of the current agrifood system, based on deep concerns about ecological degradation, economic and political imbalances, and social and ethical issues.

In the 1990s researchers began to investigate more closely various social change activities that provided a response to these commodifying, concentrating, and globalizing forces. Employing such new conceptual frameworks as food citizenship (Welsh and MacRae 1998) and civic agriculture (Lyson 2000), this research has examined responses ranging from new food production paradigms (Kirschenmann 2003) through alternative fair trade markets (Raynolds 2000) to new food and agriculture policy proposals (Benbrook 2003).

We seek to add perspective to this developing literature on change within the modern agrifood system. We draw on selected tools from the social movements literature in order to improve understanding of the dynamics of social change in the modern agrifood system. We also propose two analytical frameworks to help conceptualize more clearly the multiplicity of change activities in the modern agrifood system. We intend that these frameworks be useful in assessing the prospects of these activities in creating significant change. Finally, we highlight other chapters of this book that provide concrete examples of the work of resistance, reconstruction, and connection.

The primary audience for this chapter are fellow social change workers, a growing base of citizens concerned with developments in the agrifood system, and academic colleagues who share our concerns and professional commitments.[2] We hope that the following exploration of social movement theory; goal orientations; and warrior, builder, and weaver work will help these and others with their work and life choices and sharpen our collective understanding of social change and food citizenship.

The first analytical framework considers the goals of change activities. We distinguish change efforts based on the degree to which their goals reflect one of three orientations toward the dominant food system. The three orientations are inclusion, reformation, and transformation. With inclusion the goal is to increase participation by marginalized players in the existing agrifood system. With reformation the goal is to alter operating guidelines of the existing agrifood system. With transformation the goal is to develop qualitatively different paradigms to guide the modern agrifood system.

The second analytical framework, which is the heart of our exploration, focuses on the strategic orientation of change activities in the modern agrifood system. We refer to these as warrior, builder, and weaver work. Warrior work consciously contests many of the corporate trajectories and operates primarily, but not exclusively, in the political sector. This is the work of resistance. Builder work seeks to create alternative food initiatives and models and operates primarily (and often less contentiously) in the economic sector. This is the work of reconstruction. Weaver work focuses on developing strategic and conceptual linkages within and between warrior and builder activities. It operates in the political and economic sectors but is particularly important in mobilizing civil society.[3] This is the work of connection. Warrior, builder, and weaver strategies can be employed in any of the three goal orientations.

And, while these three strategies differ in approach and methods, they are clearly interrelated and complementary, and all challenge business as usual (Evans 2000) in the current agrifood system.

Insights from Social Movement Theory

Social movements are consciously formed associations with the goal of bringing about change in social, economic, or political sectors through collective action and the mobilization of large numbers of people. Through social movements, informal networks of individuals, groups, and organizations that share a common belief about the nature of a problem work to bring attention to the problem and then propose and advocate solutions. Within the past decade many have said that the change activities emerging in the modern agrifood system comprise a social movement (Goodman 2000; Henderson 1998; Margaronis 1999; Rosset 2000). Whether or not this is so (and we will consider this question in the final section), we find that social movement literature helps us understand the nature, limitations, and potential of contemporary change-oriented activities related to food and agriculture.

According to social movement theorists McAdam, McCarthy, and Zald (1996), the power of social movements is determined significantly by the degree to which they effectively engage three interactive elements: framing processes, mobilizing structures, and political opportunities.

Framing Processes

Framing processes refer to the discourses—the shared meanings and definitions—that describe social problems, identify the causes of those problems, suggest solutions, and mobilize adherents to action (Snow and Benford 1988, 1992). The mobilizing capacity, or power, of a frame is a measure of how strongly it resonates with citizens and compels them to action. Adapting from Snow and Benford, we identify two factors that measure the power of a frame:

- *Empirical credibility (or objective importance)* describes the degree to which a frame is testable and verifiable and reflects the fundamental significance of the contention being framed to the everyday life of potential movement adherents. Agriculture and food clearly are associated with important biophysical and socioeconomic realities, including most obviously our dependence on food for survival and the direct relationship between food and human health. In addition, agriculture is the most important form of land use on a global scale (Buttel 1997), and household food ranks behind only transportation with regard to environmental impacts per household in the United States (Brower and Leon 1999).

- *Experiential resonance (or subjective relevance)* describes the degree to which a discourse (frame) corresponds to everyday life experience and meaning. No matter how objectively important an agrifood issue may be, if it does not resonate with a substantial proportion of the population, it will be difficult to mobilize much change activity. For example, the different wartime food experiences of European countries and the United States contribute to different assumptions and attitudes about protecting small farmers and the security of national food systems.

Several framing processes predominate within current change activities in the agrifood system. We identify four examples of prominent frames; they are not exhaustive, nor do they assume distinct boundaries. We separate them to highlight their different points of emphasis. The *environmental sustainability* frame focuses on the environmental impacts of agricultural production practices and includes farmland preservation and environmental consequences of biotechnology. Historically, and still in some circles, this focus constitutes the sustainable agriculture frame. The current sustainable agriculture frame as now understood by a growing number of farmers, politicians, and the public addresses environmental impacts but adds economic and equity concerns. It does not, however, address health, food safety, or equity issues for those besides farmers and farm workers.

A second frame, *economic justice for farmers*, focuses on domestic and global economic rights and social justice. It addresses the plight of family farmers, worldwide market inequities, global trade, and land tenure, among other issues, but it limits consideration to food and farming. A third frame, *community food security*, focuses on access to food by the disenfranchised. It centers on food security as well as equity and justice issues as aligned with food-related antipoverty concerns, predominantly from the point of view of food consumers. A final frame, *health and food safety*, focuses on nutrition and diet, the nature and effects of food processing, food-borne risks and illness, the role of governmental regulation, and advertising. Within this frame some of the concerns relate to agriculture and some do not.

Frames vary in comprehensiveness, with master frames being most inclusive. In contrast to more narrowly focused frames, a master frame provides a unifying message bringing together various subissues, organizations, and networks within a social movement. A master frame both specifies the main goals of the movement and encompasses perspectives of different movement organizations. As such, it has the greatest power to resonate with potential adherents to the cause and thus to mobilize many people to action. For instance, the master frame of civil rights effectively synthesized the contentions and solutions advocated by mobilized Afro-Americans and their supporters in the 1960s and 1970s (Gerlach and Hine 1970).

As we discuss below, it is of strategic importance whether those seeking change in the modern agrifood system can forge a master frame that both resonates and coalesces, or whether change initiatives proceed under diverse frames such as those described above. The degree to which various food system frames can align with the frames of other change movements—that is, sustainable human development, fair trade, or corporate accountability—also will be important.

Mobilizing Structures

Mobilizing structures refer to the particular forms that social movement organizations take and the tactics that they engage in order to communicate a message and to press for political change (McCarthy and Wolfson 1992). Social movement organizations craft methods to highlight a problem and its solutions and to advocate for change. Movement organizations provide a venue for individual participation in a cause. Mobilizing structures also refer to the organizational capacity available to social change groups. Within the agrifood arena there are hundreds of groups and networks. Many, particularly at the grassroots level, struggle for resources, technical capacity, and specific expertise. At the same time, the multiplicity and diversity of these groups provide abundant opportunities for entry, movement resilience, and testing various tactics such as public protest (for example, against the proposed organic rule), cyclical legislative reform (for instance, the federal farm bill), or innovative market experiments (for example, value-added co-ops).

Political Opportunities

Political opportunities refer to openings for change found within political structures and processes and the likelihood of exploiting these openings for the institutionalization of long-term structural change (McAdam 1996). These openings shape the timing and outcomes of social movement activities. To achieve long-lasting change, social movements must address the issue of political change and how the movement interfaces with existing political structures. Some agrifood system social movement organizations and change networks have been successful in capitalizing on political opportunities. For instance, they have used environmental legislation, as with the Clean Water Act, to address agricultural nonpoint source pollution. Further, they have capitalized on the discrimination lawsuit brought by southern African American farmers against the USDA to successfully establish a civil rights office within the department.

Goal and Strategic Orientations for Changing the Agrifood System

Our two analytical frameworks build on the work of researchers who have examined goal and strategic orientations of contemporary agrifood change activities. Allen and Sachs (1993) observed that the sustainable agriculture movement in the early 1990s remained silent on several important factors (hunger, farm labor, race, and gender) and failed to critically examine economic structures that subordinated environmental rationality and ethical priorities to market and short-term profit-making rationalities. More recently, Allen et al. (2003) examined thirty-seven alternative agrifood initiatives in California that ranged in time from the 1970s to the 1990s. Tracing historical changes, these researchers found a decided shift over thirty years from oppositional to alternative activities that were increasingly locally focused. Their terms "oppositional" and "alternative" parallel our concepts of warrior work and builder work. However, our definition of builder work contains considerably more change potential and is not necessarily "limited to incremental erosion at the edges of political-economic structures" (Allen et al. 2003, 61). Shreck employs a goals framework similar to ours when she evaluates banana fair trade in terms of the degree to which it engages "acts of resistance, redistributive action, or radical social action" (Shreck 2005, 18).

Hinrichs (2003), Marsden (2000), and Raynolds (2000) also consider the extent to which contemporary change activities in the agrifood system result in transformation. Additionally, our categories of warrior and builder can be likened to the "politics of protest" and the "politics of proposal" (Myhre 1994), "remedy" and "alternative" (McMichael 1996b), and "resistance to capital" and "alternatives to the private sector" (Gunn and Gunn 1991). Gould, Schnaiberg, and Weinberg (1996) have emphasized the need for weaver work in their investigation of local environmental movements. Perlas' areas of resistance to elite globalization have some similarity to our notions of warrior, builder, and weaver work (Perlas 2000).

Several common elements run through these and other commentaries. One stresses the importance of challenging those elements of the prevailing food system that drive human inequality and exploitation and that lead to waste or destruction of natural resources (Magdoff, Foster, and Buttel 2000; Mann and Lawrence 2001). Said another way, these challenges mean engaging what social psychologist James Hillman (1999) calls substantial things. In the biophysical realm substantial things mean the protection and/or restoration of basic ecological systems (Hawken, Lovins, and Lovins 1999). In the political-economic realm they include values such as justice, accountability, democracy, and cultural diversity.

Another element highlights the importance of social action in engaging in a "meaningful moral discourse" (Thompson 1998) and in recognizing that "ethical and spiritual principles" should underlie strategies to rebuild local, regional, and global food systems (Mann and Lawrence 2001). According to Thompson (1998), such a discourse is most likely to occur under conditions that are relatively local, open, and democratic and in which motivations are not predominantly utilitarian and short-term. Based on what we feel is a misplaced deification of capitalist markets (Cox 1999), much of the modern agrifood system presents a seriously flawed context for meaningful moral discourse, based as it is on distance, short-term business frames, undemocratic decision-making, and a utilitarian social psychology (Thompson 1998). Because we agree that agrifood system change must involve substantial things, we advocate strategic activities that engage a meaningful moral discourse toward goals that seek substantial reformation or transformation.

Goal Orientations

Social change activities in the modern agrifood system vary considerably in the degree to which their goals and objectives challenge fundamental dimensions of the dominant economic and political paradigms (DeLind 1993; Magdoff, Foster, and Buttel 2000). Ranging from niche marketing of boutique foods that presents few challenges to the dominant agrifood system through ecolabeling that provides information that shifts consumer decision-making beyond the concerns of traditional food marketing to forms of food-based associations seeking escape from the more negative aspects of market relationships and private land ownership, change activities in the modern agrifood system come with a mix of goals and objectives. Important ideological and strategic questions are embedded in this mixture, involving varied visions of a preferred agrifood system and differences in shorter-and longer-term change strategies. We hope to deepen understanding by constructing distinctions between the following goal orientations for changing the agrifood system.

1) *Inclusion (getting marginalized players into the agrifood system)*. Here the goal is to increase participation in the existing agrifood system. Examples include helping new immigrant farmers to enter the market and enabling low-income people to obtain healthy foods through established food channels.
2) *Reformation (changing the rules of the agrifood system)*. Here the goal is to alter operating guidelines of the existing agrifood system.

Examples include regulated farmers' markets at which sellers must produce what they sell, many of the rules that undergird fair trade relations (Raynolds 2000; Shreck 2005), and getting agrifood businesses such as MacDonald's or Unilever to change their corporate practices.

3) *Transformation (changing the agrifood system)* Here the goal is to develop qualitatively different paradigms for modern agrifood systems. Examples might include some community supported agricultures (CSA) in which farmers and eaters engage through genuinely other-than-market relationships, non-traditional farmland tenure through trusts or other community-based forms, or initiatives that posit food as a human right.

However, the slope from inclusion to transformation can be a slippery one, particularly when we try to distinguish between substantial reform and true transformation. Any such distinction is relative: it depends on perspective, time frame, and ultimate vision. For instance, limiting payments in USDA conservation programs based on farm size might be seen as a substantial step away from rewarding very large livestock operations. Yet green payment policies rewarding farmers for the ecological services their farming systems provide rather than the commodities they produce could be seen over time as transforming the very foundation of national farm policy.

Strategic Orientations

Warrior, builder, and weaver activities are strategically aimed at creating social change. These orientations are not mutually exclusive. A person or an organization is not exclusively a warrior, a builder, or a weaver. Typically, these activities are intermingled and complementary and necessarily so in order to achieve maximum impact. We include intellectual work that provides the analyses and conceptual bases for each of the three strategic orientations among the important kinds of warrior, builder, and weaver activities. Table 1 summarizes some characteristics of warrior, builder, and weaver work in the modern agrifood system.

Table 1. Characteristics of warrior, builder, and weaver work

	Warrior
Activity	Resisting the corporate food trajectory
Strategic orientation	*Resistance*; public protest and legislative work; often confrontational; draws attention to issues
Goals	Change political rules; protect territory; recruit adherents from civil society; confront or thwart economic concentration or unsustainable production practices
Main target	Political; civil society
Examples of actors	Situation-specific networks of organizations for public protests; policy advocates within/outside established political structures
Sustainability challenges	Difficult to sustain mass mobilization momentum; difficult to fund policy work
Link with civil society	Recruits adherents from civil society by drawing attention to the issue; mass mobilization for protest actions
Issues and/or types of organizations typically adopting the strategic orientation	Factory farming rBGH GMOs WTO/World Bank, IMF Farm workers' rights Farm Bill Organic rule Grape boycott

Builder	Weaver
Creating new agrifood initiatives and models	Developing strategic and conceptual linkages
Reconstruction; entrepreneurial economic activities building new collaborative structures; less confrontational than warrior strategies	*Connection*; linking warriors and builders; coalition building; communicating messages to civil society
Reconstruct economic sector to include such goals as sustainability, equity, healthfulness, regionality; work within established political structures to create alternative public policies	Build a food system change movement; engage members of civil society; create and strengthen coalitions within and beyond food system change communities
Economic; political	Civil society; political
Individual and collective economic enterprises; policy advocates; agricultural researchers and producers developing alternative production systems	Nonprofit and voluntary organizations and networks; university-based extension programs; movement professionals
New business, economic, and political models are fragile; new food enterprises often need some form of protection by government or civil society	Lack of resources for grassroots and other groups; difficult to maintain food issues resources
Requests that civil society protect alternative economic spaces through consumption choices or public policies	Serves linkage function for advocates and engaged actors within the public sphere; potential to provide vehicles for participation by less engaged members of civil society
Sustainable/organic farmers Intensive rotational grazing farmers/networks Farmers' markets, on-farm operations, delivery schemes Microenterprise development Farmers' marketing cooperatives Green payment farm policies	Local and regional nonprofit organizations Food policy councils Regional and national networks and organizations Land-grant university Extension programs

Warrior Work (Resistance)

Warrior work contests and challenges aspects of the prevailing agrifood system. While it focuses primarily, although not exclusively, on the political sector, its goal is to change both political and economic structures as well as the attitudes and beliefs of civil society. Warrior work is the characteristic resistance activity. It is interventionist in the sense that it assertively initiates change and pursues reform, as with, for example, efforts to legislate limits to corporate concentration within the agrifood arena. It also resists unacceptable proposals and defends political ground that has been gained, as for instance by resisting attempts to subvert the National Organic Rule. In this sense, warrior work often is "defensive" in seeking to protect "valued and vulnerable matters" from encroachment. For instance, resistance to confined animal feeding operations (CAFOs) may stem from concern for community, defending family farming may emerge from values about ownership and land tenure, resistance to genetically modified foods may involve protecting particular conceptions of nature and health, and efforts to improve the situation of farm labor or farm animals may draw from a defense of basic rights.

Warrior work seeks to put pressure on the political system but also to recruit and mobilize adherents to the cause. It is effective to the extent that significant numbers of people become engaged. Warrior work often makes use of high profile tactics, such as public demonstrations at the World Trade Organization (WTO) meetings or the destruction of research plots evaluating genetically modified crops. Such tactics draw attention to the issues, and also open space for others to participate through less confrontational means.

While the most visible warrior work is public protest, other, more "in the trenches" forms of warrior work, particularly legislative work, are equally important. In the agrifood arena numerous organizations engage in such warrior activity.[4] Other lower profile but crucial forms of warrior work include research and analysis that logically and articulately contest prevailing political or economic structures and processes. Warrior work also can operate in the economic sector. Examples include the grape boycotts organized by the United Farm Workers Union in the 1960s and opposition to the corporate pushers of infant formula in the developing world in the 1970s and 1980s.[5]

Builder Work (Reconstruction)

Builder work seeks to create alternative approaches and models in the agrifood system. Builder activities most fully express the reconstruction orientation to change. While examples exist in the political sector, such as legislation creating green payment farm policies, the majority of builder work in the modern agrifood system occurs in the economic sector through the creation of new food production and distribution initiatives and networks. Being entrepreneurial rather than political, builder work tends to be less contentious than warrior work. Many people and organizations engaged in builder work do not see their efforts as conscious resistance to the dominant agrifood system (Allen et al. 2003; Shreck 2005).

Builder activities occur at multiple levels, none of which is business as usual. One level involves formation of individual business enterprises that exhibit inclusionary, reformative, or transformative intent (such as new generation co-ops). Examples of inclusionary initiatives are efforts to support new immigrant farmers and farmers' markets that consciously reform the guidelines for market participation. At a larger level, builder work creates new models that engage whole sectors or systems. Examples include regional networks of CSA farms or efforts by organic grain farmers to create collective marketing structures at regional or national levels.[6] Builder work at the level of research for new agricultural production systems also is important. Clearly, transformative are systems being developed that could replace current fossil-fuel-dependent agricultural technologies with ecologically based food production systems that draw resources from complex biological interactions, plant and animal diversity, and perennial plantings (Kirschenmann 2003).

Nonetheless, builder work can be precarious. As shown in several chapters in this book, starting and sustaining new agrifood businesses can be difficult for several reasons. Foremost is a frequent lack of business expertise and financial resources for building viable new food enterprises. Another reason is the harshness of many conventional agricultural and food markets. Such conditions often lead people engaged in builder activity to search for niches in conventional markets where large, more powerful food companies present little competition. This approach is exemplified by the farmers' markets described in this volume's chapter by Gilbert Gillespie and his coauthors and more protected markets such as community supported agriculture, discussed in the chapter by Marcia Ostrom. Sometimes builder activity depends on warriors to create spaces through policy reform in which these new initiatives can emerge, as when state procurement laws are reformed to allow public institutions to preferentially purchase local food products.

Many of the builder activities explored in this book involve small, direct marketing enterprises and initiatives that fall at one end of an increasingly dualistic agrifood system (Buttel and LaRamee 1991). This dualism is characterized by growth at two ends of a spectrum: smaller-scale farming and food enterprises that are successfully developing direct marketing relationships with consumers and very large farms entering into contractual supply chains with consolidated food firms that move bulk agricultural commodities around the world. However, farms and food enterprises that fall between these two very different extremes are increasingly threatened. Referred to as the "agriculture-of-the-middle," these family-based farms, together with the social and environmental benefits they provide, will likely disappear in the next decade or two if present trends continue (Kirschenmann et al. 2003). Attention to reconstruction work in the middle is beginning (Hamilton 2003; Stevenson and Born, this volume). Such builder work will likely engage new production, marketing, and policy approaches.[7]

Weaver Work (Connection)

Weaver work focuses on creating linkages that support activities promoting change. It develops networks and coalitions among groups engaged in builder and warrior work. Weaver work is performed by a wide range of players, from community-based organizations and grassroots coalitions to collaborations among food system academics and practitioners who explore new concepts and develop research and analysis. This book represents one effort to distill and disseminate such conceptual weaver work. Of the three change activities, weaver work is most explicitly oriented toward movement building. It focuses largely, but not exclusively, on civil society through outreach and organizing activities.

Weaver work takes on several tactical orientations. Intrasectoral linkages forge connections among groups in a given agrifood interest area, such as by organizing producer cooperatives into a federation to strengthen capacity and market position or by maintaining a list serve for a network of CSA farms. Intersectoral linkages connect different agrifood interests and groups with complementary agendas such as by linking farm groups with environmental groups on the issue of farmland preservation or by linking nutritionists with "buy local" campaign organizers. In their chapter in this book Kate Clancy and her coauthors explore food policy councils that serve as institutionalized forms of weaver work bringing together diverse public and private organizations having agrifood interests.

Weaver work creates horizontal linkages based on space and locality by facilitating alliances across agrifood work and complementary social change efforts within a bounded area or "foodshed" (Kloppenburg, Hendrickson, and Stevenson 1996). Vertical linkages in weaver work involve strategic connections between structural, geographic, or analytical levels. They strategically enlarge the spatial or institutional scope in which contested issues are negotiated. Our use of the terms horizontal and vertical linkages is similar to Johnston and Baker's 2003 notions of "scaling out" and "scaling up" as related to community food security work. Scaling out involves enlarging the reach and impact of successful local food security programs by multiplying them in other geographical communities. Scaling up involves engaging policies at higher institutional levels like state/provincial or national governments at which levels the underlying structural causes of inequality and food insecurity must be addressed. What emerges from combining these two kinds of strategic linkages is a promising "multi-scaled food politics" (Johnston and Baker 2003).

Integrating Warrior, Builder, and Weaver Work in Response to Elite Globalization

The term elite globalization refers to a post–World War II neocolonial strategy for global economic integration powered by and primarily benefiting a small set of transnational corporations and the wealthier industrial nations (Perlas 2000). Among the primary institutions employed by these beneficiaries has been the WTO, begun in 1995 as the replacement for the General Agreement on Tariffs and Trade (GATT). In the agrifood sector WTO delegates representing agribusiness interests in the United States and Europe have pushed through trading rules that require developing countries to remove barriers to agricultural imports. Many of these imports are agricultural commodities like grains or cotton that U.S.-and European-based corporations dump on the market at prices below the cost of production. Such artificially cheap agricultural goods drive down local prices and put severe economic pressure on many indigenous farmers. Coupled with a prohibition on government policies to subsidize or protect local farmers, these conditions push developing countries to compete in a global economy based on low wages, cheap production costs, and weak environmental laws (Rossett 2000; Shiva 2000).

U.S. agricultural policy since the institution of the WTO and the Farm Bill of 1996 has been an example of "hypocrisy and double-speak" (Ray 2003). While pressuring developing countries to reduce domestic agricultural supports, so-called emergency subsidies to agricultural commodity producers in the United States have skyrocketed, driven by the low market prices.[8] Despite these subsidies, many U.S. farmers are not better off. This is particularly true for diversified, independent, owner-operated (family) farms (Ray 2003). The primary beneficiaries of these taxpayer-funded subsidies are corporate grain traders and livestock producers who have access to agricultural commodities at below the cost of production, enabling them to consolidate further their control over the entire food production and marketing chain (Ray 2003).

Since the 1990s a growing number of people and organizations worldwide have responded to this recipe for exporting poverty internationally and jeopardizing the U.S. domestic base of diversified farms. Beginning with the third ministerial meeting of the WTO in Seattle in 1999, an increasingly sophisticated coalition of "food citizens" has challenged both the content of the rules emerging from these negotiations and the undemocratic processes involved in producing them. Illustrating how warrior, builder, and weaver work can be combined effectively, we highlight the activities of civil society organizations at the Fifth Ministerial of the WTO held in Cancun, Mexico, during September 2003. We focus on the activities of a key U.S.-based nonprofit organization, the Institute for Agriculture and Trade Policy (IATP).[9]

Though portrayed in the conventional media as a failure because they closed prematurely and with no new trade agreements, the WTO meetings in Cancun were described by IATP's founder Mark Ritchie as "one of the most successful international meetings in years because it redefined how trade can benefit the poor and how the developing world can be real players in these negotiations" (Ritchie 2003a). According to Ritchie, Cancun demonstrated that

1. equitable and effective global trade agreements cannot be negotiated when the balance of power rests exclusively with the wealthiest nations;
2. civil society has a legitimate and useful role in the discussions;
3. fair trade—trade that ensures that producers are paid a fair price and workers are paid fair wages—is the world's best hope for a sustainable trading environment (Ritchie 2003a).

Standing behind these lessons and successes are examples of creative warrior, builder, and weaver work.

Warrior work performed by the IATP and other civil society groups consisted of public demonstrations outside the Cancun meeting halls and extensive lobbying of delegates inside the ministerial. Particularly important was the intellectual warrior work of providing negotiators from the developing countries with analyses, technical information, and advice to counter initiatives by the wealthier nations. The IATP distributed a series of white papers focusing on key issues for agricultural trade, in particular highlighting the issue of agricultural dumping by U.S. and European agribusinesses.[10] In criticizing the damage that U.S. farm policy inflicts on Third World farmers, negotiators from the developing countries drew and quoted extensively from the paper on dumping (Richie 2003b). Warrior work at the Cancun ministerial also targeted the historically secretive and undemocratic nature of trade negotiations under the GATT and the WTO. Previously much negotiation took place informally among the wealthier nations with no record of discussions. Statements by the developing countries and civil society groups at the closing of the Cancun meetings made it clear that future WTO negotiations must be conducted with "clear rules, procedures, transparency and accountability" (Ritchie 2003b).

The primary builder work at the Cancun ministerial involved providing a visible and alternative model for world trade: the fair trade model described earlier in this chapter (Jaffee, Kloppenburg, and Monroy 2004; Raynolds 2000). Organized and sited within blocks of the meeting hall, a Fair Trade Fair featured more than one hundred fair trade producer groups offering fair trade items from soccer balls to coffee. WTO negotiators, nongovernmental organization delegates, and Mexican farmers and citizens attended the fair (IATP News 2003). Beyond this event the IATP facilitated several day-long discussions on trade issues that emphasized fair trade in the Americas. According to the organizers, this builder work succeeded in raising the fair trade movement to a new level of global awareness and successfully engaged trade negotiators in a dialogue over what fair trade can offer as a model for good trade rules (Ritchie 2003b).

Standing behind the warrior and builder work at Cancun was a series of coalitions made effective by skilled weaver work. The highly successful Fair Trade Fair, for example, was the product of a coalition of civil society groups from Mexico, Canada, and Switzerland, as well as participation by the IATP and Oxfam International (IATP News 2003). The shift in power away from the wealthier nations that occurred at the Fifth Ministerial resulted from the linkages and coalitions between civil society groups and developing countries begun at the Fourth Ministerial in Doha, Qatar, which matured in Cancun (Ritchie 2003b). Particularly creative weaver work was the establishment in Cancun of an Internet radio station, Radio Cancun, that provided full and up-to-date coverage of events both inside and outside the meeting hall. Linked with a range of radio and media sources worldwide, Radio Cancun provided listening and educational opportunities in the United States, Mexico, Canada, and Europe (IATP News 2003).[11]

An Agenda for Changing the Food System

Achieving substantial reformation or transformation in the contemporary agrifood system remains difficult. Transforming the system must be a long-term goal. Prevailing economic and political structures are deeply institutionalized and often provide only limited space for meaningful moral discourse. Yet such spaces are being created and widened by a spectrum of warrior, builder, and weaver initiatives. Warrior work can expose system dysfunctions and make room for builder work to explore and create models that can develop into viable alternatives. These spaces can be further developed through strategic coalitions with partners both inside and outside the community seeking to change the agrifood system.

In order to approach more substantial change, four areas stand out for us as important for building on work already undertaken. As change advocates, we must (1) strengthen our analyses of the prevailing food system and develop processes for constructing compelling alternative visions, (2) challenge or replace corporate-dominated market structures, (3) continue to translate our successes into changed public policies, and (4) build toward mobilizing master frames.

To Strengthen Our Analyses and to Construct Compelling Visions

We agree with Benbrook's (2003, 2006) observation that a key obstacle to substantial change is "a lack of consensus and clarity on what is wrong with the American food system and what steps are needed to make things 'right.'" Several authors from our community have begun fruitful analyses to understand the structural causes of problems in the prevailing agrifood system (see Bonanno et al. 1994; Magdoff, Foster, and Buttel 2000; McMichael 1996a, 1996b). Warrior work needs such critical analyses in order to identify openings for resistance, reconstruction, and connection.

Similarly, we contend that the community concerned with agrifood system change needs to develop proactive and shared visions of what should be (Allen and Sachs 1993) and a firm agreement on the steps necessary to make things right. Important work has begun to develop frameworks for preferred food systems (Kloppenburg et al. 2000; Lang 1999; Welsh and MacRae 1998). These shared visions are essential in order to produce master frames with sufficient mobilizing capacity. Their absence results from the multiplicity of issues and groups working on agrifood systems change and the multiple frames employed. Sustained builder and weaver work are required to foster dialogue toward unifying and mobilizing visions. Finally, we agree with those authors who contend that commitment to such fundamental social processes as democratic participation, organizational accountability, and philosophical as well as political pragmatism is critical to the achievement of compelling alternative visions (Hassanein 2003).

Challenge or Replace Corporate-Dominated Market Structures

We encourage the developing critique of contemporary market structures and their domination by concentrated agrifood corporations. Significant change will require combinations of challenging, reforming, and creating alternatives to these market structures. We describe three current expressions of such work. One is clearly warrior work and the others are more builder-oriented. All are relatively new and their potential for growing beyond the margins of the prevailing agrifood system remains to be developed.

The warrior work involves active challenges to the growing corporate concentration and the resulting domination and destruction of genuinely competitive markets for many agricultural commodities. Supported by a few academic voices and performed by several active groups,[12] this work involves direct court challenges to corporate mergers and takeovers in the agrifood industry as well as calls for active, innovative, long-term antioligopsony public policies (Benbrook 2003; Cochrane 2003; Stumo 2000).[13] Such warrior work needs to be supported and expanded.

Two examples of builder work involve reformative and transformative goals regarding market structures.[14] Food-based fair trade initiatives seek to reform markets by reembedding commodity circuits within ecological and social relations, thus challenging the dominance of conventional price relations in guiding production and trade conditions (Jaffee, Kloppenburg, and Monroy 2004; Raynolds 2000; Shreck 2005). Some CSA models seek to operate outside of the market paradigm altogether. Framed as "associative economies," food producers and consumers do not assume the inevitably conflicting market roles of sellers and buyers but rather see themselves as members of a community (or association) who make negotiated, collective decisions about meeting the needs of both farmers and eaters (Groh and McFadden 1997; Lamb 1996, 1997). We recommend that these and other alternative economic models be strongly encouraged and rigorously evaluated.

Extend Successes Through Public Policy Change

As important as work on market reforms and on alternative markets is, we remind ourselves that long-term sustaining change in the modern food system must be anchored by changes in public policies (Buttel 1997; Lang 1999). Such changes require moving beyond the realm of individual consumer choices in the marketplace to the realm of citizen politics in which people make positive collective decisions about the larger nature of their food system (Lang 1999; Raynolds 2000; Welsh and MacRae 1998). This need is particularly important because critical issues in the agrifood system, such as environmental impacts, farmland preservation and tenure, farm worker rights, and agribusiness concentration, are not adequately addressed by market solutions. Additionally, substantial policy changes will need to be what Benbrook (2003) terms systemic—addressing how capital, income, and other resource streams flow through the entire food system. Optimally, such policy changes also will reinforce one another, creating much greater momentum than through any single policy reform or new program (Benbrook 2003).

Build Toward Mobilizing Master Frames

Frames have been described as discourses that define social problems, identify causes of the problems, suggest solutions, and mobilize people to action. Master frames are more comprehensive, bringing together various issues and points of view. We agree with Buttel (1997, 353) that there is currently "no one unifying notion that can serve as a singular unifying focus" for the change efforts now engaged in the modern agrifood system. In other words, no coherent master frames for change initiatives in the modern agrifood system presently exist. Although there is a diversity of agrifood movements, this diversity does not mean that these groups comprise a social movement. Is this observation important? On the one hand, what we label our work is less important than understanding the strategic implications of the change activities we choose. This diversity is an asset in offering multiple doors for potential movement adherents to enter and thus resilience in facing opposing forces. On the other hand, it is not clear whether master frames with sufficient mobilizing capacity will emerge to unify the substantial diversity of issues and views within the broader community working to change the agrifood system. The absence of master frames hampers our collective efforts to effectively mobilize large numbers of people toward a unifying vision or goal.

Accepting (and acknowledging the strengths of) the diversity within the food systems change community (Hassenein 2003), we further agree with Buttel (1997, 353) that "it will only be through coalitions . . . that this social movement force can achieve the extent of meaningful impacts that are required to address" the fundamental social, political, and economic issues. We see two strategic options to push coalitions toward compelling master frames. The first involves the weaver work of creating and strengthening linkages among sectors *within* the agrifood system change community. For example, organizations associated with community food security are framing their issues in ways that create coalitions with food relocalization groups around farm-to-school programs and the acceptance of WIC vouchers at farmers' markets.[15] As a frame, sustainable agriculture has widened to enable coalitions among environmental and community development groups operating with important new concepts like civic agriculture, as discussed in Thomas Lyson's chapter in this volume.[16] However, while many agrifood system activists believe that "food and agriculture" provide a resonant master frame, there is no empirical evidence to support this. As long as food remains relatively safe, convenient, and cheap (and with its externalities not fully considered), it is likely that the claims of many agrifood activists will lack experiential resonance and remain below the emotional radar of most U.S. citizens.

The second arena for conceptual connections and coalition building involves linkages *between* the food system change community and change communities focusing on other potentially synergistic issues. These intersectoral linkages and coalitions are likely to be the most empowering of all. As discussed earlier, the contemporary anti-WTO coalition is being framed to support a powerful combination of civil society groups, including environmental, labor, antiglobalization, social justice, and family-farming organizations. This combination of interests could coalesce around a sustainable human development master frame (Ritchie 2003a). Obesity and other diet-related issues provide conceptual and potentially coalitional linkages between the agrifood sector and progressive currents in the public health community, raising the potential for health to develop into a master frame that encompasses multiple dimensions of environmental, biomedical, and sociocultural well-being (Benbrook 2003; MacRae 1997; Nestle 2002; Pollan 2003).

Such examples point to the considerable potential for agrifood system issues to serve as powerful frames of example, or frames of entry, into such larger nonfood master frames as fair trade, sustainable human development, or corporate accountability. With this orientation agrifood system change advocates will add value to other synergistic movements—joining others under a bigger tent with substantial mobilizing capacity. Ultimately, the social change we seek is about democracy, sustainability, equity, and justice. Food and agriculture point to one door; we work here to achieve these important ends.

Enhancing Warrior, Builder, and Weaver Work

The modern agrifood system is complex. Its far-reaching social, political, and economic effects have few analogues in other sectors. While various social change gains have occurred in the agrifood arena, most within our community would agree that we need to deepen and broaden our analyses, our vision, and our base of civic engagement within the context of shifts in the roles of government, the economy, and civil society.

Warrior work of the publicly confrontational variety requires a compelling, mobilizing frame and high-profile activity in order to attract adherents and to be effective. Warrior work of the low-profile trench work variety exerts pressure for change less noticeably and attracts considerably fewer adherents. Both types of warrior activity are essential to contest issues, to defend ground, and to galvanize participation. At a minimum we believe that productive warrior work in the near future should focus on a continuing strong resistance to elite globalization, including questioning the WTO agenda, an increased campaign against corporate agribusiness concentration, and a broad critique of current federal farm policy. Such warrior work could open up spaces for builder work in the areas of fair trade, competitive and alternative markets, and farm policies that would support moderately scaled, sustainable agriculture.

Builder work is an especially promising arena for activities to change the agrifood system because it can be applied at multiple levels and in multiple sectors. In part, because it is positive, less contentious, and more accessible to more people, builder work can succeed in attracting and mobilizing adherents. As the accounts of builder efforts described in other chapters of this book show, most such activity currently occurs at the local level. We believe it will be important to use locally learned lessons to both scale out geographically and scale up institutionally. Examples of scaling out include multiplying local food policy councils, farm-to-institution programs, and forms of urban agriculture. Examples of scaling up include establishing food policy councils at the state level and constructing alternative food-value-chains for an agriculture-of-the-middle.[17] We also argue that researching and designing new farming systems that shift from reliance on petroleum-based inputs to biologically based interactions will be important builder work in the agricultural research and production sector.

Weaver work involves social and conceptual organization. Framing is particularly important for conceptualizing and organizing change movements. We need to develop frames that strike a strategic balance between visions of agrifood systems that are too narrowly framed to attract significant numbers of people and frames that are too broad and abstract to resonate with specific sectors. Examples of narrower frames are food security or natural resource conservation while broader ones include democracy, sustainability, equity, and justice. Such overarching principles are important but not sufficient to galvanize people into action.

Regarding social organization, we reemphasize our earlier discussion about the importance of building coalitions, both within the community concerned with agrifood system change and with other change communities. Because of the complexity in mapping issues and groups important to agrifood systems change, weaver work must assess what role(s) various actors and concepts should play in the greater set of worldwide social change efforts. Given that transformation in the agrifood system cannot be accomplished independently of significant change in other societal sectors, it ultimately may be more effective to forge strategic linkages under broader or differently framed concerns—for example, linking food security with living wages, agribusiness domination with corporate concentration, or domestic food access with global food sovereignty issues. Finally, as a way of reaffirming the long-term nature of our work, it will be important for weaver work to focus significant energy on movement base building, particularly by creating a strong, vibrant tapestry of youth and new leadership to sustain change efforts over the long haul in the agrifood system.

Notes

1. As described by Perlas, "elite globalization" is a form of global economic integration powered by and primarily benefiting transnational corporations and the wealthier industrial nations. Based on neoliberal economic theories that favor international market forces over national governmental policies, the primary institutions carrying out elite globalization are the World Bank, the International Monetary Fund, and the World Trade Organization (Perlas 2000, 21–30).

2. Social movement theorists use the term "publics" to characterize such consciously concerned citizens (Emirbayer and Sheller 1998).

3. According to social movement theorists, "civil society" is comprised of social associations that fall outside either the economic or political sectors (Casquette 1996). Civil society provides free spaces for the formation of oppositional subcultures, and a strong civil society is viewed as a precursor to social movement formation (Emirbayer and Sheller 1998).

4. Examples include the National Campaign for Sustainable Agriculture and the Sustainable Agriculture Coalition.

5. For descriptions of the grape boycott and resistance to exporting infant formula to developing countries, see http://sunsite3.berkeley.edu/calheritage/ufw/ and http://www.babymilkaction.org/pages/boycott.html (both sites last accessed April 9, 2006).

6. For information on the Madison Area Community Supported Agriculture Coalition, see http://www.macsac.org/. For information on Ofarm, the common marketing structure of organic grains, see http://www.ofarm.org/ (both sites last accessed April 9, 2006).

7. For the goals of a national task force to address issues of renewing an agriculture-of-the-middle in the United States, see http://www.agofthemiddle.org/ (last accessed April 9, 2006).

8. By 2002 U.S. subsidies for eight commodity crops reached nearly $20 billion per year, with the bulk of these payments going to less than 10 percent of the largest farmers in program commodities such as corn, soybeans, cotton, and rice (Ray 2003).

9. For background on the IATP, see http://www.iatp.org/ (last accessed April 9, 2006).

10. The six agriculturally oriented white papers can be found at http://www.tradeobservatory.org/ (last accessed April 9, 2006).

11. For Radio Cancun, see http://www.radiocancun.org/home.cfm (last accessed May 7, 2006).

12. Examples include the Organization for Competitive Markets and the Agribusiness Accountability Initiative.

13. For further information on the warrior work done by the Organization for Competitive Markets and the Agribusiness Accountability Initiative, see http://www.competitivemarkets.com and http://www.agribusinessaccountability.org (both sites last accessed May 9, 2006).

14. For a useful discussion of the distinction between reformative and transformative strategies related to direct agricultural markets, see Hinrichs (2000).

15. See http://www.foodsecurity.org (last accessed April 9, 2006).

16. For an example of an interesting coalition between agricultural and environmental groups, see the Web site for an ecolabel collaboration between the Wisconsin Potato Growers Association and the World Wildlife Fund: http://ipcm.wisc.edu/bioipm/ (last accessed April 9, 2006).

17. For an intriguing analysis that finds regional food systems more efficient than either national/international or local ones when measured by miles that food travels, see Pirog et al. 2001. For a discussion of state-level food policy councils, see Hamilton 2002.

References

Allen, P., M. Fitzsimmons, M. Goodman, and K. Warner. 2003. "Shifting plates in the agrifood landscape: The tectonics of alternative agrifood initiatives in California." *Journal of Rural Studies* 19:61–75.

Allen, P., and C. Sachs. 1993. "Sustainable Agriculture in the United States: Engagements, Silences, and Possibilities for Transformation." In *Food for the Future: Conditions and Contradictions of Sustainability*, edited by P. Allen, 139–67. New York: Wiley.

Benbrook, C. 2003. "What Will It Take to Change the American Food System?" Paper presented at the W. K. Kellogg Foundation's Food and Society Networking Conference. http://www.biotech-info.net/kellogg.pdf (last accessed April; 9, 2006).

Bonanno, A., L. Busch, W. Friedland, L. Gouveia, and E. Mingione, eds. 1994. *From Columbus to Con Agra: The Globalization of Agriculture and Food*. Lawrence: University of Kansas Press.

Brower, M., and W. W. Leon. 1999. *The Consumer's Guide to Effective Environmental Choices: Practical Advice from the Union of Concerned Scientists.* New York: Three Rivers.

Buttel, F. 1997. "Some Observations on Agro-food Change and the Future of Agricultural Sustainability Movements." In *Globalizing Food: Agrarian Questions and Global Restructuring*, edited by D. Goodman and M. Watts, 344–65. New York: Routledge.

Buttel, F., and P. LaRamee. 1991. "The Disappearing Middle: A Sociological Perspective." In *Toward a New Political Economy of Agriculture*, edited by W. Friedland, L. Busch, F. Buttel, and A. Rudy, 151–69. Boulder CO: Westview.

Casquette, J. 1996. "The sociopolitical context of mobilization: The case of the antimilitary movement in Basque Country." *Mobilization* 1(2): 203–17.

Cochrane, W. 2003. "A food and agricultural policy for the 21st century." http://www.agobservatory.org/library.cfm?refid=29732 (last accessed April 9, 2006).

Cox, H. 1999. "The market as God," *Atlantic Monthly*, March.

DeLind, L. 1993. "Market niches, 'cul de sacs,' and social context: Alternative systems of food production." *Culture and Agriculture* 47 (Fall): 7–12.

Emirbayer, M., and M. Sheller. 1998. "Publics in history." *Theory and Society* 27:727–79.

Evans, P. 2000. "Fighting marginalization with transnational networks: Counter-hegemonic globalization." *Contemporary Sociology* 29:230–41.

Gerlach, L., and V. Hine. 1970. *People, Power, and Change: Movements of Social Transformation.* Indianapolis: Bobbs-Merrill.

Goodman, D. 2000. "Organic and conventional agriculture: Materializing discourse and agro-ecological managerialism." *Agriculture and Human Values* 17:215–19.

Gould, K., A. Schnaiberg, and A. Weinberg. 1996. *Local Environmental Struggles: Citizen Activism in the Treadmill of Production.* Cambridge UK: Cambridge University Press.

Groh, T., and S. McFadden. 1997. *Farms of Tomorrow Revisited: Community Supported Farms, Farm Supported Communities.* Kimberton PA: Biodynamic Farming and Gardening Association.

Gunn, C., and H. Gunn. 1991. *Reclaiming Capital: Democratic Initiatives and Community Development.* Ithaca NY: Cornell University Press

Hamilton, H. 2003. "Sustainable agriculture for midsized farms," *Choices*, guest editorial. Third quarter. http://www.choicesmagazine.org/2003-3/2003-3-08.htm (last accessed April 9, 2006).

Hamilton, N. D. 2002. "Putting a face on our food: How state and local food policies can promote the new agriculture." *Drake Journal of Agricultural Law* 7 (20): 408–54.

Hassanein, N. 2003. "Practicing food democracy: A pragmatic politics of transformation." *Journal of Rural Studies* 19:77–86.

Hawken, P., A. Lovins, and L. H. Lovins. 1999. *Natural Capitalism.* Boston: Little, Brown.

Heffernan, W., M. Hendrickson, and R. Gronski. 1999. *Consolidation in the Food and Agriculture System*. Research Report. Washington DC: Farmers Union.

Henderson, E. 1998. "Rebuilding local food systems from the grassroots up." *Monthly Review* 50:112–24.

Hendrickson, M., W. Heffernan, P. Howard, and J. Heffernan. 2001. "Consolidation in food retailing and dairy." *British Food Journal* 3 (10): 715–28.

Hillman, J. 1999. *The Force of Character*. New York: Random House.

Hinrichs, C. C. 2000. "Embeddedness and local food systems: Notes on two types of direct agricultural market." *Journal of Rural Studies* 16:295–303.

———. 2003. "The practice and politics of food system localization." *Journal of Rural Studies* 19:33–45.

IATP News. 2003. Vol. 7, no. 4. http://lists.iatp.org/listarchive/archive.cfm?id=84052 (last accessed May 7, 2006).

Jaffee, D., J. Kloppenburg, and M. Monroy. 2004. "Bringing the 'moral charge' home: Fair trade within the North and within the South." *Rural Sociology* 69 (2):169–96.

Johnston, J., and L. Baker. 2003. "Eating outside the box: FoodShare's good food box and the challenge of scale." *Agriculture and Human Values* 22:313–25.

Kirschenmann, F. 2003. "New Seeds and Breeds for a New Revolution in Agriculture." Paper presented at the New Seeds and Breeds Conference, Washington DC.

Kirschenmann, F., G. W. Stevenson, F. Buttel, T. Lyson, and M. Duffy. 2003. "Why Worry About the Agriculture of the Middle?" A white paper for the Agriculture of the Middle Project. http://www.agofthemiddle.org/archives/2005/08/why-worry-about.html (last accessed April 9, 2006).

Kloppenburg, J., Jr., J. Hendrickson, and G. W. Stevenson. 1996. "Coming in to the foodshed." *Agriculture and Human Values* 13:33–42.

Kloppenburg, J., Jr., S. Lezberg, K. DeMaster, G. W. Stevenson, and J. Hendrickson. 2000. "Tasting food, tasting sustainability: Defining attributes of an alternative food system with competent, ordinary people." *Human Organization* 59:177–86

Lamb, G. 1996. "Going beyond self-interest in economic life," part I. *Threefold Review* 13 (Winter/Spring): 17–26.

———. 1997. "Going beyond self-interest in economic life," part II. *Threefold Review* 15 (Summer/Fall): 6–8.

Lang, T. 1999. "Food Policy for the 21st Century: Can it be Both Radical and Reasonable?" In *For Hunger-proof Cities: Sustainable Urban Food Systems*, edited by M. Koc, R. MacRae, L. J. A. Mougeot, and J. Welsh, 216–224. Ottawa: IDRC.

Lyson, T. 2000. "Moving toward civic agriculture." *Choices* 15 (3): 42–45.

MacRae, R. 1997. "If the Healthcare System Believed 'You are What You Eat': Strategies to Integrate Our Food and Health Systems." Toronto Food Policy Council Discussion Paper. Ser. 3.

Magdoff, F., J. B. Foster, and F. H. Buttel, eds. 2000. *Hungry for Profit: The Agribusness Threat to Farmers, Food, and the Environment*. New York: Monthly Review.

Mann, P., and K. Lawrence. 2001. "Rebuilding our Food System: The Ethical and Spiritual Challenge." Inquiry In Action. No. 31. Madison WI: Consortium for Sustainable Agriculture Research and Education.

Margaronis, M. 1999. "The politics of food." *The Nation*. December 17.

Marsden, T. 2000. "Food matters and the matter of food: Toward a new food governance?" *Sociologica Ruralis* 40 (1): 20–29.

McAdam, D. 1996. "Introduction: Opportunities, Mobilizing Structures, and Framing Processes—Toward a Synthetic, Comparative Perspective on Social Movements." In *Comparative Perspectives on Social Movements: Political Opportunities, Mobilizing Structures, and Cultural Framings*, edited by D. McAdam, J. McCarthy, and M. Zald, 1–20. Cambridge UK: Cambridge University Press.

McAdam, D., J. McCarthy, and M. Zald. 1996. *Comparative Perspectives on Social Movements: Political Opportunities, Mobilizing Structures, and Cultural Framings*. Cambridge UK: Cambridge University Press.

McCarthy, J., and M. Wolfson. 1992. "Consensus Movements, Conflict Movements, and the Cooptation of Civic and State Infrastructures." In *Frontiers in Social Movement Theory*, edited by A. Morris and C. Mueller, 273–97. New Haven CT: Yale University Press

McMichael, P. 1996a. "Globalization: Myths and realities." *Rural Sociology* 61(1): 26–55.

———. 1996b. *Development and Social Change: A Global Perspective*. Thousand Oaks CA: Pine Forge.

Mehegan, S. 1999. "Merger mania: Consolidation changes the face of the North American supermarket sector." *Meat and Poultry*, September.

Myhre, D. 1994. "The Politics of Globalization in Rural Mexico: Campesino Initiatives to Restructure the Agricultural Credit System." In *The Global Restructuring of Agro-food Systems*, edited by P. McMichael, 145–67. Ithaca NY: Cornell University Press.

Nestle, M. 2002. *Food Politics: How the Food Industry Influences Nutrition and Health*. Berkeley CA: University of California Press.

Perlas, N. 2000. *Shaping Globalization: Civil Society, Cultural Power, and Threefolding*. Saratoga Springs NY: Center for Alternative Development Initiatives and GlobeNet3.

Pirog, R., T. van Pelt, K. Enshayan, and E. Cook. 2001. *Food, Fuel, and Freeways: An Iowa Perspective on How Far Food Travels, Fuel Usage, and Greenhouse Gas Emissions*. Ames IA: Leopold Center for Sustainable Agriculture. http://www.leopold.iastate.edu/pubs/staff/ppp/index.htm (last accessed May 6, 2006).

Pollan, M. 2003. "The (Agri)Cultural Contradictions of Obesity." *New York Times Magazine*. October 12. http://www.nytimes.com/2003/10/12/magazine/12wwln.html (last accessed April 9, 2006).

Ray, D. 2003. *Rethinking U.S. Agricultural Policy: Changing Course to Secure Farmer Livelihoods World Wide.* http://www.agpolicy.org/blueprint.html (last accessed April 9, 2006).

Raynolds, L. 2000. "Re-embedding global agriculture: The international organic and fair trade movements." *Agriculture and Human Values* 17:297–309.

Ritchie, M. 2003a. "A new beginning for the WTO after Cancun." IATP *News* 7 (4). http://lists.iatp.org/listarchive/archive.cfm?id=84052 (last accessed May 5, 2006).

———. 2003b. "Report from Cancun—Fair Trade and the WTO Ministerial." Institute for Agriculture and Trade Policy. http:/www.tradeobservatory.org/showFile.php?id=3015 (last accessed May 4, 2006).

Rosset, P. 2000. "New food movement comes of age in Seattle." *Community Food Security News*: 1,4.

Shiva, V. 2000. *Stolen Harvest: The Hijacking of the Global Food Supply.* Cambridge MA: South End.

Shreck, A. 2005. "Resistance, redistribution, and power in the Fair Trade Banana Initiative." *Agriculture and Human Values* 22 (1): 17–29.

Snow, D., and R. Benford. 1988. "Ideology, frame resonance, and participant mobilization." *International Social Movement Research* 1:197–217.

———. 1992. "Master Frames and Cycles of Protest." In *Frontiers in Social Movement Theory*, edited by A. Morris and C. Mueller, 133–55. New Haven CT: Yale University Press.

Stumo, M., ed. 2000. *A Food and Agriculture Policy for the 21st Century.* Lincoln NE: Organization for Competitive Markets.

Thompson, P. 1998. *Agricultural Ethics: Research, Teaching, and Public Policy.* Ames IA: Iowa State Press.

Welsh, J., and R. MacRae. 1998. "Food citizenship and community food security: Lessons from Toronto, Canada." *Canadian Journal of Development Studies* 19:237–55.

Welsh, R. 1996. *The Industrial Reorganization of U.S. Agriculture.* Policy Studies Report. No. 6. Greenbelt MD: Henry A. Wallace Institute for Alternative Agriculture.

Part II
Institutions and Practices to Remake the Food System

3. Farmers' Markets as Keystones in Rebuilding Local and Regional Food Systems

Gilbert Gillespie, Duncan L. Hilchey, C. Clare Hinrichs, and Gail Feenstra

As the opening bell rings at a farmers' market, the first customers stream in, buying tomatoes, baked goods, fresh peaches, farmhouse cheeses, and cut flowers from their preferred vendors. Particularly in summer, scenes much like this one occur in small towns and large cities throughout North America. As places where producers sell food directly to the consumers who will eat it, farmers' markets are community institutions with long histories (Tangires 2003). In the early to mid-twentieth century they declined as innovations such as refrigeration and national supermarket chains took root. However, by the 1960s and 70s farmers' markets began to reappear in a new social climate marked by interest in the healthfulness and freshness of foods. In the United States passage of the Farmer-to-Consumer Direct Marketing Act of 1976 directly supported the reemergence of farmers' markets (Hamilton 2002).

Farmers' markets seem to offer something for nearly everyone. Producers may look to farmers' markets as a profitable alternative to the low prices of commodity markets in an industrial agricultural system. Consumers seek farm-fresh food and regional specialties, and local officials hope to enliven public areas and stimulate business development. Offering these diverse attractions, it is little wonder that retail farmers' markets have increased in both number and popularity. From as few as 100 in the 1960s, U.S. farmers' markets increased to 1,755 by 1994 when the United States Department of Agriculture assembled its first national farmers' market directory. Over the next decade (1994-2004) the number increased more than 100 percent to 3,706 (USDA-

AMS 2006). Although 31 percent of farmers participating at farmers' markets only sell their products through farmers' markets, the remaining 69 percent combine farmers' market sales with other retail and wholesale markets (Payne 2002).

The popular press and many academic studies often frame farmers' markets as isolated cultural and economic phenomena. Journalists portray them as colorful, hopeful scenes, both nostalgic and *nouveau* (Hamilton 2002). Meanwhile, social scientists examine why vendors participate and what they experience at farmers' markets (Griffin and Frongillo 2003; Vaupel 1989), what consumers seek at farmers' markets and how they behave (Andreatta and Wickliffe 2002; Baber and Frongillo 2003), and how farmers' markets function socially and economically (Hilchey, Lyson, and Gillespie 1995; Holloway and Kneafsey 2000; Lyson, Gillespie, and Hilchey 1995). But as useful as these inquiries have been, many do not situate farmers' markets in a larger social context. How do farmers' markets, their vendors, and customers connect to something larger?

We maintain that farmers' markets are community social and economic institutions that can be keystones in building more localized food systems. As keystones, farmers' markets join together and support seemingly separate social and economic building blocks: the local resource bases and skills of producers, the needs and preferences of local households, and the development goals of communities. This keystone function of farmers' markets within local and regional food systems emerges through four interrelated processes: (1) making local food products and producers regularly visible in public settings, (2) encouraging and enabling producer enterprise diversification, (3) incubating small businesses, and (4) creating environments where market transactions and nonmarket social interactions are joined. Not all of these processes are present to the same degree in every farmers' market, but when they are, potential synergies abound. In this chapter, drawing on case studies of farmers' markets in New York, Iowa, and California, we show how these processes work and why they are important in making farmers' markets have wider benefits and allowing them to realize their keystone potential.

Food System Transitions, Farmers' Market Opportunities

We can think of a food system as the complex set of interrelated commercial and noncommercial activities by which people manage and move food from creation to consumption to elimination and finally to reincorporation with nature (Dahlberg 1993). Any food system has an infrastructure that facilitates certain movements of matter and energy related to agriculture and food and inhibits other movements. Food system infrastructure consists of (1) material phenomena, both natural and built; (2) organizations and enterprises; (3) individual skills and knowledge; and (4) social relations and networks. The material infrastructure includes tangible physical things that people can readily see or handle. Such tangibles include farmland, buildings, signage, roads, parking lots, vehicles, or production and processing equipment. Organizations and enterprises include input supply businesses, farms, bookkeeping businesses, wholesale buyers and brokers, food processing firms, and food retail establishments (from supermarkets to farm stands). Usually, but not always, these organizations and enterprises are commercial entities that own, operate, and manage the material resources. Food system infrastructure also includes the complex sets of skills and knowledge of people directly participating in the food system as producers, handlers, and consumers of food or indirectly through activities supporting those participating directly. Finally, food system infrastructure includes the social networks and relations, both formal and less formal, that link people within and across functional areas of the food system.

As large corporations have gained more control over agricultural production and food marketing and organized both on a more global scale, local food system infrastructures have become more fragmented and homogenized. Fragmentation occurs when a town or region specializes so that it has only some one component of food system infrastructure such as a processing plant or a supermarket rather than a mix. Homogenization occurs as national brands, like Campbell's, Coca-Cola, and Dole as well as similar generic products designed for national or global distribution, capture substantial market share, buying up or eliminating regional labels. The ubiquitous global fast food chains such as McDonald's and mass retailers such as Wal-Mart epitomize homogenization. Concentration in the supermarket sector—the main site of food acquisition for most people—has only accelerated homogenization in the food system (Hendrickson and Heffernan 2002).

As Thomas Lyson explains in this volume, these general processes of fragmentation and homogenization in food systems have eroded once thriving local and regional food production and marketing patterns. Homogenization reinforces pressures on farmers to emphasize economic efficiency and quantity of production rather than social values and product quality. Under such pressures farmers tend to specialize in particular crops or livestock in an effort to compete more successfully. Such specialization further fragments the food system and also simplifies it by reducing agroecological complexity in any given farming operation. Simplification comes to prevail across the wider landscape as those farmers remaining in a region tend to produce the same small set of crops or livestock, drawing on similar stocks of outside expert skills and knowledge and depending on the same limited set of national suppliers and buyers. In a food system less centered in localities or regions, economics, that is, closing deals, moving product, and making profit, becomes the paramount concern.

The very restructuring of the food system along more global lines, however, simultaneously exposes vulnerabilities in that system and creates opportunities for relocalization (Hendrickson and Heffernan 2002). The current renaissance of farmers' markets can be seen as a logical response by food producers and consumers to such globalizing trends in the food system. With increasing consolidation and coordination across the conventional food supply chain, smaller farms often lose access to mainstream globally-oriented markets for their commodities. Similarly, small-scale food processors generally experience difficulties in getting their products into large conventional supermarket chains. Consequently, lacking opportunities, both small-scale and beginning farmers have sought an alternative in farmers' markets, which offer a relatively low-risk market outlet that can yield a steady, though typically modest, income stream (Feenstra et al. 2003).

As farmers' markets support the economic prospects of smaller farms and other food producers, they maintain crucial infrastructure for local and regional food systems—independent farms and regional working landscapes. Farmers' markets can provide this maintenance by serving as informal business incubators that nurture entrepreneurship, diversification, and expansion of small farms and food enterprises (Feenstra et al. 2003; Hilchey, Lyson, and Gillespie 1995; Hinrichs, Gillespie, and Feenstra 2004). In short, as venues for more direct producer-consumer relations, farmers' markets contribute to the viability of small farms and food enterprises. Some of these farms and enterprises that flourish will develop food system links beyond their farmers' markets.

Farmers' markets also have thrived due to their ability to offer precisely what many consumers now feel is missing from huge, conventional supermarkets—"food with a face." They attract customers who care about product quality and market ambiance as well as convenience and price. They usually provide foods that are grown locally and these foods appear only when they are in season locally. Many of the processed foods sold at farmers' markets are unique to the particular market or local area, or for common products like jams and sausages, they have labels and distinctive recipes connected to the families and the small businesses of their producers. Shoppers often can find varieties of produce (for example, gooseberries or heirloom tomatoes) that lack characteristics like long shelf life or mass-market appeal and therefore rarely appear in mainstream produce channels. For customers jaded by the dizzying choices, but ultimately standardized fare, of slick supermarkets, farmers' markets offer a different, appealing food shopping experience (see Sommer 1989; Sommer, Herrick, and Sommer 1981; Tyburczy and Sommer 1983).

Farmers' markets then hold potential to build and rebuild local and regional food systems by offering different possibilities than other forms of market exchange such as grocery stores, roadside stands, or community supported agriculture farms. In this chapter we draw upon farmers' market case studies conducted as part of a larger study on farmers' markets and rural entrepreneurship in order to examine processes influencing how farmers' markets contribute to the development of local food systems.

Study Methods

The empirical material for this chapter comes from case studies of farmers' markets conducted between 1999 and 2002 in New York, Iowa, and California. These three states have led the United States in their number of farmers' markets; by 2006 California had 444, New York 278, and Iowa 178 (USDA-AMS 2006). The three states represent distinct regional contexts for farmers' markets, given their different agricultural, social, and economic histories. This chapter emphasizes not the distinctions among the three states but rather the experiences common to the organizational context and operation of farmers' markets.

Following a telephone survey of farmers' market managers and a mail survey of vendors in all three states (see Feenstra et al. 2003 and Hinrichs, Gillespie, and Feenstra 2004 for a fuller account of the larger study), we conducted case studies of fifteen farmers' markets (six in New York, six in Iowa, and three in California). The farmers' markets were selected to capture varying patterns of entrepreneurial development and innovation representative of farmers' markets in urban, small town, and rural places in each state. The data came from observation at farmers' markets and from interviews with managers, selected vendors, and in some cases, customers; the results were used to develop comprehensive pictures of the development of the respective markets.

Making Local Food More Visible

As public venues farmers' markets are places where colorfully displayed fresh and seasonal produce is visible both to market shoppers and passersby. Creating this visibility is the most fundamental process by which farmers' markets become keystone institutions in rebuilding local food systems. Whether markets occupy landmark sites or nearly hidden locations, they are grounded in public life very differently than is a closed commercial venue like a supermarket. Although a new market pavilion on the edge of a downtown park may attract considerable foot traffic, while a cluster of trucks on the edge of a highway may not, farmers' markets, regardless of location, generally occupy public spaces that are open to anyone. This more open, public aspect of most farmers' markets distinguishes them from grocery stores and supermarkets, which are dedicated commercial spaces. Like farmers' markets, community supported agriculture (CSA) farms feature locally grown, and sometimes specialty processed, foods, but CSA farms usually lack the clearly public location and accessibility of farmers' markets. CSA members may go to their farm to pick up their shares, but these farms are rarely in well-trafficked public locations that by virtue of their openness might serendipitously attract new people into the experience.

As public spaces, particularly when visibility and accessibility are good, farmers' markets stimulate community awareness about the possibility of locally sourced foods. The Downtown Des Moines Farmers' Market (Iowa), for example, takes place in the historic, redeveloped Court Avenue District, and its management consciously promotes a special events atmosphere for the market. As its fame has grown, this market has drawn in a wider cross-section of citizens from the greater Des Moines area as well as tourists and visitors from beyond. People may enjoy the music and the mimes, but they also learn through sights, smells, sounds, and taste that Iowa produces excellent Asian vegetables, delectable morel mushrooms, and choice cuts of lamb. The visibility of local food to passersby then extends to the wider Des Moines metro area as the local media disseminate images and stories about the market opening in spring, special events through the season, and peak harvest in late summer.

The periodic character of farmers' markets also figures in making high-quality, local food visible and valued by consumers. A farmers' market punctuates the flow of public life once or twice a week. Because farmers' markets are periodic and seasonal rather than 24-hour everyday constants, people notice them and are less likely to take their products or services for granted. The regular weekly rhythm of a farmers' market is itself marked by fare that shifts with the progression of the seasons. These patterns educate farmers' market observers and customers to the possibilities—and also the limits—of local food supplies. In this way the regular and public visibility of high-quality, diverse local food in farmers' markets restores consideration of the provenance of food. Such awareness is foundational if food systems are to be relocalized.

Encouraging Diversification

Farmers markets' also help rebuild local food systems by encouraging local farmers and food producers to diversify their products and services. Such diversification is a keystone process because it enhances the economic viability of small agricultural and food businesses while also developing consumer demand for local food products and services. For producers, diversifying into new crops or products or new varieties of familiar crops or products can lengthen the market season, add value to products, attract more or different customers, and better utilize resources, including labor and equipment. Diversification is a time-honored way to reduce the risks of production failures and market price fluctuations, and it remains important for farmers and food producers marketing locally. Beyond the ability to stabilize local food enterprises, diversification both builds and responds to consumer demand. As a wider, more attractive and interesting array of locally produced foods entices more customers to a farmers' market, the resulting larger customer base attracts additional producers to participate in the market. Farmers' markets support diversification simply by providing opportunities for vendors to experiment with different crops and value-added products at low cost and low risk (Hilchey, Lyson, and Gillespie 1995). Indicators of diversification include appearance of new produce categories and value-added products in vendors' lineups. By encouraging diversification, farmers' markets "prepare the fields" so that other potential farmers and food producers will not only notice but enter and experiment with local markets.

Farmers' markets in many parts of the country have helped traditional commodity farmers to diversify and reinvent themselves as specialty and niche marketers and thereby contribute to relocalizing local food systems. In 1991 in New York State the Middletown Chamber of Commerce established a farmers' market with the goal of revitalizing a struggling downtown and making fresh food more available to local residents. For the Bialas family, for example, those goals were achieved through diversification out of their longstanding pattern of commodity production. They had grown only onions for years and, like other onion growers in the area, struggled because of low wholesale onion prices. When the Middletown Farmers' Market opened, Mrs. Bialas took a few things from the family's home vegetable garden to the market. She sold everything she took that day, as she did on every succeeding market day that season. Over time, Mrs. Bialas' market garden business grew as she joined new farmers' markets that opened in the lower Hudson Valley along with the Greenmarkets in New York City. The Bialases came to grow more than seventy varieties of vegetables, with most of their sixty-acre farm in market garden crops. This diversification has led to a more economically viable farm and in the process has created opportunities for the Bialas children, now adults, to stay on the farm. In short, farmers' markets facilitated the Bialas' enterprise diversification and connection to the local food system.

The downtown Des Moines Farmers' Market (DMFM) in Iowa also demonstrates how farmers' markets can support diversification by providing new marketing opportunities to replace unprofitable conventional commodity markets. The Des Moines Chamber of Commerce started the DMFM in 1975 to help revitalize the downtown business district. Since then the market has evolved from a few vendors and a trickle of customers to more than 140 vendors and thousands of customers on an average market day. The Terpstra and Mast families each have businesses that illustrate how diversification can build connections to local food systems. During the farm crisis in the early 1980s, the DMFM provided the Terpstras a crucial alternative when continuing to produce hogs was no longer economically feasible. Since then the Terpstras have sold vegetables at the DMFM, and nearly all of their family income comes from the vegetables produced on fifty acres and sold at the DMFM and two other farmers' markets.

Similarly, the Mast family began selling strawberries at the DMFM in 1988 as a strategy for dealing with the growing instability of income from their conventional farm operation. Since the berry season was short, they tried supplementing that enterprise by selling home-baked bread. Successfully selling the bread, they tried other kinds of baked goods. They now have an on-farm bakery, and although they continue to farm, this bakery and the farmers' market sales it generates account for 80 percent of their net family income. Nurtured by the farmers' market, their food enterprise is now an established provider of local foods in the Des Moines metro area.

Diversification does not only occur in larger urban farmers' markets. The Laytonville Certified Farmers Market (LCFM) in rural Mendocino County, California, also has encouraged enterprise diversification and thereby helped to develop the local food system and economy. In this region depletion of timber resources and environmental concerns have heightened the need to create alternative economic opportunities. Located on a highway traveled by many tourists, the LCFM has evolved to take advantage of that customer pool while continuing to contribute to local needs. During a three-year period, 20 percent of the farmer-vendors diversified into value-added products and crafts sold through their farmers' market businesses. The creation of a market directory describing each vendor's products and services has stimulated mail order sales with travelers and tourists, but vendors also have found that the directory has promoted their businesses locally. Here enterprise diversification has helped to tap into a wider customer base. This strengthens local farm and food enterprises and ensures their longer-term viability as components of the local food system.

Enterprise diversification is a keystone process not only because it strengthens businesses but because it also can draw new and different consumer groups into local food system relationships. In parts of the Midwest, for example, new immigrant populations are changing the demographic composition of traditionally European-American communities. The Marshalltown Farmers' Market in Iowa offers an example of how product diversification targeted to different consumer segments can build the local food system. Home to a large meatpacking plant, Marshalltown experienced an influx of people of Mexican origin in the 1990s. Many are now settling permanently in Marshalltown and the surrounding rural area. Some Anglo vendors have responded to these changing demographics by rethinking their product lines. Several now grow and sell specific foods used in Mexican cuisine—for example, tomatillos, cilantro, and more varieties of both sweet and hot peppers. This diversification of products in turn lures new community residents to shop at the farmers' market, improves vendors' sales, and piques the interest of other prospective vendors. Here diversification by individual vendors helps the farmers' market improve as a provider of culturally appropriate foods for a wider range of local residents.

Supporting Business Incubation

A third way in which farmers' markets rebuild localized food systems is by incubating small businesses that then may expand beyond farmers' markets. A setting that incubates small businesses helps these enterprises address the particular challenges of start-up, expansion, and continuation. Business incubation involves identifying and providing market opportunities, building motivation to undertake needed business tasks, improving business skills, cultivating entrepreneurial attitudes, and providing information or technical assistance, all of which increase the likelihood of business success for entrepreneurs. As low-cost marketplaces with few barriers to entry for new and existing small businesses, farmers' markets can work as business incubators (Feenstra et al. 2003; Sommer 1989). Vendors at farmers' markets can avoid exacting grading and packing standards common in many wholesale markets, and farmers' markets often eliminate nonlocal competition (Feenstra and Lewis 1999). By observing and interacting with more established, successful vendors, vendors learn both what does and does not work (Hinrichs, Gillespie, and Feenstra 2004). Some market managers coach vendors informally, and some markets sponsor more formal training or educational programs for vendors. Indeed, useful strategies for supporting vendor businesses may come to light through the collaborative research techniques described in the chapter by Larry Lev and colleagues in this volume. Supporting business incubation is a keystone process because it nurtures and develops new local food enterprises within farmers' markets to the point where some then can move into other direct and local wholesale markets. Such transitions increase the density of local food networks and relations.

The processes of small business incubation at farmers' markets occur both through individual initiative and through more formalized training sometimes offered by markets. Larry Cleverly, a garlic and vegetable grower at the downtown Des Moines Farmers' Market, used visibility and sales at the farmers' market to develop special marketing relationships with chefs at white tablecloth restaurants in the city. Wholesale direct sales to restaurants were reinforced by his farmers' market customers, who sought Cleverly products prepared by noted chefs at their favorite restaurants. Similarly Panna Putnam, a producer of samosas and other Indian foods at the Davis Farmers' Market, found that her success at the farmers' market afforded both the contacts and confidence to pursue catering jobs for parties and festivals and for selling wholesale to area grocery stores. Putnam's exposure at the farmers' market gave her the credibility to market her products through other local channels.

Some entrepreneurs develop new markets outside their farmers' markets more or less on their own. Others, however, benefit from opportunities more formally structured by their farmers' markets. Under the leadership of Joan Petzen, a Cooperative Extension educator who works on technical support for alternative agricultural enterprises, the Rural Enterprise Association of Proprietors (REAP) serves several counties of western New York. REAP fosters microenterprises by providing education and mutual support for new and retooling entrepreneurs. Recognizing the critical need for marketing opportunities, REAP establishes and operates area farmers' markets, which become proving grounds for new local food system businesses.

The Saturday Stockton Certified Farmers' Market (SSCFM) in California offers an example of a more urban-focused business incubation that strengthens the local food system. The market serves a part of Stockton, a city populated with many Southeast Asian immigrants. It provides market opportunities for vendors, 90 percent of whom are Southeast Asian and generally lacking the English language and occupational skills needed to be competitive in the U.S. job market. Market organizers also have emphasized building the skills needed to produce the Asian vegetables and other culturally appropriate food products demanded by the market customers. Community organizations, such as Southeast Asian Farm Development, which supported production, marketing, and English language training, and Cooperative Extension, which provided other training and information, have been critical in helping ethnic minority farmers run more profitable enterprises while providing desired foods to local Southeast Asian communities. For some of these enterprises the farmers' market serves as the launching pad for more extensive business operations.

Facilitating Social and Economic Interaction

Farmers' markets also serve as keystones in rebuilding food systems through the wider social exchange that they facilitate alongside market transactions. In this respect they carry on the historical legacy of public markets as "civic spaces—the common ground where citizens and government struggled to define the shared values of community" (Tangires 2003, xvi). While foodstuffs and money change hands, so, too, do ideas, enthusiasms, reservations, rebuffs, and more. The range and density of interactions revitalize civic life, representing a deeper set of considerations for building a more food-based local economy. As Dan Kemmis, the former mayor of Missoula, Montana, writes in *The Good City and the Good Life*: "During the course of the market's two hours, the conversations will run a gamut from Little League to potholes to events in Eastern Europe, but what I have come to be as attentive to are the unspoken conversations. As Steve weighs my broccoli and Lucy counts out my change, the whole history of their farm and of our friendship is part and parcel of what we exchange (Kemmis 1995, 5)."

Because the first business of farmers' markets is selling food, a life necessity, they attract a broad spectrum of people who might not under other circumstances meet or interact (Sommer 1989; Sommer, Herrick, and Sommer 1981). Many vendors value the social and recreational aspects of selling among friends at their market and learning how to be better marketers from the example and mentoring of other vendors. While some vendors at farmers markets may be inclined to compete with each other, most recognize and prioritize their common interest in having well-organized, interesting farmers' markets.

In the direct marketing encounter itself, vendors and customers interact at the point of sale, an economic exchange suffused with social information. Conversation about preferences and products informs what vendors sell and the loyalties and trust customers develop. Beyond this interchange customers are more likely to interact with other customers at the farmers' market since shopping takes place alongside taking in the scene, encountering friends and acquaintances, and lingering and talking.

Without minimizing the importance of cash flow and profitability at farmers' markets, this entwining of the economic and the social is a crucial keystone process of farmers' markets. By recasting food marketing relationships as broader human relationships, farmers' markets offer a different mix of criteria to consider in purchasing food than do supermarkets. Beyond just making local food supply visible, farmers' markets make local food *suppliers* visible so that they can be more valued by community members. As social ties develop through farmers' markets, consumers might well ask whether they also can obtain food from those or other local producers at other outlets, like farm stands or retail shops.

For more than twenty-five years the Davis Farmers' Market has been building just such relationships in Davis, California, a progressive community that is also home to a major university. The market's first alliance was with a local food co-op, which agreed to buy all of the market vendors' leftover produce at the end of each market day. This safety net convinced vendors to participate during the critical start-up phase of the market. The market's longtime manager, Randii MacNear, and its board of directors have worked to ensure the market is connected to the Davis community in diverse ways. MacNear has been active in local civic associations such as the Davis Chamber of Commerce, the Downtown Davis Business Association, the school district, and a variety of nonprofit organizations. The market supports civic development by being available as a forum for community groups, political organizations, and candidates for public office. Monthly market events showcase different community organizations and a business fair called Chamber Day sports displays by more than one hundred businesses and chamber organizations. Vendors donate substantial quantities of food for community meals and emergency food assistance (contributions to the food system). The market pays no rent to the city, but it has contributed very significantly to park maintenance, renovations, and improvements. For its vendor members the market facilitates connections with restaurants, grocery stores, and other vendors. Through these many and varied relationships, the market has become a highly valued local institution for many people in the community.

The Williamsburg Farmers' Market in a low-to-moderate-income area in Brooklyn, New York, further illustrates the importance of social connections within the economic sphere of the farmers' market. Known locally as *La Marqueta Communitaria*, the Williamsburg Farmers' Market has become a hub for building relationships among people in the neighborhood's diverse ethnic enclaves as well as better connecting urban people with nearby farmers. The market was established in 1998 through the work of two Hispanic community groups, *Los Sures* and *El Puente*. Although small, the market is at the core of a constellation of programs supporting food production and enterprise development, including a community garden, a shared-use kitchen, and a CSA enterprise, the latter established in collaboration with Just Food and Cornell Cooperative Extension. In short, the social connections developed in and through the market have been foundational in the success of these other local food system initiatives.

As social and economic institutions farmers' markets have the potential to be keystones in rebuilding local and regional food systems. Most fundamentally, they make local food visible in public spaces on a regular basis. This trait puts local food on the map and cultivates public awareness of local food production and producers. Beyond this function they can encourage enterprise diversification and business incubation, crucial processes for developing viable small businesses and building more broad-based consumer interest and commitment. Finally, farmers' markets facilitate economic interactions that are counterbalanced by shared social information and concern, which many would say is a hallmark of local and regional food systems. Through the twin processes of fragmentation and homogenization, the conventional globalizing food system actually creates opportunities for relocalization to succeed, with disenchanted consumers and disenfranchised small producers finding a coincidence of interest (Hendrickson and Heffernan 2002). Farmers' markets address such opportunities particularly well.

The four keystone processes discussed in this chapter clearly overlap. Since most farmers' markets are open-air venues, they make local food more visible, although the effect may be stronger or weaker depending on the market's physical location. The fulfillment of the other three processes may vary in other ways. Direct marketing tends to be more established in urban and urbanizing areas due to the attractive numbers and demographic profiles of the potential consumer base (Gale 1997).

For similar reasons keystone processes of enterprise diversification and business incubation overall find more impetus in farmers' markets in urban and urbanizing areas. The support and training for these processes often are associated with having a paid market manager, something more common in farmers' markets in larger places and those with a longer history (Oberholtzer and Grow 2003). Some rural farmers' markets we studied, particularly in areas with little or no tourism, demonstrated only limited enterprise diversification or business incubation, suggesting those markets' weaker role in catalyzing beleaguered local food systems. While such markets do provide some fresh foods on a seasonal basis in communities that may no longer even have a local grocery, the ongoing basic struggle with the twin shortages of vendors and customers can preclude more focused attention to small business development through the market. In this respect, as much as political will or organizational skill, location and demographics influence the vigor of farmers' markets, the success of participating enterprises, and the role they play in rebuilding local and regional food systems.

An important outcome of a successful farmers' market is its potential contribution to community food security—a community's capacity to feed itself. Community food security is "a situation in which all community residents obtain a safe, culturally acceptable, nutritionally adequate diet through a sustainable food system that maximizes community self-reliance and social justice" (Hamm and Bellows 2003, 38). Some farmers' markets play a central role in providing local residents with safe, culturally acceptable, and nutritious food at the same time that they stimulate new and stronger local food and farming enterprises. In these cases eaters, farmers, and food businesses rely on and support each other through the institution of the farmers' market.

This complementary relationship is particularly evident at the colorful Saturday Stockton Farmers Market, discussed earlier, which takes place beneath a crosstown freeway (Lewis 2001). On a busy day 10,000 customers might shop at the market—most of them from a nearby low-income immigrant neighborhood that is largely Asian. Many of these neighborhood families rely on such government food assistance programs as food stamps and the Women, Infants, and Children (WIC) supplemental food program. In 2000 families could use both their food stamps and Farmers Market Nutrition Program (FMNP) coupons at the market. In much the same way as Carro-Figueroa and Guptill describe for Puerto Rico in their chapter in this volume, government nutritional assistance programs help the Stockton Farmers' Market to build a more localized food system. Indeed, the Stockton market sold more than $600,000 in food through food stamps and FMNP coupons in its first year. Most of this food was farm fresh Asian vegetables (such as bitter melon, taro, daikon radishes, winter melon, Chinese long beans, bok choy, and specialty mustards), fish, eggs, and chicken—foods neither likely to be of the same quality nor for some specialties even available at local supermarkets. Here the enhancement of community food security is entwined with development of the local food system through the farmers' market.

In general, the farmers' markets that contribute the most to local food system development are those organized and conducted with more deliberate community development intent. Their success stems from the strategic dedicated actions of individuals and groups who have identified needs and recognized challenges, which they have worked steadily to overcome. However, the practical details of how this happens, noting both accomplishments and setbacks, tend to be downplayed in most of the celebratory accounts of farmers' markets. This bias may create undue optimism about the ease of starting or improving farmers' markets and the prospects for their success.

Nonetheless, building on the strong historical tradition of public markets in North America and with modest governmental support for their development and promotion today, farmers' markets tend to be one of the first manifestations of a relocalizing food system. Their benefits, though sometimes small, for small-scale agricultural producers and food customers participating in the farmers' market are well documented. Organizers and promoters of farmers' markets as well as researchers should attend to their potential synergistic effects (Knickel and Renting 2000) within local and regional food systems. Understanding the processes by which farmers' markets operate as keystones in local and regional food systems can support more integrated planning and development, tailored to the circumstances, strengths, and needs of particular communities.

Acknowledgments

This study was supported by the USDA Fund for Rural America.

References

Andreatta, S., and W. Wickliffe. 2002. "Managing farmer and consumer expectations: A study of a North Carolina farmers market." *Human Organization* 61:167–76.

Baber, L. M., and E. A. Frongillo. 2003. "Family and seller interactions in farmers' markets in upstate New York." *American Journal of Alternative Agriculture* 18:87–94.

Dahlberg, K. A. 1993. "Regenerative Food Systems: Broadening the Scope and Agenda of Sustainability." In *Food for the Future: Conditions and Contradictions of Sustainability*, edited by P. Allen, 75–102. New York: Wiley.

Feenstra, G., and C. Lewis. 1999. "Farmers' markets offer new business opportunities for farmers. *California Agriculture* 53 (November-December): 25–29.

Feenstra, G. W., C. Lewis, C. C. Hinrichs, G. W. Gillespie, and D. L. Hilchey. 2003. "Entrepreneurial outcomes and enterprise size in U.S. retail farmers' markets." *American Journal of Alternative Agriculture* 18:46–55.

Gale, F. 1997. "Direct farm marketing as a rural development tool." *Rural Development Perspectives* 12 (2): 19–25.

Griffin, M. R., and E. A. Frongillo. 2003. "Experiences and perspectives of farmers from upstate New York farmers' markets." *Agriculture and Human Values* 20:189–203.

Hamilton, L. 2002. "The American farmers market." *Gastronomica* 2 (3): 73–77.

Hamm, M. W., and A. C. Bellows. 2003. "Community food security and nutrition educators." *Journal of Nutrition Education and Behavior* 35:37–43.

Hendrickson, M. K., and W. D. Heffernan. 2002. "Opening spaces through relocalization: Locating potential resistance in the weaknesses of the global food system." *Sociologia Ruralis* 42:347–69.

Hilchey, D., T. Lyson, and G. W. Gillespie Jr. 1995. *Farmers' Markets and Rural Economic Development: Entrepreneurship, Business Incubation, and Job Creation in the Northeast.* Community Agriculture Development Series Bulletin. Farming Alternatives Program, Rural Sociology. Ithaca NY: College of Agriculture and Life Sciences, Cornell University.

Hinrichs, C. C., G. W. Gillespie, and G. W. Feenstra. 2004. "Social learning and innovation at retail farmers' markets." *Rural Sociology* 69:31–58.

Holloway, L., and M. Kneafsey. 2000. "Reading the space of the farmers' market: A case study from the United Kingdom." *Sociologia Ruralis* 40:285–99.

Kemmis, D. 1995. *The Good City and the Good Life*. Boston: Houghton Mifflin.

Knickel, K., and H. Renting. 2000. "Methodological and conceptual issues in the study of multifunctionality and rural development." *Sociologia Ruralis* 40:512–28.

Lewis, C. 2001. *The Stockton Farmers' Market: An Urban Community Market*. Davis CA: University of California Sustainable Agriculture Research and Education Program. http://sarep.ucdavis.edu/cdpp/stockton.htm (last accessed May 7, 2006).

Lyson, T. A., G. W. Gillespie Jr., and D. Hilchey. 1995. "Farmers' markets and the local community: Bridging the formal and informal economy." *American Journal of Alternative Agriculture* 10:108–13.

Oberholtzer, L., and S. Grow. 2003. *Producer-Only Farmers' Markets in the Mid-Atlantic Region*. Arlington VA: Henry A. Wallace Center for Agricultural and Environmental Policy at Winrock International.

Payne, T. 2002. *U.S. Farmers Markets—2000: A Study of Emerging Trends*. Washington DC: Agricultural Marketing Service, USDA. http://www.ams.usda.gov/directmarketing/farmmark.pdf (last accessed May 7, 2006).

Sommer, R. 1989. "Farmers' Markets as Community Events." In *Human Behavior and Environment*, edited by I. Altman and E. H. Zube, 57–82. New York: Plenum.

Sommer, R., J. Herrick, and T. Sommer. 1981. "The behavioral ecology of supermarkets and farmers' markets." *Journal of Environmental Psychology* 1:13–19.

Tangires, H. 2003. *Public Markets and Civic Culture in Nineteenth Century America*. Baltimore: Johns Hopkins University Press.

Tyburczy, J., and R. Sommer. 1983. "Farmers' markets are good for downtown." *California Agriculture* 37 (May-June): 30–32.

USDA-AMS. 2006. *AMS Farmers Markets*. http://www.ams.usda.gov/farmersmarkets/(last accessed May 7, 2006).

Vaupel, S. 1989. "The farmers of farmers' markets." *California Agriculture* 43 (January-February): 28–30.

4. Practical Research Methods to Enhance Farmers' Markets

Larry Lev, Garry Stephenson, and Linda Brewer

Farmers' markets play a significant and expanding role in ensuring the viability of small farms and the vitality of towns and communities. They have grown in popularity because they provide communities, producers, and consumers an alternative to a mainstream food distribution system dominated by large-scale firms focused on efficient buying and selling within the global marketplace. Yet farmers' markets remain poorly understood and underappreciated. Most individual markets lack both the resources and skills for documenting their role in the community and to make effective changes and improvements.

In a world in which Wal-Mart has real time data on all of its operations, farmers' markets languish at the other end of the information spectrum. Although the markets provide a vibrant meeting ground for independent businesses (Hinrichs, Gillespie, and Feenstra 2004), the limited market staffs are hard pressed to do anything beyond simply getting the markets up and functioning each week. In many instances the market organizations collect no information beyond the number of vendors and the stall fees paid. They do little to develop themselves or their vendors so as to achieve greater successes in the future. Still there is little doubt that an improved understanding of how these markets function would enable vendors to better meet customer demands and would help markets gain increased public and business community support.

We have designed three quick, inexpensive, and reliable methods that address both the information needs and the severe financial and personnel constraints faced by markets.[1] We also achieved our most ambitious goal of developing data collection methods that add to, rather than detract from, market atmosphere. As such, our efforts can be seen as tools that are useful for the construction of a civic agriculture—locally based systems that are linked to social and economic development, as presented in the chapter by Thomas Lyson in this volume (see also Lyson 2000).

The first two methods, attendance counts and dot surveys, were designed to be used independently by individual markets. The third more integrated evaluation approach, the Rapid Market Assessment (RMA), adds qualitative assessments to the first two. In our experience an RMA works best as a participatory learning process in which a team consisting of external market managers (and others) visits and studies a host market. The learning is two-way: the host market and the visiting RMA team members both gain knowledge and experience.

Our approach draws from a broad spectrum of prior work on social and collaborative learning both within the agricultural community (Pretty 2002; Röling and Jiggins 1998; Uphoff 2002) and outside of it (Bruffee 1993; Whyte 1991). This type of learning requires the expertise and the active participation of the "audience" and represents a departure from the approaches most commonly used by the Extension Services in the United States. The set of techniques can be adapted for use by any organization with an interest in fostering education and learning in a specific community.

Counting Customers

Knowing how many people shop in a market provides diverse benefits. Prospective vendors gain a basis for estimating their potential sales, while existing vendors can better understand the exposure they achieve through selling in the market. The market organization can use attendance data to estimate potential spillover sales for neighboring businesses and to document the market's role as a social center. Few markets, however, collect attendance data on a routine basis because they lack the personnel to tackle the complex task of counting people in an open-air market without clearly defined entrances.[2]

How then do markets estimate their attendance? Many just throw out a number and see what people think. Some use the run-through-the-market count system. On a given market day a person runs through the market (usually once an hour) and counts customers, and then this number is multiplied by some estimate of how long customers stay in the market. Only a few markets are ambitious enough to attempt comprehensive counts. Once completed, any of these attendance counts may be updated periodically by multiplying by an estimated growth percentage.

We developed a simple and inexpensive counting method so that markets would have a standardized way of estimating their attendance. To minimize personnel requirements, we use a sampling process of counting all *entering* customers for only ten minutes at the midpoint of every hour. Multiplying these ten-minute counts by six provides an hourly average. In advance we carefully determine all possible entry points and assign a counter to each. Although greater accuracy would be achieved by counting every person entering the market, that approach demands much more labor. In our experience counting for ten minutes per hour provides an acceptable estimate. When the attendance count is part of a broader RMA, team members are available for other RMA activities during the balance of the hour. If conducted as a stand-alone activity, volunteers or market workers can conduct the counts and still be free most of the time.

When counts are conducted periodically, they document how market attendance varies both within a season and between seasons. Because markets have found the approach to be simple, doable, and useful, many more now collect this attendance information. At the state level the standardization of the approach makes comparisons and collation more meaningful.

When the city of Corvallis, Oregon, began to redesign its riverfront, the credible attendance data that the farmers' market presented allowed it to demonstrate the importance of the market as a gathering place and ultimately to influence the design of the riverfront. The first attendance count that the Oregon State University Extension Service conducted took place at this market in 1998. Before the count the very knowledgeable manager guessed that attendance would be 800. The actual count that day of 1650 convinced us that it would be worthwhile to train other markets to collect this elusive, but valuable data.

Dot Surveys: A New Method for Collecting Customer Information

All businesses need to collect information from their customers and many tried and tested data collection techniques exist for this purpose (Salant and Dillman 1994). While some markets make use of these techniques, many find them too expensive and complicated to use.

Random mail or phone surveys are valuable in providing information on starting markets and in comparing attitudes and behaviors between market shoppers and nonmarket shoppers (Gallons et al. 1997; Rhodus, Schwartz, and Hoskins 1994; Stephenson and Lev 1998). Response rates of 60 percent or higher can be achieved so there can be reasonable confidence in the validity of the data. But in many instances the specific needs of farmers' markets are not well met by these two types of surveys because such surveys require considerable expertise to do properly, are fairly expensive if conducted by paid consultants, and are not adequately focused if trying to target market shoppers.

In-market intercept interviews provide an excellent means of collecting detailed market specific data (Lockretz 1986; Sommer, Herrick, and Sommer 1981; Wolf 1997). However, this approach generally results in a small sample size (or requires a large crew of skilled interviewers to increase sample size) and therefore needs to be conducted carefully to attain a representative sample.[3]

Handout/mail-back surveys address both the targeting weakness of phone and mail surveys and the sample size concerns of interviews (Brooker et al. 1987; Eastwood, Brooker, and Gray 1995; Govindasamy, Italia, and Adelaja 2002; Kerr Center 2002). Mail-back surveys raise their own concerns. They are characterized by wide variations in response rates (from 20 to 66 percent in the surveys cited). This response rate may raise concerns about bias in the sample.[4] In fact, even the proper definition of response rates continues to be surrounded by uncertainty—is it the percent of surveys handed out or the percent of potential respondents who were approached (CASRO 1982)? Finally, in many instances markets contract with outside experts for survey design and analysis.

We designed the dot survey approach to supplement existing data collection methods and to better address the unique problems and constraints facing farmers' markets (Lev and Stephenson 1999). It is a self-service research approach that asks a limited number of close-ended questions (generally four) that are displayed on easels in a central location in the market. Consumers indicate their response to each question by placing a colorful, round, self-stick dot label on the poster in the category where it makes the most sense. Answering all four questions completes the dot survey and takes each participant only one or two minutes (although some choose to stay and discuss the research). Dot surveys achieve far greater participation and fewer refusals than other methods. They also fit more easily within the financial and expertise limitations faced by most markets.

The most striking difference between this and other survey techniques is that respondents can see how others have responded. This visibility is both a weakness and a strength. It is a weakness because respondents may be influenced by what they observe on the posters. However, this is not a concern for the majority of questions. For example, people do not change their answer to questions such as "Where do you live?" or "How old are you" based on how others have answered. In fact, in our experience the response rates on difficult questions like age are higher than through other survey approaches because it is clear to the respondent that anonymity is preserved.

Because early responders may influence later responders for certain questions, it is necessary to keep the influence of previous respondents in mind when crafting all questions. Several strategies can help reduce this influence concern. We generally seed the posters with scattered dot responses and later remove the seeded dots. We also replace the posters with fresh sheets at regular intervals so that new respondents will place their dots with less prior information.

A second major concern with the approach is that only a limited number of close-ended questions can be asked. This is certainly a major drawback and should not be underestimated. More positively, though, it does force the market to focus on what information is most important and also enables the data to be analyzed in a timely fashion. In addition, because this is a low-cost approach, it is feasible (and preferable) for the market to collect information on multiple occasions.

A third concern with the approach is the difficulty (but not impossibility) of conducting cross-tabulation analysis of the data. Ordinarily answers to question one cannot be related to answers to question two. While it is possible to code the dots so that this information is available, this addition considerably increases the time needed to conduct data analysis.

In our view the advantages of dot surveys in farmers' markets far outweigh the disadvantages. The approach is both simple and inexpensive. The transparency of the research process in which everyone sees everything turns out to be one of the most appreciated aspects of the approach. In this sense it makes the research interactive rather than extractive. Participants often stop back later in the market to see how the responses become arrayed on the posters. Overall, this data-gathering approach actually adds to rather than detracts from the overall atmosphere of the market and thereby allows us to achieve our most ambitious goal.

Dot surveys are a high-volume research approach. We have had as many as 1,000 participants in five hours (200 per hour). When using the approach, we have kept careful records and have documented that 90 percent of the consumers that we approach are willing to participate in the surveys. This rate compares very favorably with response rates for all the other survey methods discussed above. The superior quality of the sample should improve the accuracy of the data collected (Salant and Dillman 1994). We also have specifically asked consumers whether they prefer this approach or more traditional written surveys, and 94 percent favor the dots.

The results of the dot survey can be quickly tabulated. When Oregon State University Extension conducts such a survey as part of an RMA, we typically provide the results back to the market in forty-eight hours. Finally, as mentioned above, these surveys are not intimidating to conduct. Therefore markets can easily repeat questions and gain greater confidence in results.

Our assertions were confirmed by Suzanne Briggs, one of the organizers of Portland's Hollywood Farmers' Market. More than any other market, Hollywood has embraced the data gathering possibilities of the RMA techniques. Briggs says that "the dot research method is very valuable. First and foremost, the process is fun and engaging for

Table 2. Responses to "Was the farmers' market your primary reason for coming downtown this morning?"

	Corvallis Saturday farmers' market (%)	Albany Saturday farmers' market (%)
Yes	78	88
No	12	8
Partially	10	4

both the volunteers and our customers. Our survey results are currently posted in the market every week in front of our volunteer booth. Each week people stop and read the results, then share how we could use this information to build a stronger market in our neighborhood. The process is so simple that we are continuing to do the dot survey in the market this year."

The information obtained with these surveys has been eye-opening and useful to the markets that we have studied. We always let the market being studied select the questions to be asked. As a result our statewide and regional efforts suffer a bit by not having the same questions asked at all markets, but the gain in relevance to the specific market more than makes up for this lack of uniformity.

To show the usefulness of the dot survey, we offer the following examples of information collected along with comments explaining why the information was useful. To obtain that information, we asked these three questions:

1. Was the farmers' market your primary reason for coming downtown this morning?
2. Will you be doing additional shopping in this area on this trip? If yes, how much do you anticipate spending?
3. What stopped you from buying more at the market today?

The responses are tabulated in tables 2, 3, and 4.

Many downtown business communities are lukewarm supporters of farmers' markets. One reason is that they don't believe that markets really attract people downtown. In conjunction with the attendance counts, question one provides data that directly addresses that concern

Table 3. Responses to "Will you be doing additional shopping in this area on this trip? If yes, how much do you anticipate spending?"

	Albany (1998 average)	Corvallis (1998 average)	Hollywood district (2000)
Percent doing additional shopping	38	63	47
Average amount spent	$4.32	$10.13	$8.98

(see table 2). In both Corvallis and Albany this data has served to increase business support for the markets because the markets were able to show that additional shoppers were attracted downtown.

Question two examines whether people attracted to the market actually do any shopping at neighboring businesses (table 3). The extent of spillover sales depends on the attractiveness of the adjoining businesses (it must be noted that the downtown Corvallis district has more to offer than downtown Albany). In several markets we have found that spillover sales have been as high as 80 percent of in-market sales (Lev, Brewer, and Stephenson 2003).

Question three was prompted by repeated comments by some dot survey participants that prices in farmers' markets are too high (table 4). The question assesses whether price or other factors really do constrain sales. The results are remarkably similar across markets, and the overall conclusion is clear—most consumers do not regard prices as a major factor limiting their purchases. This result shows the tremendous advantage of collecting responses from hundreds of shoppers rather than simply listening to the opinions of a vocal few.

The markets have used dot surveys to collect a variety of data that can be used in different ways. Some examples include

1. attendance counts and market sales that
   a. demonstrate to community leaders the value and popularity that residents place on markets;
   b. provide potential vendors with market size information;
2. drawing power of market and spillover sales that
   a. quantify for community leaders the broader economic impact of markets;

Table 4. Responses to "What stopped you from buying more at the market today?"

	Corvallis Saturday farmers' market (%)	Corvallis Wednesday farmers' market (%)	Hollywood district farmers' market (%)
Nothing else wanted	67	60	55
Couldn't carry more	13	11	17
Out of money	10	14	14
Prices too high	6	8	5
Ran out of time	4	7	8

b. demonstrate to neighboring businesses the value that the market can provide to them;

3. information on where customers live, why they come to the market, and what stops them from making additional purchases that
   a. assists market boards in allocating their budget (advertising, entertainment, and so on);
   b. assists vendors in selecting and pricing their products.

Rapid Market Assessments: Developing a Participatory Action Research Model

The data collection methods described above proved useful and productive. We studied several markets and distributed widely both the results and the explanation of the research methods used. While we hoped that other markets would be inspired to replicate the methods (a few did), the more common result was that the other markets simply made use of these initial results with the comment that their own market was "sort of the same." Most managers viewed the data collection methods as too great a hurdle to undertake.

As a result we refocused our attention on developing a way to encourage the spread of these research methods. One intriguing idea was to enlist market managers as data collectors outside of their own markets. What better way to learn than by actually doing? But once the managers had put their time and effort into getting to the target market, there had to be a better means for taking advantage of both their expertise and their eagerness to learn about market management. Bit by bit, we moved toward a *participatory action research* (PAR) model in which the practitioners (market managers) were fully involved in the research process (Krasny and Doyle 2002; Whyte 1991).

We developed a system of recruiting teams of four to five managers or board members from markets around the state and bringing them together to study a market on a specific day. During the RMA the team members both collect quantitative information, using the methods discussed above, and also conduct a qualitative assessment of the market. Two sets of people gain from the RMA: those associated with the market studied and the outsiders who participate in the RMA. Over time, we discovered that we could greatly expand the total value of study team members' benefits by increasing the team size (we commonly use teams of eight and have used teams with as many as thirty people) and by incorporating a more formal classroom training and discussion session just before the market study begins.

The process we use follows a very precise format. The night before the market, all of the team members gather for dinner with the market manager and market board. The manager and the board discuss the history and current circumstances of the market. We ask them to focus on the specific issues of how they define success and where they want the market to be in five years. The dinner both provides context and sets up relationships for the next day.

At the market the team members learn firsthand how to conduct the attendance counts and the dot survey. Once they have experienced the simplicity and effectiveness of these methods, they are much more likely to use them in their own markets. The team members also complete a series of three worksheets on physical setting of the market, vendors and products, and market atmosphere over the course of the market. This part of the process particularly captures their interest and involvement. Each person brings a fresh set of eyes to the task and notices different things. They are encouraged to do whatever it takes to understand the market—follow shoppers around, talk to them, watch vendors, or explore the neighborhood.

At the end of the market we come back together for a debriefing of both the qualitative and quantitative research activities. During the debriefing we discuss the team members' main observations as well as their suggestions for improving the research process. Rebecca Landis, manager of the Corvallis-Albany Farmers' Markets, offered this feedback: "Participation in Rapid Market Assessments offers a rare opportunity to carefully observe other markets in action. It's especially helpful to examine the strengths and weaknesses of other policies, practices and site designs before experimenting in my own markets."

Within two days of the study the Oregon State University Extension Service, as the group facilitating the RMA effort, e-mails the quantitative results to the studied market and the RMA team members. Providing the information so quickly is very useful. The process of collating the qualitative reports requires more time. All of the market worksheets are typed, and then one of the researchers takes on the job of compressing the five or six different accounts into a single short report. This report follows the format of the three worksheets and has separate sections. One section details what the observers appreciated about the market and another lists either questions they had on what they saw or suggestions for improvement. This report is also distributed by e-mail. After the distribution of the reports a conference call is scheduled to discuss the top three to five issues to come out of the overall process.

The Hillsdale Market RMA study (Lev, Stephenson, and Brewer 2002) provides an example of the value of the overall process. As described in *Better Together: Restoring the American Community* (Putnam and Feldstein 2003), a group of Hillsdale residents decided to organize a farmers' market to serve as a community meeting place. Since nearby Portland area markets already operated on Saturday, thus competing for both customers and vendors, the Hillsdale group chose to operate on Sunday. The RMA study was conducted two months into the first marketing season.

Although the Hillsdale market management knew that both customers and vendors were happy with the first two months of operation, the count data provided the market with its first objective measure of success (2,132 adult customers). The responses to one of the dot survey question showed that 84 percent of the customers lived within two miles of the market, demonstrating that the market had achieved the objective of attracting local residents. A second question provided detailed information on how this new market interacted at the consumer level with existing markets. Team member qualitative comments helped the market to focus on key issues including determining vendor location within the market and making effective use of volunteers. The participating market managers, several of whom represented neighboring markets, learned valuable lessons about community participation in market development and demonstrated that farmers' markets in the Portland area continue to be collaborators rather than competitors.

These research techniques provide something unique and needed for strengthening community food systems. As a whole they represent an excellent example of what the public sector—or any organization with an interest in working on social issues—can provide to organizations such as farmers' markets (Abel, Thomson, and Maretzki 1999). The attendance counts are simple but very useful. The dot surveys add to the market atmosphere and supply critical data that most markets have been too intimidated to collect. The overall RMA process has proven to be an exhilarating and empowering means of strengthening the statewide network of farmers' market managers. During an RMA all of the involved markets gain new insights and ideas through a two-way learning process. The studied markets receive detailed quantitative and qualitative information on important issues. The RMA team members become better connected with their peers, more analytical, and more confident about conducting research in their own markets. Based on their experience of intensively studying another market in a short period of time, they go back to their own markets with many new management ideas and options. As Dianne Stefani-Ruff of the Portland Farmers' Market observed: "My experience with the assessment project was wonderful, both as a team participant in Ashland and having a team visit our Wednesday market. The information and insights were wonderful. It was time well spent and enjoyable, even in the middle of a busy market season! What a good way to meet my peers. Sometimes I feel isolated (even though I am surrounded by vendors, customers and board members!). No one really understands like another market manager. It is great to work with such a sharing, caring, dedicated bunch of folks!"

Acknowledgments

This research was supported by a USDA Initiative for Future Agriculture and Food Systems (IFAFS) grant. While the results reported here represent Oregon studies, related market studies have been conducted in Idaho and Washington.

Notes

1. Detailed descriptions of all three research techniques are in Lev, Brewer, and Stephenson (2004).

2. Small indoor markets are the most likely ones to keep track of attendance because for them it is an easy task.

3. The willingness to be interviewed (response rate) is often not reported for this survey process.

4. Potential participants who refuse to accept the survey are generally ignored rather than counted as nonrespondents for the purpose of calculating the response rate.

References

Abel, J., J. Thomson, and A. Maretzki. 1999. "Extension's role with farmers' markets: Working with farmers, consumers, and communities." *Journal of Extension* 37 (5) (October). http://www.joe.org/joe/1999october/a4.html (last accessed April 9, 2006).

Brooker, J. R., C. L. Stout, D. B. Eastwood, and R. H. Orr. 1987. *Analysis of In-store Experiments Regarding Sales of Locally Grown Tomatoes*. Bulletin 654. Knoxville TN: University of Tennessee Institute of Agriculture.

Bruffee, K. A. 1993. *Collaborative Learning: Higher Education, Interdependence, and the Authority of Knowledge*. Baltimore: Johns Hopkins University Press.

Council of American Survey Research Organizations. 1982. *On the Definition of Response Rates: A Special Report of the CASRO Task Force on Completion Rates*. http://www.casro.org/resprates.cfm (last accessed April 9, 2006).

Eastwood, D. B., J. R. Brooker, and M. D. Gray. 1995. *An Intrastate Comparison of Consumer Patronage of Farmers' Markets in Knox, Madison, and Shelby Counties*. Research Report 95–03. Knoxville TN: University of Tennessee Institute of Agriculture.

Gallons, J., U. C. Toensmeyer, J. R. Bacon, and C. L. German. 1997. "An analysis of consumer characteristics concerning direct marketing of fresh produce in Delaware: A case study." *Journal of Food Distribution Research* 28 (1): 98–106.

Govindasamy, R., J. Italia, and A. Adelaja. 2002. "Farmers' markets: Consumer trends, preferences, and characteristics." *Journal of Extension* 40 (1) (February). http://www.joe.org/joe/2002february/rb6.html (last accessed April 9, 2006).

Hinrichs, C. C., G. Gillespie, and G. W. Feenstra. 2004. "Social learning and innovation at retail farmers' markets." *Rural Sociology* 69:31–58.

Kerr Center. 2002. *Creating A Successful Farmers' Market*. http://www.kerrcenter.com/farmers–market/index.html (last accessed April 22, 2006).

Krasny, M., and R. Doyle. 2002. "Participatory approaches to program development and engaging youth in research: The case of an inter-generational urban community gardening program." *Journal of Extension* 40 (5) (October). http://www.joe.org/joe/2002october/a3.shtml (last accessed April 9, 2006).

Lev, L., L. Brewer, and G. Stephenson. 2003. *Research Brief: How Do Farmers' Markets Affect Neighboring Businesses*. Oregon Small Farms Technical Report. No. 16. Corvallis OR: Oregon State University Extension Service. http://smallfarms.oregonstate.edu/publications/techreports/techreport16.pdf (last accessed April 9, 2006).

———. 2004. *Tools for Rapid Market Assessments*. Oregon Small Farms Technical Report. No. 6. Corvallis OR: Oregon State University Extension Service. http://smallfarms.oregonstate.edu/publications/techreports/techreport6.pdf (last accessed April 9, 2006).

Lev, L., and G. Stephenson. 1999. "Dot posters: A practical alternative to written questionnaires and oral interviews." *Journal of Extension* 37 (5) (October). http://www.joe.org/joe/1999october/tt1.html (last accessed April 9, 2006).

Lev, L., G. Stephenson, and L. Brewer. 2002. *Hillsdale Farmers' Market Rapid Market Assessment*. Oregon Small Farms Technical Report. No. 13. Corvallis OR: Oregon State University Extension Service. http://smallfarms.oregonstate.edu/publications/techreports/techreport13.pdf (last accessed April 9, 2006).

Lockretz, W. 1986. "Urban consumers' attitudes toward locally grown produce." *American Journal of Alternative Agriculture* 1 (2): 83–88.

Lyson, T. A. 2000. "Moving toward civic agriculture." *Choices* 15 (3): 42–45.

Pretty, J. 2002. "Social and Human Capital for Sustainable Agriculture." In *Agroecological Innovations: Increasing Food Production with Participatory Development*, edited by N. Uphoff, 47–57. Sterling VA: Earthscan.

Putnam, R.D., and L. Feldstein. 2003. *Better Together: Restoring the American Community*. New York. Simon & Schuster.

Rhodus, T., J. Schwartz, and J. Hoskins. 1994. *Ohio Consumer Opinions of Roadside Markets and Farmers' Markets*. Report to the Ohio Rural Rehabilitation Program, Ohio Department of Agriculture. Columbus OH: Department of Horticulture, Ohio State University. http://hcs.osu.edu/hcs/em/rfmarket/rfmarket.html (last accessed April 9, 2006).

Röling, N., and J. Jiggins. 1998. "The Ecological Knowledge System." In *Facilitating Sustainable Agriculture*, edited by N. G. Röling and M. A. E. Wagemakers, 283–306. New York: Cambridge University Press.

Salant, P., and D. A. Dillman. 1994. *How to Conduct Your Own Survey*. New York: Wiley.

Sommer, R., J. Herrick, and T. R. Sommer. 1981. "The behavioral ecology of supermarkets and farmers' markets." *Journal of Environmental Psychology* 1:13–19.

Stephenson, G., and L. Lev. 1998. "Common Support for Local Agriculture in Two Contrasting Oregon Cities." Paper presented at Sixty-First Annual Meeting of the Rural Sociological Society, Portland OR.

Uphoff, N. 2002. "The Agricultural Development Challenges We Face." In *Agroecological Innovations: Increasing Food Production with Participatory Development*, edited by N. Uphoff, 3–20. Sterling VA: Earthscan.

Whyte, W. F., ed. 1991. *Participatory Action Research*. Newbury Park CA: Sage.

Wolf, M. M. 1997. "A target consumer profile and positioning for promotion of the direct marketing of fresh produce: A case study." *Journal of Food Distribution Research* 28 (3): 11–17.

5. Community Supported Agriculture as an Agent of Change

Is It Working?

Marcia Ruth Ostrom

Increased recognition of the negative impacts of global-level economic restructuring on social welfare and the environment has prompted social theorists and activists to look for alternatives at a grassroots level. Within this context community supported agriculture (CSA), in which consumers form direct connections with local farmers to obtain their food, has been proposed as a strategy for revitalizing local agricultural economies, preserving farmland, enhancing community food security, and educating consumers about farming and the environment. The CSA approach is unique in that it seeks to change how agricultural goods are bought and sold by forming alliances between farmers and consumers. The goal is to cover the true costs of production by dividing them fairly among the end consumers of the products, factoring in the costs of environmental stewardship and fair returns for labor. Because consumers who join a CSA farm make a payment in advance of the growing season in exchange for a share of whatever the farm produces each week, the farmers' production and marketing risks are minimized. The CSA farm is a prominent contemporary example of a grass-roots effort to protect land and farmers from the volatility of a globally organized, corporate-driven system of commodity food production and distribution.

Gathering momentum in the United States after its introduction from Japan and Europe in the mid-1980s, the CSA idea has spread from a few farms in the Northeast to hundreds of farms across North America. Today, tens of thousands of consumers, known as farm *members*, or *shareholders*, are eating food produced and distributed by these farms. Regarded by some as a cornerstone of an emergent new agriculture, various authors have envisioned CSA as the basis for the "farms of tomorrow" (Groh and McFadden 1997) or the "catalyst for a new economy" (Lamb 1994). The CSA concept simultaneously celebrates land stewardship, the community, the small farmer, and the spirit of urban-rural cooperation. The goal is to recreate the local connections between food production and consumption based on a new kind of civic-minded, economic contract. And, unlike efforts to forge environmental, agricultural, or food system change at a policy level, CSA participants can reap immediate, practical rewards for their efforts. Consumers can run their hands through the soil that produces their food; savor the taste of fresh melons, crisp greens, tomatoes, and herbs; and tend to their own health—even as they make their stand for social and environmental improvements. Farmers can collect payments for their crops in advance, regardless of the uncertainties of weather and commodity markets. It is no wonder that the CSA concept generates such enthusiasm.

Two decades after CSA was first introduced in the United States, we still have little understanding of its composition or long-term potential. To address this knowledge gap, this study investigates the strengths and weaknesses of the organizational configurations, tactics, and outcomes of CSA as a social movement for change in the food and farming system. Twenty-four CSA farms serving major metropolitan areas in Minnesota and Wisconsin are tracked over a ten-year period. Critical topics of investigation include consumer and farmer participation frameworks, the conditions needed for farms to succeed, and the challenges encountered. Emphasis is placed on the individual understandings and action rationales of the participants themselves as well as the potential for them to achieve their goals. Based on this regional study, questions emerge about the broader implications of CSA. Can such localized, grassroots efforts offer a meaningful challenge to a globally organized corporate system of food production and distribution, and can they ultimately move us any closer to more democratic control of food and farming systems?

Our Current Understanding of CSA

Identified with such popular ideals as saving family farms and protecting the environment, CSA has received generous media coverage. For nearly two decades feature stories on CSA have appeared regularly in the mainstream press as well as in the leading periodicals of the environmental and sustainable agriculture movements. Featuring titles such as "Vegetables for All" (VanderTuin 1987), "Share the Land" (Sugarman 1991), "Sharing the Harvest" (O'Neill 1997), and "Mission Possible" (Maxim 2006), these articles offer photogenic and inspiring accounts of farmers and consumers joining together to preserve farming. A sympathetic press has been instrumental in piquing consumer and farmer interest.

On a practical level several excellent field manuals or handbooks have been developed by farmers and educators as resources for starting and managing CSA farms (see Blake et al. 1995; Gregson and Gregson 1996; Henderson and Van En 1999; Rowley and Beeman 1994; Wilson College 1998). *Farms of Tomorrow* by Trauger Groh and Stephen McFadden (1997) provides an important overview of the philosophical and historical foundations of CSA and the unique social and environmental values that set it apart from other forms of direct marketing. Dynamic conferences and e-mail discussion groups have further facilitated knowledge transfer and resource sharing among farmers.

Even though journalistic, promotional, and how-to references on CSA are becoming common, accounts that offer a more critical and balanced analysis of the overall strengths and weaknesses of CSA as a movement remain rare. Most references to CSA are extrapolated from case studies of a single farm or tend to be primarily conceptual or anecdotal in nature. While CSA is regularly cited as a core element of an emergent consumer-based resistance to dominant agrifood paradigms (Buttel 2000; Imhoff 1996; Lacy 2000), little systematic research is referenced.

Nonetheless, a slowly evolving body of academic theses, dissertations, survey reports, and journal articles, grounded in empirical evidence from diverse regions of the country, has begun to address this gap. Critical questions have been raised about various aspects of CSA, including the challenging economic and quality of life issues encountered by farmers (Ostrom 1997a), problems with member retention (Kane and Lohr 1997), and its failure to achieve goals of social justice and inclusiveness (Allen et al. 2003; DeLind 1993; Hinrichs and Kremer 2002). Hinrichs (2000) examines the extent to which CSA as it is currently practiced succeeds at creating an alternative to conventional market relationships by reembedding economic transactions within the fabric of civil society (see also Hendrickson 1996; O'Hara and Stagl 2001). DeLind and Ferguson (1999), along with Cone and Myhre (2000), have begun to explore the gender dynamics of CSA, and studies from diverse regional contexts have identified consistent profiles of consumer participation (Cone and Myhre 2000; DeLind 1993; Hendrickson 1996; Kane and Lohr 1997; Laird 1995; O'Hara and Stagl 2001) and farmer participation (Lass et al. 2003; Tegtmeier and Duffy 2005). Elsewhere, Ostrom (1997a) has made the case for regarding CSA as a social movement.

After reviewing this emerging literature, however, several important questions remain unanswered. Can CSA become a significant and lasting force in remaking the food system? While the basics—consumers pay in advance for regular portions of seasonal farm produce—are consistent across farms, the model is being adapted and applied in increasingly diverse forms. Which organizational forms and tactics are proving to be successful over the long-term? The upper Midwest provides an important region for examining these questions. Farmers here were early innovators, with several CSA farms already in operation by the late 1980s. This two-decades presence allows consideration of how CSA has commenced and unfolded over time. Furthermore, farms exist in sufficient numbers to allow for investigations and comparisons across farms and interactions among farms.

Participatory Research Methods

Research with CSA farms in Minnesota and Wisconsin was first initiated by the farmers at Philadelphia Community Farm (PCF) in Osceola, Wisconsin. The widespread local and national interest attracted by their farming model prompted the PCF farmers to assemble a team of farmers, university specialists, and sustainable agriculture organizations to conduct research and outreach on CSA. Initial research goals included analyzing management and production systems on area CSA farms, identifying the organizational and philosophical foundations of successful farms, and determining the economic and social impacts of

CSA. In 1993 PCF secured the first of two USDA grants to lead a study that eventually grew to include twenty-four farms serving the vicinities of Minneapolis-St. Paul and Madison (Kragnes and Hall 1993).

A regional case study integrating a combination of qualitative and quantitative research methods seemed best suited for exploring the research questions. This project utilized firsthand, ongoing interaction and participant observation with CSA farmers, farm coalitions, and farm members as well as formal interviews and surveys to investigate the collective and interactive activities surrounding CSA. Because access to knowledge dramatically affects the success of CSA projects, we wanted local farmers, farm members, and other organizers, to participate in and to learn from the research process as much as possible. As university researchers we viewed ourselves as committed participants and co-learners rather than as detached observers in the manner detailed by Maguire (1987) and Fals-Borda and Rhaman (1991).

Research with farmers and members from twenty-four Minneapolis and Madison area CSA farms took place at various times from 1993 to 2003. Data collection was designed to capture the four analytical dimensions of (1) farmer participation, (2) member participation, (3) member/farmer relationships, and (4) farm-to-farm interactions. Research with members was based on year-end surveys, focus groups, and telephone interviews. A total of 642 year-end member survey responses were analyzed from 1993-94. In 1995 farmers decided to administer their surveys independently. From 1993 to 1995 six focus groups were held with 55 members, and telephone interviews were completed with 75 nonrenewing members. Farm level data were collected using in-depth personal interviews (in 1995 and 2003) and on-farm research by farmers. Our team attended and participated in everyday activities (harvesting and share pickups), special events (field days, potlucks, and organizational meetings), and the functions of the two regional CSA farm coalitions. In addition, we gathered and analyzed written materials such as newsletters, farm brochures, farm directories, meeting minutes, press releases, and media stories. In contrast to a strictly macro or microlevel of analysis, our approach recognized that attempts to resist and reshape the existing agrifood system are unfolding at different locations and scales. While we did not focus on global-level forces and actors, we explored CSA with reference to the dominant agrifood paradigm. We sought to integrate detailed analysis of local movement centers with a broader awareness of the ways macrolevel conditions and opportunity structures affect local actors, as well as the reverse (see Buroway 1991, 9).

Successes and Challenges at the Farm Level

The Face of the Farmer

In a Japanese version of CSA, known as *Teikei*, organizers stress the importance of seeing "the farmer's face on the vegetables" (Getz 1991). The face of CSA farmers in the upper Midwest is female and male, young and old, gay and straight, and white. These farmers share a uniquely challenging occupation. Unlike conventional farms with one or two primary commodities, CSA farms commonly produce more than forty different types of fruits and vegetables on an extended basis, perhaps in addition to supplemental meat, egg, herb, or flower shares. Typically, the ecologically based production practices employed are labor and management intensive. In addition to complex production, harvesting, and distribution schedules, CSA as currently configured requires farmers to develop and coordinate a social network of members.

Our farmers closely fit the profile of other CSA farmers reported from national surveys (Lass et al. 2003; Tegtmeier and Duffy 2005). The twenty-four farmers in the study had high levels of formal education, including graduate degrees in agricultural, as well as nonagricultural fields such as nutrition, education, business, and social work. On average these CSA farmers were younger than conventional farm operators. A few were raised on farms, but most had acquired their practical experience in other ways such as home and market gardening, apprenticeships, and commercial nursery work. Even those with farm backgrounds felt that CSA demanded a whole new set of knowledge and skills.

The farmers in our study were as likely to be female as male. We did not find typical patterns for the way labor and decision making were allocated on farms. Women farmers and apprentices appeared to have the same tasks and decision-making roles as their male counterparts. The division of labor on the farm and in the household appeared to be more strongly influenced by who worked off-farm than by prescribed gender roles, and unlike the typical U.S. farm, men were as likely to hold outside jobs as women. In most cases the partner without the off-farm job had primary responsibility for running the farm.

In contrast to the conventional construction of the nuclear family, and hence the family farm, CSA farm families were also breaking new ground. Three farms in our study group were owned and operated exclusively by female couples. Another farm was jointly owned and operated by two women and a man who shared a household. One family provided assisted living for two developmentally disabled adults who were integral to the farm. Another farm was run by two male friends and yet another by a single father. Finally, one farm was owned and managed by a collective. Thus while CSA farms often are referred to as family farms, this form of agriculture, at least in Minnesota and Wisconsin, is not reinscribing the gender codes associated with traditional constructions of the nuclear family. Indeed, CSA farmers are breaking down and reconstructing conventionally held notions of farmer and family. While it should in no way be inferred that male and female farmers experience their involvement in CSA in the same way, it is worth noting the new social space that may be opening up for people to define their own roles on the farm in accordance with their unique skills, interests, and goals rather than their gender.

Farmer Goals and Objectives

Farmers started out with very idealistic visions for the CSA movement. Those interviewed were nearly united in their assessment of the problems with the conventional agrifood system, their conviction that change was a necessity, and their desire to contribute to a larger social cause. Asked why they had chosen to become CSA growers, the answers reflected the multifaceted nature of the movement. Some found their calling in the simple act of feeding and nourishing others, while others found it in working outdoors or educating urban youth. The farmers universally emphasized their commitment to protecting and restoring the environment. Although not everyone was certified, all claimed to use organic or biodynamic farming methods. Many farmers were committed to addressing food security issues and had taken steps to provide food to low-incomes households. Issues of economics and lifestyle also were strong motivators. As one farmer said, "This is a way of redoing the whole economic structure so that the producer gets a fair share and doesn't take all the risk."

Where is the Community?

Despite holding strong ideals about CSA as a vehicle for agrifood system change, the logistics of managing a CSA farm turned out to be more challenging than most had initially realized. As originally conceived, the core of CSA is "the open support of households whose members are not actively farming but who share the responsibility, the costs, and the produce with the active farmers" (Groh and McFadden 1997, 31). While the original CSA farms in the United States may have been largely consumer driven, today's farms are primarily started, administered, and sustained by the farmers. Indeed, most farmers put their livelihoods on the line in order to establish and operate a farm, and yet they continually struggle to get their members to become invested. The need to attract and retain members to sustain their income leaves the farmers more dependent on the consumers than vice versa since the latter have plenty of other options. As one farmer put it, "We're still peddling vegetables." Another problem is that the social ties that form around the farm tend to radiate outward from the farmer to the members, rather than developing among the members. While some farms had a supportive community of members that stuck with the farm from year to year and worked together to support it, many farms struggled with high turnover and member apathy. Some have given up hope that their farm will ever generate the social capital they envisioned.

While most CSA farmers said that the CSA model provides distinct economic advantages in comparison with other farming options, they still do not feel that they are earning adequate returns (see also Tegtmeier and Duffy 2005). There is an obvious gap between the income levels of the farmers and those of the consumers they feed. While most farmers were willing to accept a lower income in order to pursue their farming dream, they generally lacked provisions for health care or retirement and found it challenging to make large capital expenditures such as acquiring land, a house, or equipment. Like conventional farm-

Table 5. Selected characteristics of CSA study farms, 2001

Farm	No. shares[a] 1996	No. shares 2001	Year started[b]	Still active? (last active year)	Estimated return rate (%)
A	13	0	1995	1999	50
B	15	10.5	1994	Yes	88
C	20	NA	1990	Yes	85
D	21	0	1993	1993	na
E	25	0	1993	1993	na
F	25	0	1995	1997	50
G	26	0	1994	2000	na
H	30	60	1995	Yes	30
I	30	0	1993	1995	50
J	33	33	1995	Yes	63
K	35	33	1994	Yes	70
L	40	65	1993	Yes	40
M	47	0	1993	1998	55
N	50	0	1994	2000	60
O	83	156	1994	Yes	90
P	85	20	1994	Yes[c]	75
Q	95	500	1995	Yes	60
R	100	55	1989	Yes	75
S	120	158	1992	Yes	94
T	125	75	1992	Yes[d]	50
U	130	24	1993	1997	75
V	146	220	1989	Yes	86
W	200	0	1993	1999	80
X	250	410	1993	Yes	73

[a]A full share equivalent is designed to feed a family of four with two adults and two children.

[b]No new farms were added to the study after 1995, although several new farms started after this time.

[c]All shares are u-pick.

[d]Former apprentice has taken over farm.

ers, most CSA farmers find that to some extent their farm income is still circumscribed by the cheap food policies of the marketplace. Rather than basing their prices on the real costs of production in partnership with a knowledgeable core group of members as envisioned (Lamb 1994; Groh and McFadden 1997), most farmers said they set their prices according to the limits of what local consumers would accept. In addition, most have had to diversify into other direct and wholesale markets as a way of generating added revenue. Consequently, CSA farmers often find themselves caught up in the same problems as conventional farmers, with their products and labor chronically undervalued. Given these tensions, it is not surprising that many farmers seemed uncertain about their future and turnover was substantial during the ten-year study period.

From our original group of twenty-four study farms, only fourteen are still involved with CSA, as shown in table 5. Of those fourteen one farmer has cut back to U-pick shares and another has turned the farm over to a former apprentice. While the total number of CSA farms in the region expanded rapidly after the first farms started in 1988, growth appears to have leveled off. The overall number of farms in the region has remained fairly constant due to a steady influx of start-up farms, but many of the CSA programs fold after only one or two years. When farmers were interviewed about their reasons for leaving CSA, the number one reason given was economics, followed by health and quality of life issues.

CSA in the upper Midwest appears to have reached a crossroads. Will farmer participation level off or even decrease in future years or will we see continued expansion? The various options for confronting the problems encountered by farmers need to be carefully evaluated. Already dismissed by some as yuppie chow, CSA is frequently criticized for its lack of affordability and accessibility. At the same time, however, from the farmer's standpoint it is clear that current share prices, ranging from $300 to $575, still are not high enough to reflect the actual costs of providing farmers and farm workers with a living wage. In most cases labor from volunteers or social capital has not been sufficient to compensate for this deficiency. On most farms members have yet to shoulder the true costs of producing food in a sustainable manner.

Consumers and CSA: Moving Beyond Vegetable Anxiety

The extent and quality of consumer support that can be mobilized for CSA will be important in determining its reach and long-term viability. What will be required to transform passive food consumers into farm members who are active, cognizant partners in creating new farming models that protect and nurture the environment and the farmer? To better understand what motivates people to join farms and become invested in the well-being of the farm over time, we analyzed the general characteristics of CSA members, their rationales for joining and leaving farms, their levels of participation in farm activities, and the changes they experience.

Farm Member Profile: Who Belongs to CSA Farms and Why

Like other CSA studies, we found limited socioeconomic diversity among members, with most being middle-class, urban, white, and highly educated. A growing minority of members came from rural towns located near the farms. Although most member households had two adults and around half had children, member composition varied by farm, with various farms attracting higher proportions of single, older, younger, female, or low-income members.

The reasons given for participation on member surveys were quite diverse. The top motivations for joining a farm, in declining order of importance, were obtaining fresh, nutritious produce, buying local produce, supporting small-scale farmers, obtaining a source of organic produce, and caring for the environment. Building community and a desire to learn about agriculture were ranked near the bottom of possible reasons for CSA participation. These findings—in which an interest in the personal benefits of alternative food streams such as health and taste take precedence over public concerns such as community and the environment—correspond to results of consumer-related research from other parts of the country (Hartman Group 2000, 2001; Ostrom and Jussaume in this volume). The lack of emphasis placed on *community* illustrates the divergence between farmer and member expectations.

The focus group format, which facilitates more indepth exploration of motivational frames, revealed a more complex interplay between self-interest and social values on the part of many members. While getting fresh, top-quality vegetables appeared critical to member satisfaction, this reason was seldom identified as a primary impetus for getting involved. Instead, many members offered well-developed critiques of the conventional agrifood system to explain their participation. They complained that most food was "trucked in" from great distances, was "too manufactured," and that you "can't know where your food is coming from or how it was produced." They expressed a belief that food was different from other commodities: "It's easy to start thinking

Table 6. Why members leave CSA farms, 1995

Reason	Percentage responding
Wrong vegetables/limited seasons	25
Out of town or moved away	16
Changing farms	11
Too much food	11
Have own garden	11
Poor quality	11
No time	8
Not affordable	6

Note: N = 75

about food as just another product. You can go to the store and buy a tomato or you can go to the store and buy a toaster and they feel kind of the same. I really like the idea of being acquainted with not only where your food comes from, but with the idea that it comes from the earth. Tomatoes aren't toasters. You can live without a toaster, but you can't live without food." Thus members had a continuum, or in many cases, a combination of rationales for participating in CSA, ranging from improving personal health to solving community and global problems.

Why Members Leave CSA Farms

Telephone interviews with a sample of nonrenewers revealed a variety of reasons for discontinuing membership. Table 6 shows that 11 percent of the nonrenewers contacted were actually just changing CSA farms. Another 36 percent of respondents were positive but reported circumstantial constraints such as a change in location, frequent traveling, time challenges, or focusing on gardening themselves. Often these respondents said they might join a farm again in the future if their circumstances changed.

Another significant cause of member attrition (36 percent) was supermarket withdrawal, a problem we characterized as "receiving the wrong vegetables in the wrong quantities at the wrong times." CSA clearly cannot compete with supermarkets when it comes to providing the staples people are accustomed to having on demand. The unprepared found it onerous to adapt their menus to the vagaries of seasonality and the midwestern weather instead of seeing each week's share as the "wonderful surprise offered up by the soil" referred to by more positive members. Many of the root crops well-suited for the region—like rutabagas, parsnips, celeriac, beets, and Jerusalem artichokes—were unfamiliar and unappreciated. Similarly, while such leafy greens as arugula and chard thrive in the heat of a midwestern summer, lettuce does poorly. Consequently, ex-members complained of too many strange root crops and unknown leafy greens and not enough of their favorites such as potatoes, head lettuce, tomatoes, corn, and broccoli. As one ex-member remarked, "The veggies were too weird. We ended up replacing things we didn't like by shopping at the farmers' market. Some people no doubt love purple potatoes, mystery greens and guess-a-squash as a staple. We weren't expecting this. We won't participate next year: it's another one of those theoretically good ideas, but it suited the farmer's convenience, not ours."

This issue also had an economic component. While a minority said they could not afford the payment (6 percent), many people felt that other shopping choices were more economical because they did not have to pay for food they did not want.

Other complaints about the vegetable quality, quantity, or selection—such as wormy corn, rotten melons, dirty carrots, unripe fruit, wilted greens, or a lack of variety—could be traced to farmer inexperience. Some former members clearly understood and agreed with the larger principles behind CSA, but a belief in such ideals was insufficient to sustain their participation if the quality was lacking or they could not cope with eating and preparing the vegetables.

Levels of Participation in the Farm

As described previously, a challenge for nearly every farmer was figuring out how to induce greater member investment in the farm. While many members seemed to enjoy their vegetables and farm activities, most did not initially see themselves as integral to farm operations. The least invested members saw themselves as buying a service or a product. If something went wrong with a crop, a share box, or a drop-off site, they logically concluded that it was the farmer's responsibility and sometimes even requested a monetary refund. More involved members often developed a personal connection with the farmer and were more

Table 7. Willingness to participate in farm activities, Madison CSA farms, 1994

	Strongly agree (%)	Agree (%)	Disagree (%)	Strongly disagree (%)	No answer (%)
Would like to actively participate more in my farm	5	40	36	3	16
Would like to attend more festivals and field days	17	34	3	1	45
Would like to help more with farm work	8	19	12	4	57
Would attend planning meetings	8	16	16	2	58
Would help with phoning and newsletters	2	14	20	4	60

Note: N = 274

tolerant of inconveniences or problems with particular crops or shares. Finally, at the most involved end of the spectrum were the minority who took active responsibility for problems as they arose and volunteered for specific farm tasks such as planting, weeding, harvesting, packing boxes, organizing drop-offs, writing newsletters, or serving on core groups (a committee of members that helps make farm management decisions).

This spectrum of willingness to become involved in farm activities is evident from the survey responses shown in table 7. While 45 percent of respondents indicated they would like to participate in their farms more, this interest centered on fun activities like festivals and field days. When queried about helping out with specific farming, organizational, or administrative tasks, interest declined sharply. There was little recognition of the farm as an organization that required planning, communication, and effort. One comment scribbled in the survey margins captures this perspective: "I hate meetings! I certainly do not want a meeting to discuss the family vegetables." Although most members appeared content to remain peripherally involved in the farm, a significant minority on each farm expressed willingness to help out with the farm tasks listed, indicating clear opportunities for strengthening member involvement.

Do Members Change as a Result of Participation?

Understanding whether ideological, behavioral, or knowledge shifts occur as a result of CSA participation is central to assessing its potential as a movement as well as for improving the educational and organizational tactics employed. It is the vegetables that force the most obvious changes in members' everyday routines. Because they had to adapt their menus and diets to the contents of their weekly share, members were induced to experiment with new recipes and to eat foods that they would normally never have tried. Changes typically noted included eating more and fresher vegetables, eating a greater variety of vegetables, having a healthier diet, and shopping less. Variations on this theme were reiterated throughout the surveys and focus groups: "It changed our eating habits. . . . We eat better with more variety. We eat things we never would have before like squash and kale that are very nutritious. We shop a lot less." Around 90 percent of survey respondents said that their household eating and shopping habits had changed in positive ways as a result of CSA membership.

The changes were not always easy. Even enthusiastic members conceded that, although they liked "having to do something with what they got," at times their lives were too busy, and they ended up wasting vegetables. One woman coined the term "vegetable anxiety" to describe the way she felt when it was time for another CSA delivery before she had used up the vegetables from the last one. Whether or not people experienced the changes prompted by the vegetables as positive or negative depended upon their ideological frames and their practical knowledge of cooking and storage options.

Some participants had clearly joined a farm out of prior concern about the conventional agrifood system; CSA just provided them with an opportunity to "put their money where their mouth was," as one participant explained. Others commented that getting to know a farm had made them more knowledgeable eaters: "There is much more consciousness in the house about where food comes from and how its growing, transporting, and processing impacts the earth. Responsibility to eating sanely has grown a great deal." Members said that CSA had changed "what they talked about." As food had begun to take on expanded meaning for them, it had become a more frequent topic of conversation at home and at work.

Thus changes took place on various interrelated levels. Even though people joined CSA farms with interests ranging from personal health to new social, economic, and environmental relationships to complex philosophical and symbolic statements, they were unified by the common thread of the lifestyle shifts that they experienced. Even the least involved members had to make some lifestyle adjustments in order to cope with the vegetables. Some social movement theorists would argue that it is through doing, through such small changes in everyday life habits, that evolution in meanings eventually occur (Melucci 1985). Accordingly, part of the power of CSA as a social movement lies with its ability to gradually forge a new understanding of what it means to eat.

What Distinguishes Successful CSA Farms?

The farms that have persisted over time have found successful ways to bridge the divide between member and farmer expectations, thereby reducing member turnover (see table 5). They also have improved their overall efficiencies, devising ingenious systems for producing, harvesting, and distributing quality products in labor-saving ways. Many have addressed the issue of supermarket withdrawal by providing clear introductory information to prospective members before they join, tips and recipes for coping with excess and unfamiliar vegetables, and choices wherever possible. Others are addressing the problem of member apathy by requiring participation in tasks like harvesting and share distribution. As capital expenses were being paid off, knowledge efficiencies improved, and stable member communities evolved, many farmers became increasingly optimistic.

As partially illustrated by table 5, each farm is organized in a unique way. Our study identified three generalized management strategies that appear to result in stable farm enterprises with a committed consumer base:

1. The *classical* approach, based on the original model of CSA described by Groh and McFadden (1997), demands the most from members, engaging them directly in decision making,

operational tasks, and financial management through a core group and clear-cut volunteer expectations. Three of the study farms were close to achieving this ideal type even though they were initiated by farmers instead of consumers. This type of farm was the most successful at disengaging and insulating farmers from the conventional market system.

2. The second management strategy, the *nonprofit*, is an innovative variant of the first. One farm has incorporated as a nonprofit, formed a board of directors, and taken on an educational mission. It successfully organized an expansive capital campaign to purchase the farm and bordering natural areas and preserve them in a land trust.
3. The *entrepreneurial* approach was business oriented and farmer directed. This management strategy was based on improved efficiencies, hired labor, and customer service. While at first glance this model might appear contradictory to the original CSA ideals, the entrepreneurial farmers had a strong ideological commitment to the environmental and social principles of CSA, worked hard to educate their members, and played key leadership roles in the movement. Their farms had among the highest quality produce, the highest member return rates, and they successfully supported the farmers without off-farm income.

While the three management strategies challenged conventional marketing relationships to varying degrees, they each provided important educational opportunities for members and security for farmers. Contrary to many opinions expressed within the CSA movement, the classical ideal of CSA farming, based on community building, shared leadership, and volunteer labor, may not be the most appropriate or realistic model for every farmer or consumer.

The Future: Stronger Ties among Farms

Some have argued that resolving the discrepancy between what members want to pay and what farmers need to earn will inevitably lead to vast increases in efficiency and scale. Such a result could enable farmers to earn a decent living while keeping the food affordable and accessible. Trends toward large CSA farms are already occurring here and in many other parts of the country. In addition, some farms are finding it more economical to act as brokers, collecting products from specialty growers or wholesale houses and assembling them into customized CSA shares for distribution on a large scale. Whether a farm can deliver more than two thousand shares a week and still uphold the idealistic CSA values and principles, such as building a personal farmer and a place, remains to be seen.

In a considerably different approach, many farmers in the upper Midwest and elsewhere see networking with one another as a way to build efficiencies and to overcome the obstacles encountered in CSA. Farmer coalitions in Minnesota and Wisconsin have long been important for sharing knowledge and equipment, coordinating outreach and building public awareness, developing resources such as cookbooks and directories, and addressing the needs of low-income and special needs families (Ostrom 1997b). According to recent interviews, many farmers are ready to take farmer-to-farmer cooperation to new levels. One farmer said, "I'd like to see a growing network of CSAs around the state or around the country serving local communities. It would be important to bring a lot of other farmers into the CSA loop with other products, like meat, eggs, or honey, or whatever, so that CSA becomes part of a much bigger process of people eating locally and with the seasons." Another suggested that it could be more efficient to have some farms specialize in certain products for distribution by multiple farms. This kind of cooperation was already happening to some extent for certain crops, and there are successful national examples of multiple farmers contributing specialized crops to comprise a complete CSA share, for instance, as does the Pike Place Market Basket CSA in Seattle. However they may evolve, farmer networks offer an exciting opportunity to complement and strengthen the efforts of individual farms.

Implications for the Democratization of Agrifood Systems

CSA is by no means a complete model for effecting change, but it does illustrate the astonishing rapidity with which new ideas and lifestyle shifts can take hold. The CSA movement shows how practical solutions to global-level problems can be effectively implemented and spread from one grassroots setting to another when change is conceived within a positive and achievable framework. Rather than focusing on confrontational protests or the political process, this movement has focused more on the builder and weaver work discussed in part 1 of this volume. Among the most interesting features of CSA is the potential over time for participants themselves to become transformed as a result of their involvement. Thousands of new consumers around the country are now refashioning their daily eating, cooking, and shopping routines around the seasonal output of local agroecosystems. Hundreds of new people are now farming or are farming differently than they were fifteen years ago. Our research indicates that rather than going along on a whim or a fad or to make a profit, significant numbers of farmers and consumers in the upper Midwest are passionately committed to the underlying ideals and principles of CSA. Many participants in the movement are convinced that by reorienting their everyday habits and lifestyles in accordance with their values they can effect change at a wider level. Where the practical needs of the members and farmers can be met, the experience of CSA invites movement across the barriers that separate such issues of personal concern as health from issues of wider social and political significance. Melucci has commented that contemporary social movements derive much of their power from their ability to integrate both "private life in which new meanings are produced and experienced and publicly expressed commitments" (1989, 206). While many CSA members are a long way from making a public or a personal commitment to changing the agrifood system, let alone the economic system, at a minimum thousands of them are literally chewing on the roots of a new agriculture. As they eat, they gain opportunities to increase their understanding of food, the challenges faced by farmers, the needs of the environment, and the potential role informed citizens can play in reshaping food and economic systems.

Rather than evaluating the success of this movement by traditional measures—sales volume, degree of institutionalization, or ability to influence the political process, among others—it may be most pertinent to evaluate the transformative potential of the CSA movement with reference to the dynamic new ideas it has generated. While there is tremendous potential to expand the amounts and types of food that can be supplied through a CSA model, these farms are unlikely to ever become the major producers of food in the United States. Indeed, it is not clear at this point whether the tensions between the contrasting expectations of the farmers and the consumers in the movement can be resolved using existing organizational forms. Regardless of whether CSA persists in its current configurations, its lasting legacy may turn out to be the ideas it has set in motion. The very concept of CSA restores a sense of agency to local communities and begins to suggest the elementary outlines of what an economic system driven by local needs rather than international markets might look like. Linking the producers and consumers of goods and services at a local level has the potential to return certain aspects of economic and environmental decision making to communities, thereby restoring some degree of local control over the material conditions of everyday life.

Acknowledgments

The author would like to acknowledge all the members of the research team for their tireless work on this project over many years. This study would not have happened without the leadership and guidance of Verna Kragnes and Rick Hall at Philadelphia Community Farm and John Greenler at Zephyr Community Farm, as well as the patience of all the participating CSA farms. Further, John Hendrickson and Steve Stevenson at the Center for Integrated Agricultural Systems, University of Wisconis–Madison, have been invaluable partners in this research effort for more than a decade. Finally, Cynthia Cone of the Department of Anthropology, Hamline University, has played a key role.

References

Allen, P., M. FitzSimmons, M. Goodman, and K. Warner. 2003. "Shifting plates in the agrifood landscape: The tectonics of alternative agrifood initiatives in California." *Journal of Rural Studies* 19:61–75.

Blake, B., S. Junge, R. Ingram, G. Veerkamp, and M. Rosenzweig. 1995. *Community Supported Agriculture-Making the Connection: A Handbook for Producers*. Auburn CA: University of California Cooperative Extension Placer County.

Buroway, M. 1991. "The Extended Case Method." In *Ethnography Unbound: Power and Resistance in the Modern Metropolis*, edited by M. Burowoy. 271–81. Berkeley CA: University of California Press.

Buttel, F. 2000. "The recombinant BGH controversy in the United States: Toward a new consumption politics of food?" *Agriculture and Human Values* 17:5–20.

Cone, C., and A. Myhre. 2000. "Community-supported agriculture: A sustainable alternative to industrial agriculture?" *Human Organization* 59 (2):187–99.

DeLind, L. 1993. "Market niches, 'cul de sacs,' and social context: Alternative systems of food production." *Culture and Agriculture* 47 (Fall) :7–12.

DeLind, L., and A. Ferguson. 1999. "Is this a women's movement? The relationship of gender to community-supported agriculture in Michigan." *Human Organization* 58(2):190–200.

Fals-Borda, O., and M. A. Rahman, eds. 1991. *Action and Knowledge: Breaking the Monopoly with Participatory Action-Research*. New York: Apex.

Getz, A. 1991. "Design for community: Consumer-producer co-partnerships, a direct marketing approach." *The Permaculture Activist* 25 (Winter): 1–10.

Gregson, R., and B. Gregson. 1996. *Rebirth of the Small Family Farm: A Handbook for Starting a Successful Organic Farm Based on the Community Supported Agriculture Concept*. Vashon Island WA: IMF Associates.

Groh, T., and S. McFadden. 1997. *Farms of Tomorrow Revisited: Community Supported Farms, Farm Supported Communities*. Kimberton PA: Biodynamic Farming and Gardening Association.

Hartman Group. 2000. *The Organic Consumer Profile*. An Industry Series Report. Bellevue WA: Hartman Group; Surrey BC: Hartman Group CANADA. http://www.hartman-group.com/products/reportorganicprofile.html (last accessed April 20, 2006).

———. 2001. *Food and the Environment Update*. An Industry Series Report. Bellevue WA: Hartman Group; Surrey BC: Hartman Group CANADA. http://www.hartman-group.com/products/reportfoodupdate.html (last accessed April 20, 2006).

Henderson, E., and R. Van En. 1999. *Sharing the Harvest: A Guide to Community-Supported Agriculture*. White River Junction VT: Chelsea Green.

Hendrickson, J. 1996. Providing practice in the art of commensality, community, and the moral economy: A case study of Zephyr Community Farm. Master's thesis, University of Wisconsin–Madison.

Hinrichs, C. 2000. "Embeddedness and local food systems: Notes on two types of direct agricultural market." *Journal of Rural Studies* 16:295–303.

Hinrichs, C., and K. Kremer. 2002. "Social inclusion in a Midwest local food system project." *Journal of Poverty* 6 (1): 65–90.

Imhoff, D. 1996. Community Supported Agriculture: Farming with a Face on it. In *The Case against the Global Economy: And for a Turn Toward the Local*, edited by J. Mander and E. Goldsmith, 425–33. San Francisco: Sierra Club.

Kane, D., and L. Lohr. 1997. "The Dangers of Space Turnips and Blind Dates: Bridging the Gap Between CSA Shareholders' Expectations and Reality." Stillwater NY: CSA Farm Network: 2.

Kragnes, V., and R. Hall. 1993. "Sustainable Community Values Project." North Central Region 1993 LISA Program Proposal. USDA.

Lacy, W. 2000. "Empowering communities through public work, science, and local food systems: Revisiting democracies and globalization." *Rural Sociology* 65 (1):3–26.

Laird, T. 1995. CSA: A study of an emerging agricultural alternative. Master's thesis, University of Vermont.

Lamb, G. 1994. "CSA: Can it become the basis for a new associative economy?" *Threefold Review* 11:39–44.

Lass, D., G. W. Stevenson, J. Hendrickson, and K. Ruhf. 2003. "CSA across the Nation: Findings from the 1999 Survey." Madison: Center for Integrated Agricultural Systems, University of Wisconsin–Madison.

Maguire, P. 1987. *Doing Participatory Research*. Amherst MA: University of Massachusetts Press.

Maxim, N. 2006. "Mission Possible." *Gourmet*, June.

Melucci, A. 1985. "The symbolic challenge of contemporary movements." *Social Research* 52(4): 789–816.

———. 1989. *Nomads of the Present*. Philadelphia: Temple University Press.

O'Hara, S., and S. Stagl. 2001. "Global food markets and their local alternatives: A socio-ecological economic perspective." *Population and Environment* 22(6): 533–54.

O'Neill, M. 1997, "Sharing the harvest: Urban living off the land." *New York Times*, July 9.

Ostrom, M. 1997a. Toward a community supported agriculture: A case study of resistance and change in the modern food system. PhD diss., University of Wisconsin–Madison.

———. 1997b. "Community Farm Coalitions." In *Farms of Tomorrow Revisited: Community Supported Farms, Farm Supported Communities*, edited by T. Groh and S. McFadden, 87–102. White River Junction CT: Chelsea Green.

Rowley T., and C. Beeman. 1994. *Our Field: A Manual for Community Shared Agriculture*. Guelph, Ontario: University of Guelph.

Sugarman, C. 1991, "Share the land: An Innovative Way to Shoulder the Burden and Save the Family Farm." *Washington Post*, May 15.

Tegtmeier, E., and M. Duffy. 2005. "Community Supported Agriculture in the Midwest United States: A regional characterization." Ames IA: Leopold Center for Sustainable Agriculture, Iowa State University.

VanderTuin, J. 1987. "Vegetables for all." *Organic Gardening* 34 (9): 72,75–78.

Wilson College Center for Sustainable Living. 1998. *The Community Supported Agriculture Handbook: A Guide to Starting, Operating, or Joining a Successful CSA*. Chambersburg PA: Center for Sustainable Living, Wilson College.

6. Food Policy Councils

Past, Present, and Future

Kate Clancy, Janet Hammer, and Debra Lippoldt

State or provincial and local policies greatly influence the sustainability of food systems. For example, land use and transportation decisions affect farm viability and food access; education and public health programs influence the ability of citizens to effectively participate in the food system; institutional purchases shape the local economy and environment; various programs influence resource efficiency in farm and nonfarm businesses and homes; and economic development strategies inhibit or promote a more sustainable food economy.

The food policy council (FPC) has emerged over the last two decades as a potentially useful tool in shaping state and local policy agendas to support sustainable food system goals. FPCs are institutions that can bring a broad array of people together to consider and respond to connections among diverse but interrelated facets of the food system. The values and visions that underline the development of these bodies are myriad. They revolve around a desire that local and state governments assume responsibility for the food needs of their citizens. Councils hope to ensure that an adequate and nutritious food supply is available to all citizens, that they can strengthen the economic vitality of the local food industry, that they can improve citizen food choices, that they can increase local food production, and that they can minimize food-related activities that degrade the natural environment.

This chapter reviews the history and performance of government-sanctioned food policy councils in North America. Cases examined include a range of FPCs—enduring, foundering and failed. Attention will be given to what has worked and what has not, as well as to intended and unintended outcomes. Lessons will be drawn regarding the potential of FPCs to play a leading role in developing local and regional food systems. The analysis is based on interviews with key leaders from FPCs as well as from a review of relevant literature and archival material.

It must be emphasized that this chapter examines only FPCs that are government sanctioned and have a minimum three-year history of operation. We recognize that many different organizations work on food policy in their communities; however, a comprehensive analysis of all these efforts is beyond the scope of this chapter. One can find descriptions of many of these entities elsewhere (Biehler et al. 1999; Borron 2003; Dahlberg et al. 1997; Hamilton 2002).

A Short History of Food Policy Councils

Local Councils

Bob Wilson could not have known what he started when he assigned his University of Tennessee landscape architecture class to study food-related planning issues in Knoxville in 1977 (Blakey et al. 1982). The students' findings and Bob's vision triggered the development of a food policy council in the city, formed soon after the passage of a resolution by the city council in October 1981. The Council's resolution stated that food was a matter of governmental concern and encouraged the formation of a group to "continually monitor Knoxville's food supply system and to recommend appropriate action to improve the system as needed" (Knoxville Food Policy Council 1988).

Knoxville's council was followed by at least five local food policy Councils or commissions that were officially sanctioned by city or county governments and by more than thirty ad hoc committees and coalitions that have offered recommendations to various city and county agencies and policy bodies over the last twenty years. Onondaga County, New York, formed the second local FPC in the country. Preceded by a formal food and nutrition policy for the city of Syracuse signed by the mayor in 1976, the county executive and county legislature jointly issued a resolution in March 1984 that established the Onondaga Food System Council that was mandated to study and discuss local food issues and to advise the Planning, Research, and Development Committee of the county legislature.

Also in 1984 the U.S. Conference of Mayors (USCM), struck by the "emergence of hunger as a serious urban problem," recruited the entrepreneurial mayors of five cities, including Knoxville, to examine both the issues related to food and nutrition and the feasibility of establishing a municipal food policy (U.S. Council of Mayors 1985). Two of the cities, Charleston, South Carolina, and Kansas City, Missouri, did not accomplish much after the adoption of food policies by their respective city councils. Charleston's efforts fell victim to recovery demands imposed by Hurricane Hugo, and Kansas City's efforts suffered from the domination of the council by the food bank that effectively limited the scope of the council's efforts.

Philadelphia's Food and Agriculture Task Force met for several years, but a change in the job duties of the chair and problems securing funds made it difficult to proceed. In St. Paul, building on earlier work of the Minnesota Food Association and the Mayor's Homegrown Economy Strategy, an ad hoc task force was established to develop a food policy. The policy was adopted in 1985 by the city council, and the St. Paul Food and Nutrition Commission was charged to carry out its program.

In 1990 and 1991, respectively, food policy councils in Toronto and Hartford were formed (City of Hartford 2002; Toronto Food Policy Council 1995), making six councils or commissions that were brought into existence through a resolution of a local governmental body (the definition of a food policy council used in this chapter) between 1981 and 1991. Of those six, three were still active in 2003. An analysis of the durability or demise of these councils is presented below.

State Councils

The history of state food policy councils differs from those at the local level and begins earlier. State nutrition councils existed in the 1960s, those in California and Illinois being perhaps the best known. In the early 1970s in reaction to food price and energy crises, the governor of Massachusetts appointed a Commission on Food that did a thorough study of the food situation in the commonwealth and that pointed out that there was "no central focus in the state government for the coordination and implementation of policies and programs necessary for the food system to operate efficiently and equitably in providing a wholesome and dependable supply of food to Massachusetts consumers"

(Harvard Business School 1974). The commission recommended that an office of food policy be established in the governor's office. No such office was created, but the commission did continue to engage with food policy issues for a period of time.

In 1975 the New York State Assembly created a Task Force on Farm and Food Policy, composed of assembly members, that was quite active. After more than a decade and many conferences, hearings, and other machinations, a State Council on Nutrition and Food Policy finally was established in 1986. The council was housed in the Department of Health and took as its principal task the development of a five-year food and nutrition plan that contained recommendations regarding the coordination of food and nutrition programs, food access, diet-related disease, monitoring, farmland preservation, agricultural development, infrastructure, food safety, marketing, transportation, and food affordability (New York State Council on Food and Nutrition Policy, 1988). It discontinued operation several years later.

In 1997 a third northeastern state, Connecticut, established a food policy council within its Department of Agriculture. Its charge was to promote the development of a food policy for Connecticut and the coordination of state agencies that affect food security (Connecticut Food Policy Council 2001).

More recently, councils in Iowa, New Mexico, North Carolina, Oklahoma, Utah, and Michigan have been created, several through the assistance of the USDA Risk Management Agency. In Iowa the FPC developed from efforts to focus on local foods and the desire to "expand the state's food system." This council is the only one that is administered by an academic unit, the Agricultural Law Center at Drake University in the state capital Des Moines, in cooperation with the office of the governor (Hamilton 2002). The New Mexico Food and Agriculture Policy Council came into existence through a memorial of the legislature and is coordinated by a nonprofit organization, Farm to Table. The councils in North Carolina, Oklahoma, and Utah are housed in their state departments of agriculture.

The initiation and demise of FPCs around the country is dynamic. Information can be found at several Web sites listed in the references (see Drake Agricultural Law Center 2005; World Hunger Year 2007).

Why Councils Emerged

Over the thirty-year history of food councils and commissions in North America, different clusters of local, national, and global crises and concerns have prompted arguments for their establishment. In the early years key drivers included rising food prices, the oil crisis, and a fresh sensibility to food engendered by the back-to-the-land enthusiasts of the 1960s that eventually combined with an interest in sustainable agriculture that emerged in the 1980s. The early 1970s also brought efforts by consumer groups to influence the agricultural policy agenda. Later, the Reagan administration's assault on nutrition programs during a recession and the alarming growth in urban hunger provided additional impetus. Then the farm bankruptcies in the mid-1980s, along with expanding public environmental consciousness, helped more people to recognize the need for a systems perspective on food. In the last decade an emphasis on participatory democracy, the growing attraction of local and regional food concepts, and the problems of globalization and concentration in farming, ranching, and agribusiness have provided more arguments for directing the attention of policymakers to the importance of the food system and to their failures thus far in assuring its safety and sustainability.

While these factors have shaped the larger context, each local council, as expected, has its own idiosyncratic history. Most revolve around a key leader or core group: in Knoxville an enterprising professor; in Onondaga County a very persistent community activist; in St. Paul a city employee who would not let the farmers' market be demolished; and in Toronto a group of determined food systems visionaries. The fortuitous availability of the USCM grants added resources to the efforts in Knoxville and St. Paul.

In states the larger context was and is the same, but as suggested, the specific impetus has come from a larger number of leaders and more frequently than in the case of local councils from top political leaders rather than from grassroots efforts. In addition, while both local and state councils face the same larger national context, the broader scope of state responsibilities as well as the strong presence of agricultural commodity organizations often sets a very different political context for the operation of state councils.

What FPCs Look Like and Do

We define Food Policy Councils as officially sanctioned bodies of representatives from various segments of a state or local food system (Hamilton 2002). Dahlberg (1994) describes the structure of six local councils (not all so sanctioned), including Knoxville, Onondaga County, and St. Paul (see his report for much greater detail). In most councils the members were or are appointed by the mayor or county or state legislatures, and most have ex officio or special representatives from government agencies. The types of groups represented on councils vary. Some have membership that focuses on urban food access, while others include a wider range of farming, health, and environmental interests (Borron 2003; Dahlberg 1994). Most of the groups have either paid or in-kind staff from agencies such as a department of health or Cooperative Extension or from nongovernmental organizations (NGOs; for example, The Hartford Food System) (Borron 2003). Most, except Toronto, have little or no consistent funding.

Examining their own words as taken from various resolutions, council goal statements, bylaws, and so on, one can see what FPCs are created to do:

- Advise and make recommendations to state, city, and county government on food policy issues
- Monitor and evaluate the performance of the local food system
- Foster better communications among all actors in the food system, including policymakers and the public
- Assist residents in understanding the food system and food policy
- Act as a forum for discussions on improving the food supply
- Educate FPC members about each others' roles and concerns
- Plan and oversee food system projects
- Facilitate research on food issues

The topical issues that FPCs address run the gamut from hunger to farmland protection, from community development to composting (Biehler et al. 1999). In the following sections we describe in detail the activities of local and state councils as reported by key council actors.

Research Protocol

Because it has been almost ten years since Dahlberg (1994) conducted his study of FPCs, it is important to examine what has happened in the interim and also to consider councils formed at the state level. Through an analysis of interviews with key actors in two state food policy councils (Connecticut and Iowa) and in six regional or local councils that are either currently or were previously in operation, we can report on successes and failures. From the interviews and other literature we also draw lessons learned regarding the potential that FPCs have to shape the development of local and regional food systems.

From a review of the literature, Web sites, and informal conversations, we identified more than thirty possible food policy councils. We identified contacts for each entity through knowledge of one of the authors or by consulting publications. Whenever possible, e-mail or telephone contact permitted us to determine whether the council (1) had official government sanction, (2) had been in existence for at least three years, and (3) included a multi-issue food system focus rather than a single issue focus such as antihunger or promotion of agriculture. We selected these review criteria in order to make the most meaningful comparisons between councils and their outcomes.

Government sanction has been suggested by many as important for success of councils (Clancy 1988; Hamilton 2002). Further, food policy councils typically are distinguished from issue-oriented councils or task forces. The latter, for example, concentrate on combating hunger or ensuring farmland protection, while FPCs are mandated to consider a range of food system issues from production to access to consumption. Many other entities also have legitimate stakes and roles in influencing food policy, including NGOs, advisory groups, trade groups, and so on; these have been described elsewhere (Biehler et al. 1999; Borron 2003; Dahlberg 1994; Hamilton 2002). Although much could be learned from recently formed FPCs, such as the Portland/Multnomah Food Policy Council, the New Mexico Food Policy Council, and others now in the early stages of development, the experiences of these councils are still emerging. Inclusion of all potentially related policy efforts remained beyond the scope of this chapter.

Table 8. Summary of studied food policy councils

	Start/End	Paid staff	Funded	Self-rating of success[a]	Reason for demise	Future outlook
Local/Regional						
Knoxville	1981	City (In-kind)	In-kind	Was 4, now 2	—	Same structure; more focus
Onondaga	1984–95	For 2 yrs	In-kind/One 2 yr grant	3	Funding politics	—
St. Paul	1985–99	For 10 yrs	For 10 yrs	3	Funding	—
Toronto	1990	Yes	Yes	5	—	Sustainability of council; compensation of members
Hartford	1991	No	Yes and in-kind	4	—	Little change; need for vigilance and intention remains
Los Angeles	1997–2001	Yes	Yes	3	Politics	—
State						
Connecticut	1997	Contract	State funds 1997–2002; grant 2003	4	—	50/50 chance of survival, depending on politics and leadership
Iowa	2000	Yes	In-kind and grant	3	—	Uncertain; depends on governor

[a]Based on a scale of 1 to 5, with 1 being "low" and 5 "high."

Eight FPCs met the criteria for our study and were selected for interviews: Connecticut Food Policy Council, City of Hartford Advisory Commission on Food Policy, Iowa Food Policy Council, Knoxville Food Policy Council, Los Angeles Food Security and Hunger Partnership, Onondaga Food System Council, St. Paul Food and Nutrition Commission, and Toronto Food Policy Council. In order to remove some potential biases or vested interests, we identified key actors who had at least a three-year history with the council but who were not in most cases its founder.

Using the elements for a successful FPC that Clancy described in 1988, we developed an interview instrument (see the appendix to this chapter). We contacted each respondent to schedule a time to talk and requested any relevant written information in advance. Telephone interviews lasted from thirty minutes to an hour and took place in July and August 2003.

Findings

The following sections present the interview findings based on the responses of the key actors. Table 8 provides summary information for each council. We offer some analytical reflections about the results in the final section of this chapter.

Longevity

Of the eight councils studied, one (Iowa) was returning from a one-year hiatus (due to failure of the governor to renew its sanction as required every two years), and three (Los Angeles, Onondaga, and St. Paul) no longer exist. The Knoxville Food Policy Council has the greatest longevity, having been established more than twenty years ago. Onondaga and St. Paul, both established in the mid-1980s, existed for at least ten years before disbanding. Hartford and Toronto, both established in 1990/91, are still functioning as of this writing. It is possible that Los Angeles will resume a council structure in the future. The state food policy councils are the most recently established councils in our study.

Government Sanction and Funding

As described earlier, these councils were all created by state legislation or city/county resolution. Two of the councils (Knoxville and St. Paul) made a transition some years after their founding from city to city/county councils, incorporating additional resolutions by the county government at the time of expansion. After six years Onondaga County became a nonprofit in order to apply for foundation grants. Several of the councils were created following a public forum or formal government process such as a task force on hunger. A champion such as a legislator or mayor was integral to the establishment of several councils.

Of the different types of support (administrative, staffing, and projects), only Toronto has received adequate and generally consistent funding for all three needs. Its support comes from the city's Board of Health. Three of the eight councils (Hartford, St. Paul, and Toronto) receive(d) funding from the government sponsor. A fourth (Connecticut) had state funding until the current year. Five councils have been supported by short-term grants. All councils relied on some form of in-kind support, typically as administrative support from a government sponsor (for help with minutes, meeting notices, and so on), but only a few have had support for actual staffing. Council members often contribute(d) significant amounts of in-kind support to agendas and projects. Two of the three councils (Onondaga and St. Paul) that are no longer active cite lack of funding as the key factor responsible for their demise.

Food Policy Council Membership and Leadership

Council membership ranges from 9 to 24 individuals with an average of 12 to 14 members. Councilors are officially appointed by city, county, state, or other government department leaders; in practice, most additional appointments often are made by recommendation of the incumbent councilors. Most councils try to maintain a mix of specific food system stakeholders from such areas as farming and agriculture, anti-hunger, health, food industry (such as processors and retailers), government agencies (such as departments of health and social services), and nonprofit organizations. Several councils include city or county councilors or commissioners, while others mandate representation by specific government departments, either as official or ex officio members.

In nearly all of the councils the chairperson is elected by the council members themselves for a one-to three-year term. One council (Iowa)

Table 9. Typical activities of food policy councils

	Legislation	Policy	Programs	Research	Education	Other
Local/Regional						
Knoxville	—	√	√	√	√	—
Onondaga	√	—	√	√	√	—
St. Paul	—	√	√	√	√	Networking
Toronto	√	√	√	—	√	Facilitation and brokering
Hartford	√	√	√	√	√	—
Los Angeles	—	—	√	—	—	Monitoring/Assessment
State						
Connecticut	√	√	√	√	√	—
Iowa	√	√	—	√	√	—

has an appointed chair. Most councils also report an informal leadership that fills a variety of roles from activist to visionary.

Activities of Food Policy Councils

We asked each respondent whether or not his or her council is or was engaged in certain activities. The responses are listed in table 9. The most frequently identified activities are programs and education, which are necessary if the councils are to fulfill their function of getting members and others engaged in the council itself and other food system activities.

Six of the eight councils engage(d) in policy activities of various kinds, although how frequently or visibly depends on the specific council. Especially in the local councils, policy activities have to be done carefully to keep all the members comfortable. For this reason one council engages with administrative processes but not legislative proposals per se. Another made forays into the policy realm but was not successful and has not tried again. In the state councils it has been more of a given that policy recommendations will be on the agenda.

Although only St. Paul explicitly mentioned its role as a network facilitator, many councils engage in such activity as they bring diverse stakeholders together with the result of increased understanding of food system issues and/or new or better programming. One respondent said "the connections made around the table can lead to individual actions by the departments. There is a degree of coordination and new ideas that emerge from getting together monthly."

Since September 11, 2001, several councils also have become active in emergency preparedness efforts that local and state governments have undertaken. This activity includes attention to food supply and transportation.

Documentation of Food System

All but two of the councils (Los Angeles and Toronto) report producing regular or episodic documentation on the local, regional, or state food system. Many of these reports are produced in collaboration with a council member's agency or organization. For example, the Onondaga FPC produced a directory of the food system through a council member's office at a local academic institution. The Knoxville FPC benefited from a council members' receipt of a USDA Community Food Projects grant to develop a food security indicator system, although lack of continuing funds has prevented updating of the indicators.

Engaging Government and Community and Visibility of a Council

Councils are mixed in their interest and ability to garner public attention. Some councils see a lack of public visibility as a liability. However, others do not have visibility as a priority. In fact, those councils see a low public profile as creating a stronger ability to work behind the scenes in less threatening, but more effective, interaction and networking with government agencies. Most of the councils that would like to create a stronger public image report limited success. They try to engage the public through programs, events, and public forums but have not often attracted large audiences or local media (Connecticut and Iowa being the major exceptions). Two councils (Knoxville and Hartford) have received some public attention through community awards programs. Ironically, councils with prominent members may find it difficult to gain public recognition as it is "challenging to clarify when the council is being represented as opposed to the individual with overlapping duties and affiliations."

Not surprisingly, councils whose membership includes specific government agencies and departments or that have a particular city councilor or county commissioner on the council report more collaboration and effective interaction with those agencies than councils that have no formal government representatives. Many councils report a period of orientation, or learning, during which a particular government agency representative initially does not understand the role or purpose of the FPC; however, over time, the relevance and opportunity become more apparent and the relationship proves to be very fruitful. For example, in Connecticut a representative of the state's Department of Transportation (CDOT) was appointed by the department commissioner to represent the CDOT on the FPC. Not until the FPC completed creation of the Connecticut Agriculture Map, targeted to consumers and the general public, did the CDOT representative begin to appreciate the value of the council's work.

Key Successes and Factors Leading to Success

As shown in Table 8, the respondents as a whole display variability in how they report the success of their councils—both across councils and over time individually. There are peaks and valleys for most councils, especially those that have no stable funding. All the respondents claim that networking at the table and the education that FPC members receive are the starting point for successes. Most report increased interaction with government agencies. All can point to successful projects: for example, the agricultural map of Connecticut, a nutrition education supervision position in the school system, farmers' market and coupon promotion programs, school breakfast promotion, transportation to improve food access, access to land for urban gardens, and institutional purchasing to emphasize local and regional foods. Some less concrete but still important accomplishments also have occurred, such as in Toronto where the term "food security" is now an accepted phrase among professional groups in the city.

The respondents listed a dozen factors that contribute(d) to their success: strong leadership and champions, vision, offering win-win solutions, very experienced people, member commitment, persistence, staff support from a key agency (Cooperative Extension or health department), diversity in membership, government engagement, connections to the community, running below the radar, and media exposure. Obviously, given the last two items, although each council may take a different approach in order to reach its goals, the need for leadership and commitment were clear and consistent themes for all FPCs.

Key Challenges

Without question financial support and leadership were the most frequently identified challenges facing these FPCs. The Toronto Food Policy Council is in the enviable position of enjoying relatively stable staffing and funding as compared to others that we studied. However, the Toronto FPC sees the need to find funding to compensate councilors for their service in the near future.

Lack of staffing or inconsistent staffing, a direct outcome of inadequate funding, is also seen as a problem for FPCs. Even the in-kind support they receive for staffing and administration is insecure and dependent on the director of the organization or on the department supplying the support understanding what a council is trying to accomplish.

The limited support from existing institutions (especially from farm and agriculture groups in some conservative areas), the lack of land-grant institutional support in the United States (this assistance varies from council to council), and the potential breadth of issues makes it difficult to focus on only a few issues and to accomplish any one thing. Further, as reported by the respondents, councils constantly need to bridge the gap between members who do and do not understand the concepts of food systems and food policy.

Politics and the reality of changing political perspectives, the whims of current administrations, financial decisions, and communicating the value of the FPC work remain ongoing frustrations. For example, the Toronto FPC faces the problem of getting the government to understand the value of funding an entity that will criticize it; yet one of its greatest contributions is as an advocate within the city structure.

Five Years from Now

When asked where the council was likely to be five years from now, two respondents representing councils (Onondaga and St. Paul) that no longer exist anticipated little chance for reinstatement. Factors that would influence future FPC formation include fiscal and political changes. The remaining respondents describe varying degrees of optimism for the future. Regarding Hartford, there is a sense that its council will look about the same since some of the problems facing food systems are perennial and since going over the same ground regularly may actually support the "vigilance and intention that is a good part of what the Commission is all about." However, the future of the Connecticut FPC is seen as more questionable, with the respondent giving it a fifty-fifty chance of its surviving, depending on the political, financial/economic, and leadership influences. The future of the Iowa FPC is also seen as somewhat uncertain, depending on politics and eventual shifts in who is state governor. The Knoxville council's future may depend on the quality of leadership within the council and the ability of the council to establish a clear plan and goals. The Toronto FPC is exploring options for future sustainability and hopes to secure funds to compensate the volunteers who currently serve on the council.

Advice for New Councils

"Get money, staff and government support" was the main advice to new councils offered by one seasoned FPC participant and echoed by other respondents. Government sanction is important so as to "get government on record as seeing their role in food planning." While no respondent claimed to have *the* model legislative language for the creation of FPCs, government sanction was seen as critical to establishing the legitimacy of the councils. Even with the sanction FPCs still find they struggle to establish their own place alongside other better-established commissions, such as those for water or the environment.

Unquestionably, funding is a particularly critical requirement. All councils struggle with the challenge of inadequate or nonexistent funding; as one respondent from a now-defunct council said, "[We were] always trying to survive and it took away from what we wanted to do." But current political and economic climates make it difficult to rely on public funding. For example, the Connecticut FPC at one point had its entire budget cut by the state legislature. Grants provide support, but such assistance is sporadic. Several councils (Hartford and St. Paul) received some support from city health departments, but it did not continue.

Staffing is another critical component. Only the Toronto Food Policy Council has had reliable government support for its director position. The council with the greatest longevity (Knoxville) interestingly has neither paid staff nor operating funds. Staffing has been provided from the Community Action Committee as an in-kind contribution since the inception of the council, and the Knoxville FPC relies on strong organizational support from members of the council for projects.

A final point of advice is inclusion of broad representation from across the food system in the development and operation of the FPC. Respondents cited diversity of food system stakeholders as key for starting a FPC, even though such broad representation may make narrowing the council focus to a manageable agenda more challenging.

Lessons Learned

Our research and the small body of literature that exists on this topic offer many lessons learned regarding the potential of FPCs to lead or participate in the development and sustainability of local and regional food systems. One of those lessons is that the development and successful institutionalization of FPCs requires overcoming the general lack of awareness of the food system as encompassing anything more than agriculture or hunger (Dahlberg 1994). Most authors of the literature referenced here have mentioned the low visibility of the food system to policy officials and residents. Pothukuchi and Kaufman (2000) and Abel and Thomson (2001) also have documented the low level of involvement by planners in food systems. A number of our respondents mentioned that many members of councils, including the chairs sometimes, just do not "get" the concept. Despite its indisputable place as a basic requirement for human survival, food is not accepted as the domain of government responsibility in the same way as other basic human needs such as water and housing. Food usually is taken for granted, is conceived as a rural and agricultural issue, and has a very strong free market aspect (Clancy 2004; Pothukuchi and Kaufman 1999; Roberts 2001). Given this backdrop, it is not surprising that none of the respondents reported that their FPC had reached the level of integration or institutionalization within the government that it desired.

It is curious that food policy councils have not enjoyed the same influence and importance as other citizen commissions, such as Water Quality, Planning, or Air Quality. A review of the similarities and differences between these public entities might shed light on the potential for increasing the legitimacy and influence of FPCs relative to these institutional counterparts. We believe this area is an important one for further research and one that could benefit from greater attention to framing, that is, thinking about the conceptual constructs that relate to people's values and beliefs (see the Frameworks Institute 2006; and Stevenson et al., this volume).

Another lesson relates to defining the appropriate institutional structure for an FPC. When advocates attempt to translate a vague concept into a new institution, inside or just outside of traditional government bureaucracies and agencies, it is easy to understand the tensions that arise. Only two local councils have had the distinction of being formal government bodies: an ordinance gave the St. Paul Commission formal status, and the Toronto FPC has always been a subcommittee of the Board of Health. Yet even with its designation St. Paul's council was not able to garner the credibility and resources of other commissions within the city's bureaucracy. The Toronto FPC succeeded for two major reasons. First, the city has a history and leadership role in the Healthy City movement. Second, the council has found a way to link all of its activities to public health so that its members are always fulfilling their mandate even if it might seem that they have strayed from the Board's interests.

The other councils provide examples of a hybrid institutional structure that is both pragmatic and problematic and suggests several more lessons. Internally—in the interaction inside an FPC among its members—there are many challenges and opportunities. The diverse representation on a council—a visual and interactive reminder of the complexity of the food system—stands for about as many different agendas as people. These people also may be more interested in process than action or in projects than policy or in a simple focus rather than multiple foci (see Biehler et al. 1999; Dahlberg et al. 1997). These differences can lead to divisiveness and inefficiencies and require consistent and strong leadership to overcome. Councils that have survived have likely profited from leadership that can capitalize on the diversity at the table. On the positive side diverse representation of food system stakeholders in FPCs results in important dialogue that is unlikely to occur elsewhere. The value of hearing perspectives from different sectors of the food system is difficult to measure but likely invaluable.

In general, FPCs also contribute to more effective monitoring of the food system, which encourages a more holistic view of government's role. In many cases the FPC is the only vehicle for an annual accounting of food system activities (Hartford, Iowa, Knoxville, Connecticut). Annual reports on the state of its local, regional, or state food system assist in holding governments accountable for their responsibility in food assistance, agriculture viability, local markets, and more.

Finally, FPCs initiate or carry out food system research that is unlikely to occur within conventional government structures. Often this research is presented in public forums that raise public or government awareness of an issue. For example, the Connecticut Food Policy Council's first public forum brought attention to the threats to farmland preservation in that state. This event resulted in renewed grassroots and government efforts to preserve Connecticut farmland from development.

Externally—in the relations between the FPC and the governmental and political entities with which it interacts—there are also pitfalls and rewards. The history we have reviewed suggests that it can be important to have a government champion in the early stages of setting up a council. Yet this was still not sufficient in the cases of St. Paul, Onondaga County, and Los Angeles. According to Rod MacRae, the first director of the Toronto Food Policy Council, the ideal situation for a FPC is to be tied to a government department (Borron 2003). This connection provides the most direct access to government, the opportunity to affect specific policies, and accountability. But as Biehler et al. (1999) point out, after speaking to many FPC leaders, local politics and bureaucracy can be difficult to navigate, political will can change, and politicians will retire or lose elections. When such changes happen, as in St. Paul, FPCs can lose previous support and resources. Local FPCs can also experience tension when political and legal separation of a town or a city from its school districts affects opportunities for food system activities such as school gardens, control of soda "pouring rights" and vending machines, and local sourcing for school breakfast and lunch, among other issues. (Kenneth Dahlberg, personal communication).

We agree with Dahlberg (1994) that "the more institutionalized a council, the more likely it is to have budget and staff support as well as perhaps some review and/or planning powers" and with MacRae (2002) that it is best to "try to avoid fund-raising." Yet Webb et al. (1998) found in their research that there were relatively few opportunities for FPCs to secure funding, and government budgets at the present time only continue to tighten. For this reason and others, nonprofit organizations may be a logical institutional structure from which to tackle food policy. They can more easily bring grassroots public pressure to issues than can FPCs and just as importantly apply for grants if government funding is not forthcoming. However, at least one respondent reported that local and national foundations would not support a council because FPCs can be seen as the local government's responsibility.

Given the varying situations facing currently operating councils, we present the recommendation offered by several respondents: that groups interested in a food policy council take small steps toward its creation, perhaps first through a representative task force that takes the time to understand the local context and to educate one another on relevant issues. There may be other institutional structures (for instance, ad hoc committees, study groups, and coalitions) that can accomplish specific tasks more efficiently (Biehler et al. 1999). This step also can be important for building both grassroots and administrative support.

A further lesson is that expanding interest in local food economies and renewed attention to emergency preparedness and the safety of the food supply from natural and terrorist threats may create opportunities to involve FPCs and potentially to enhance their role in government service. So far, however, even in this endeavor government agencies are not readily turning to FPCs for input, and FPCs have had to work to create a place at the table. Working with established emergency preparedness officials is likely to increase the profile and appreciation for what FPCs bring to the conversation. But it can also overwhelm the message regarding other critical food system issues and at the state level be controlled by powerful commodity groups (Dahlberg 2003).

A final lesson is that food policy councils clearly have a role to play in helping to shape a more sustainable local and regional food system. We have shown that food policy councils have a unique role to play as quasi-governmental bodies in putting food topics on politicians' radar, elevating discussions about food, making connections, and getting useful projects implemented. In these endeavors they exemplify weaver work, as described by G. W. Stevenson and colleagues in this volume. But success does not come easily. Few local and state governments appear willing or able to take on more responsibility for the long-term food security of their citizens. Therefore sustainable food systems proponents need to advocate for long-term visions and policies that promote the public good. Although there have been few evaluations of the outcomes of FPCs (see Webb et al. 1998), we heard from our respondents that they require patience, focused efforts within a systems viewpoint, a clear definition of success, an effective leader/champion, continuous leadership support, and last, but definitely not least, funding and in-kind support.

Food policy councils have been an underutilized tool for reshaping the food system. We note that our findings reflect only the past and present of government-sanctioned FPCs that endured for over three years. The environment and the concerns about the sustainability of the U.S. food supply may be quite different in the future. We trust that the councils that have recently started up and those being contemplated for the future can learn from the successes and failures of the ones that have pioneered this innovative institution.

Acknowledgments

The authors would like to thank Ken Dahlberg for his thoughtful review of an earlier version of this chapter and for his helpful suggestions. We also appreciate the time that our respondents gave in order to answer our questions and to reflect on their efforts to improve the food system in their home places.

References

Abel, J., and J. Thomson. 2001. *Food System Planning: A Guide for County and Municipal Planners*. State College PA: College of Agricultural Sciences, Pennsylvania State University. http://pubs.cas.psu.edu/freepubs/pdfs/ui353.pdf (last accessed April 22, 2006).

Biehler, D., A. Fisher, K. Siedenburg, M. Winne, and J. Zachary. 1999. *Putting Food on the Table: An Action Guide to Local Food Policy*. Venice CA: Community Food Security Coalition and California Sustainable Agriculture Working Group.

Blakey, R. C., E. H. Cole, K. Haygood, S. L. Hebert, P. B. King, F. D. Luce, M. T. McGrane, J. W. Trombly, M. I. Williams, and R. L. Wilson. 1982. *Food Distribution and Consumption in Knoxville: Exploring Food-Related Local Planning Issues*. Emmaus PA: Cornucopia Project of Rodale Press.

Borron, S. M. 2003. *Food Policy Councils: Practice and Possibility*. Hunger-Free Community Report. Eugene OR: Congressional Hunger Center.

City of Hartford Advisory Commission on Food Policy. 2002. Annual report. Hartford CT: Hartford Food System.

Clancy, K. 1988. "Eight Elements Critical to the Success of Food System Councils." Paper presented at Cornell Nutrition Update, New York.

———. 2004. "Potential contributions of planning to community food systems." *J Planning Ed and Res* 23 (4): 435–38.

Connecticut Food Policy Council. 2001. *Food Security in Connecticut: The 2001 Indicators and Annual Report of the Connecticut Food Policy Council*. Hartford CT: Connecticut Department of Agriculture.

Dahlberg, K. 1994. "Food Policy Councils: The Experience of Five Cities and One County." Paper presented at the annual meetings of the Agriculture, Food and Human Values Society and the Society for the Study of Food and Society. Tucson AZ.

———. 2003. "Homeland security: Alternative approaches needed." *Michigan Organic Connections* 10:4.

Dahlberg, K., K. Clancy, R. L. Wilson, and J. O'Donnell. 1997. *Strategies, Policy Approaches, and Resources for Local Food System Planning and Organizing*. Minneapolis MN: Minnesota Food Association.

Drake Agricultural Law Center. 2005. http://www.statefoodpolicy.org/profiles.htm (last accessed April 15, 2007.

Frameworks Institute. 2006. *Strategic Frame Analysis*. http://www.frameworksinstitute.org/strategicanalysis/index.shtml (last accessed April 22, 2006).

Hamilton, N. D. 2002. "Putting a face on our food: How state and local food policies can promote the new agriculture." *Drake Journal of Agricultural Law* 7(20):408–54.

Harvard Business School. 1974. *The Commonwealth of Massachusetts: Final Report of the Governor's Commission on Food*. 4-576–129. Harvard Business School–Harvard School of Public Health, Department of Nutrition.

Knoxville Food Policy Council. 1988. *Food Policy Council of the City of Knoxville*. Brochure. Knoxville TN: City of Knoxville.

MacRae, R. 2002. "Food Policy Councils." Paper presented at Community Food Security Coalition Conference, Seattle.

New York State Council on Food and Nutrition Policy. 1988. Five-year Food and Nutrition Plan 1988–1992. NYS Department of Health.

Pothukuchi, K., and J. L. Kaufman. 1999. "Placing the food system on the urban agenda: The role of municipal institutions in food system planning." *Agriculture and Human Values* 16:213–24.

———. 2000. "The food system: A stranger to the planning field." *Journal of the American Planning Association* 66 (2): 113–24.

Roberts, W. 2001. "The way to a city's heart is through its stomach: Putting food security on the urban planning menu." Toronto Food Policy Council. http://www.city.toronto.on.ca/health/tfpc–hs–report.pdf (last accessed April 9, 2006).

Toronto Food Policy Council Policy Manual. 1995. Toronto: Toronto Food Policy Council.

U.S. Conference of Mayors. 1985. *Municipal Food Policies: How Five Cities Are Improving the Availability of Food for Those In Need*. Washington DC.

Webb, K. L., D. Pelletier, A. N. Maretzki, and J. Wilkins. 1998. "Local food policy coalitions: Evaluation issues as seen by academics, project organizers, and funders." *Agriculture and Human Values* 15:65–75.

World Hunger Year. 2007. http://www.worldhungeryear.org/fslc/faqs/ri9-043.asp?section=88click=3 (last accessed April 15, 2007).

Appendix: Survey Instrument

1. What year was your FPC established? Is it still active? (If not, what happened? Why did it cease operation?)

2. Does/did your FPC have official government sanction? If so, what type of government sanction do they have?

3. Does/did your FPC have paid staff? If so, how many FTES?

4. Does/did your FPC have designated funding for staffing and projects? If so, how much? What was the source? How stable is/was the funding?

5. How many Council members are/were there? Who selects the Council members (e.g., appointed by mayor)? Do the Council members represent specific stakeholder groups (e.g., elements of the food sector or ethnic/socioeconomic diversity?).

6. Does your FPC have documentation about the local Food System? If so, what information do they have? Who collected this information? When and how was it gathered?

7. How, if at all, does your FPC engage community members?

8. How visible is the FPC in the community (do the people in your community know you exist?) e.g., is there coverage in the newspaper and elsewhere.

9. How if at all, does your FPC engage diverse government departments/agencies?

10. How would you characterize the leadership of the FPC?

11. Does your FPC have a Vision-Mission statement? What is it?

12. What activities and policies has the FPC undertaken over the last three years.

13. Which of the following roles are filled by the FPC? review and respond to proposed legislation, develop/propose food system policies, develop/propose food system programs, conduct local food system research, provide education/outreach on food system issues to policymakers and staff, public, other?

14. On a scale of 1 to 5, with 1 being low and 5 being high, how successful would you say your FPC is?

15. In your opinion, what have been the FPC's primary successes?

16. In your opinion, what have been the key factors contributing to the success of your FPC?

17. In your opinion, what have been the key challenges facing your FPC?

18. What do you think your FPC will look like five years from now?

19. If another community/state were establishing a FPC, what advice or recommendations would you give to them?

7. The "Red Label" Poultry System in France

Lessons for Renewing an Agriculture-of-the-Middle in the United States

G. W. Stevenson and Holly Born

In recent decades the United States agrifood system has become increasingly dualistic. On the one hand, in many regions small-scale farming and food enterprises have successfully defined niches and developed direct marketing relationships that allow them to thrive and increase in numbers. This trend is encouraging and offers benefits to the communities in which these new markets exist. On the other hand, larger farms have increasingly entered contractual supply chains with consolidated food firms that move bulk commodities around the globe, often at the expense of local communities and environments.

Midsized farm and food enterprises that fall between the large supply chains that move vast quantities of agricultural commodities and the small niche businesses that market food directly to consumers tend to be left out as this dualistic food system evolves. In this chapter we refer to these enterprises as the agriculture-of-the-middle. Analysts using data from the 1997 USDA Census of Agriculture estimated that in the late 1990s 575,000 farms, or 30 percent of U.S. farms, fell into this family-size middle and accounted for approximately 30 percent of total farm sales while owning more than one-half of U.S. farmland (Cochrane 2003; Newton and Hoppe 2002).

As considerable research has revealed, such farms are very important to many rural communities (see Goldschmidt 1978; Strange 1988; Welsh and Lyson 2001). The polarizing forces threaten to hollow out rural America in many regions by moving out many of the agricultural economic activities that have long sustained rural communities, weakening local agribusiness viability, job creation, and the maintenance of local tax bases. These farms are mostly those that have been in their families for several generations; their farmers value good land stewardship as part of the family's heritage. If present trends continue, these farms, together with the socioeconomic and environmental benefits they provide, will likely disappear in the next decade or two.

The poultry broiler industry stands out as the first sector of the U.S. agrifood system to develop a strongly dualistic structure. At one extreme large industrialized poultry firms have consolidated sharply over the past twenty years. In 2000 the top four U.S. poultry companies owned and processed more than 50 percent of the nation's broilers, and 95 percent of these broilers were produced under contract with fewer than forty firms (Heffernan and Hendrickson 2002). The other extreme is characterized by the emergence over the past ten years of a number of small poultry producers (500–5,000 birds annually). These farmers often employ a free-range or pasture-based production system and use some form of direct marketing to consumers.[1]

To find models that might renew the middle segment of the poultry sector, one must look beyond U.S. borders. Particularly important lessons can be learned from France. The "red label" sector of the French poultry industry demonstrates how public policy and private agribusiness strategy are joined to result in high-quality poultry products for French consumers and sustainable economic returns to midsize French farmers. The remainder of this chapter examines the red label poultry sector in France, draws lessons from this analysis, and applies these lessons to two alternative poultry enterprises of the middle being developed in the United States.

The Red Label Poultry System in France

The red label, or "Label Rouge," poultry system created over the past thirty-five years in France provides an important model for developing an agrifood sector that can function successfully between large industrialized food firms on the one hand, and small direct marketing enterprises on the other. The Label Rouge concept emerged in the late 1950s when French farmers who used traditional methods of raising poultry faced a new wave of industrial chicken production techniques. These farmers rejected the idea of relying on chemical feed additives and high-growth genetics to raise poultry inside large buildings. Support came from French consumers who for generations had supported artisanal farming, baking, wine making, and cheese making and were unwilling to give up access to regionally produced, high-quality food (Paybou 2000).

The French government responded with the development of quality seals that could be attached to a range of local food products and brands. The first and most demanding of these seals was the Label Rouge, which was attached to a particular product with a clearly circumscribed geographical origin. The traditional French poultry sector aggressively pursued possibilities with the Label Rouge concept and was among the first sectors adopting the label. In 1965 the first two poultry products awarded the red label originated with a farmers association in southwestern France (Westgren 1999).

The strong performance of the Label Rouge poultry sector is captured by several indicators, foremost being the increase in birds sold under the seal from less than 10 million in the mid-1970s to more than 130 million in the year 2000.[2] Accounting today for approximately one-half of France's poultry farmers and more than one-half of the fresh poultry sold in French supermarkets, the success of Label Rouge has two main explanations: consumer, cultural, and government support as well as creative farmer-centered economic organizations with considerable market power.

Consumer, Cultural, and Government Support

For more than thirty-five years the sophisticated French system of private sector certification coupled with public sector oversight has built consumer confidence in poultry products bearing the red label. To obtain the Label Rouge, an organization called a "quality group" must request the seal from a joint commission of the agricultural and commerce ministries of the French government. The quality group typically consists of poultry farmers located in a given geographical region. Quality groups must present a formal document called a *cahier des charges*, an elaborate business plan that gives full details of the poultry supply chain (called a *filière*) from the genetic selection and rearing of chicks, through production and processing practices, and to delivery of products to retail stores. The *cahier des charges* designates a series of quality control tests organized around the principles of Hazard Analysis Critical Control Points (HACCP). A minimum of sixty-five tests along the supply chain is required of quality groups seeking the red label (Paybou 2000). The *cahier des charges* also names a third party certifying organization in the private sector that will be paid by the quality group to oversee its performance with regards to food quality and safety.

Consumer support for Label Rouge poultry is based on understandings about taste, safety, type and scale of farming, and locality.[3] Specific standards required of all Label Rouge quality groups are associated with each of these criteria. Bird genetics and age, along with feeding and processing regimes, all influence the taste of poultry. For poultry to carry the red label, birds are limited to one of five genetic crosses specified in the *cahier des charges* and must be raised to a minimum of eighty-one days, versus forty-five days for conventionally raised chicken. Feed rations must consist of at least 75 percent cereals and cannot contain animal products or growth stimulants. Air chilled poultry processing systems used throughout France and Europe result in better tasting meat than the water chilled systems that dominate in the United States. Finally, certifying organizations regularly perform taste tests on Label Rouge poultry using both expert and consumer panels, and supermarket shelf life for Label Rouge poultry cannot exceed nine days.[4]

Food safety standards are upheld by a series of HACCP inspections and bacteriology tests performed throughout the food supply chain. A minimum sanitation ("clean out") period of twenty-one days is required between flocks, and any dead birds on the farm must be frozen for analysis by the certifying organization. In addition to enhancing the taste of poultry, the air-chilled systems used to process Label Rouge birds reduce bacterial contamination. The combination of HACCP-based food safety tests results in extremely low incidence of *Salmonella* among Label Rouge flocks (3 percent) compared to industrial poultry flocks in France (70 percent) (Westgren 1999). Label Rouge standards are closely associated with traceability mechanisms that give consumers the ability to know information about a given flock's history. These systems reassure consumers because a food taste or safety problem can be readily traced, located, and solved (Paybou 2000).

Maintaining high food quality and safety standards is at the heart of the Label Rouge sector's strategy and success. As one observer put it, the red label poultry supply chains effectively deliver products that are "vividly distinguishable" from industrial poultry products (Westgren 1999, 1107). French consumers are willing to pay extra for such high quality and safe food. On average, fresh Label Rouge poultry products sold in French supermarkets command a 100 percent premium over industrially-raised poultry.[5] Data from a French national research institute that works on food quality issues indicate that Label Rouge poultry is purchased by a wide demographic of young and old urban consumers.[6] In the earlier years of the label, consumers purchased Label Rouge poultry primarily for Sunday and holiday meals. Since the revelation in 1996 that human deaths could be linked to the presence of Bovine Spongiform Encephalopathy (BSE) in the English beef herd, many French consumers eat only Label Rouge poultry.[7]

As will be explored in more detail below, thirty-five regional, farmer-based quality groups produce poultry under the Label Rouge seal. Recognizing the value that French culture places on place of origin, or *terroir*, related to food, these quality groups seek to differentiate themselves in the marketplace by emphasizing their geographical distinctions and adding standards to their *cahier des charges* that go beyond the minimum standards required of all Label Rouge quality groups. For instance, the quality group in Landes, a region in southwestern France where Label Rouge poultry originated in the early 1960s, emphasizes that the birds are to be provided with complete free range in the region's extensive pine forests (Paybou 2000). France's largest and most powerful quality group is the Loué farmers' cooperative that represents more than one thousand poultry producers in a region near Le Mans. In addition to the required Label Rouge standards, Loué farmers use no pesticides where their poultry range, exceed the standards regarding the proportion of cereals in the poultry feed, and have contracted with Brazilian farmers to source GMO-free soybeans for use in feed rations (Paybou 2000; Westgren 1999).

Finally, farming standards include those pertaining to bird welfare and farm enterprise scale. The Label Rouge seal imposes a maximum bird density in production houses and minimum areas for open range. After six weeks of age all birds must have access to the outdoors from

9 a.m. until dusk. To assure that Label Rouge poultry production cannot be concentrated on a few large sites but rather is done on small and midsize farms, a maximum of four poultry rearing houses is allowed per farm. These maximum density standards translate into no more than about 50,000 birds raised annually per farm. Such standards effectively regulate the scale of farming enterprise and are pivotal for maintaining an agriculture-of-the-middle in France's poultry industry.

The French government's support for the red label extends beyond initial certification requirements in two important ways. The first involves protecting the integrity and legitimacy of the seal. First, these functions are delegated to another government agency that is charged with the maintenance of strict industry standards throughout the agrifood sector regarding the use of quality labels such as the Label Rouge. Examples of this agency's activities include protecting the red label from being copied by unauthorized supermarket store brands or private labels and defending the legitimacy of the Label Rouge during political challenges posed by the World Trade Organization (Westgren 1999).

Second, the original rationale for establishing the Label Rouge involved a conscious strategy on the part of the French government in the early 1960s to maintain and support economic activity in poorer rural regions of France. At that time traditional French poultry farmers faced the growing industrialization of the poultry industry and the movement of chicken production to northern French regions with cheaper labor.[8] As seen above, significant sections of the Label Rouge standards were put in place to ensure that the red label would support small and midsize farms in several of France's most rural regions. In general, French agricultural policy has focused on support for differentiating food products through quality certification and marketing. In contrast, U.S. policy more typically has focused on support for the production of undifferentiated bulk agricultural commodities like corn, soybeans, wheat, and cotton.

Strong Creative Farmer-Centered Economic Organizations

Label Rouge incorporates an organizational model designed to maintain farmer power while linking quality groups with key affiliates in the red label poultry supply chain. This model, known in French as the *fil-*

ière, complements cultural, government, and consumer support as key factors for the success of the Label Rouge poultry sector. The *filière* positions farmer-based quality groups at the center of a set of strategic alliances with upstream affiliates (hatcheries and feed mills) and downstream affiliates (processors and distributors). The result is a vertically coordinated supply and value chain that differs significantly in its locus of power from the vertically integrated industrial poultry sectors in both the United States and France (Born and Stevenson 2002). Additionally, supermarkets are consciously not included among the partners in Label Rouge *filières* for strategic reasons related to the growth of supermarket power in the French food system.[9]

While *filières* generally can be described as farmer-centered supply networks, considerable variation exists among the twenty-four networks operating under the red label. In some cases a quality group may consist of a single farmers' cooperative, while in others there may be three or more co-ops or associations of farmers, as is true in the Landes *filière* (Paybou 2000). In other instances, such as the Janzé *filière*, the processor shares significant decision-making power with farmers (Paybou 2000). In still other networks there is a straightforward contractual relationship between farmers and processors (Westgren 1999). Sometimes the growers' cooperative owns all or parts of the upstream or downstream assets. An example is the partial joint ownership by both the Loué and Landes farmers of France's leading poultry genetics enterprise, Sasso.[10] Clearly, the Label Rouge *filière* model allows for varying patterns of strategic alliance that may shift the power dynamics.

Farmers retain power within the Label Rouge supply chains through internal discipline within producer cooperatives, strong communication with consumers, red label designation, relationships with processors and supermarkets, and high-quality food products. Given the potential conflicts of interest within the *filières*, it is particularly important that farmer associations or co-ops maintain strong internal discipline regarding issues like selling prices and profit redistribution. The Loué and Gers cooperatives are examples of highly disciplined farmer groups that work together and respect group goals, thus strengthening their *filières* and brand names (Paybou 2000). In commanding the highest supermarket prices, Loué is arguably the strongest red label brand.[11]

Consumer education to reinforce the "vivid differences" between Label Rouge and industrial poultry products remains critical for *filière* success (Paybou 2000). Under the Label Rouge system such education is conducted at two levels. Individual *filières* educate consumers in the course of strongly promoting their own brands. Consumer education at the national level is conducted by *Syndicat national des labels avicoles de France* (SYNALAF), the national association of poultry labels to which all *filières* belong. SYNALAF collects a check-off fee per bird for national education campaigns about the benefits of Label Rouge poultry.[12] All enterprises in the quality group contribute to SYNALAF's educational efforts, which often are conducted in cooperation with French consumer groups.

The French supermarket sector is highly concentrated, with five chains controlling nearly 80 percent of the French food retail market (Born and Stevenson 2002). Farmer-led quality groups employ several strategies to counteract this kind of power. The first strategy, employed by the larger farmer cooperatives like Loué and Gers, establishes strategic alliances with large, national processors that can better negotiate with supermarket chains. The Gers co-op negotiated such an alliance with the large processor, Bourgoin in the late 1980s. Loué, the largest red label quality group, actually facilitated the merger of three small local poultry processing plants in the late 1980s in order to create the third largest processing company in France. This deal yielded national negotiating power for the *filière*. During the 1990s three Label Rouge brands (Loué, Gers, and to a lesser extent Landes) became capable of supplying the largest supermarket chains in France. The smaller *filières* have continued selling to the smaller regional supermarket chains in France.[13]

As indicated above, Label Rouge poultry products command, on average, a 100 percent retail premium over industrially raised poultry. Throughout the 1980s and 1990s this margin was retained at the processing and farm levels (Westgren 1999). Our interviews with French government researchers and farmer members of the Loué and the Landes *filières* reveal the following indicators of Label Rouge economics at the farm enterprise level:

1. Poultry enterprises represent from 20 to 50 percent of farm income for French farmers associated with a red label quality group. A Loué farmer told us that poultry represented 50 percent of his farm income, beef 35 percent, and grain sales 15 percent. A large-scale Landes farmer said that poultry represented half of his farm income, while sale of corn accounted for the other half.
2. The annual number of birds raised ranged from less than 15,000 to 52,000 per year, the maximum allowed under Label Rouge certification rules. The Loué farmer raised more than 26,000 chickens and 2,400 turkeys per year. The production of Landes farmers we interviewed ranged from 12,600 chickens per year to the maximum of 52,000.
3. Capital investments for physical structures varied widely—from $3,000 to $200,000—depending on the scale of poultry enterprise and the type of poultry houses utilized. Landes poultry farmers, who let their birds range freely in pine forests, spend only $500 per house, while Loué farmers use more expensive houses that can cost as much as $50,000 each.
4. Estimates of net annual incomes from Label Rouge poultry enterprises suggest median incomes of around $14,000 per year, with maximum incomes of approximately $21,000 per year. These figures, taken from Unité de recherche sur l'économie des qualifications agro-alimentaires (UREQUA) research, indicate that Label Rouge chicken farmers receive net incomes per bird ranging from 0.75 French Francs (FF)/kilogram to 1.5 FF/kilogram. In U.S. dollars these amounts translate to from $0.20/bird to $0.40/bird. A good manager associated with a strong *filière* like Loué can expect to net $0.35/bird.)[14]

The *filière* system is predicated upon fair margins being negotiated throughout the supply network (Born and Stevenson 2002). In most *filières* farmer margins are evaluated each year to account for changes in input costs or consumer demand (Paybou 2000). Throughout the 1980s and 1990s margins involving Label Rouge poultry processors and farmers remained on a par with those earned by the supermarkets. Leading national brands like Loué, however, commanded higher prices, resulting in larger gains for those farmers.[15]

The sales of Label Rouge poultry in France dropped for the first time in 2001 (Born and Stevenson 2002).[16] Additionally, the margins received by supermarkets exceeded those received by Label Rouge poultry processors and farmers, representing an important divergence from the equitable supply chain model developed over the previous thirty-five years. Farmer margins fell 10 percent as retail prices increased (Born and Stevenson 2002).

Analysts believe that the drop in sales occurred for several reasons. First, by the late 1990s the Label Rouge poultry sector was mature, with growth coming primarily from new products such as "feast birds"—guinea fowl, turkeys, and capons—and from processed products like cutup, precooked, frozen, and microwaveable chicken (Paybou 2000). Second, demand for Label Rouge poultry increased following the 1996 BSE scare and then contracted as French consumers returned to eating more beef, resulting in overproduction of Label Rouge chicken on a national scale (Born and Stevenson 2002). Finally, the higher retail prices resulting from unilateral supermarket margin increases likely helped to lower demand.

There have been several responses from the Label Rouge *filières* to these challenges. Importantly, the various brands experience such problems to different degrees. For example, the nationally recognized Loué brand that historically has brought higher margins has not been significantly affected (Born and Stevenson 2002). The smaller, more affected *filières*, however, have put in place response strategies that include longer sanitation periods between flocks to reduce supply; diversification into new markets such as food service, export, and convenience poultry products; and intensified consumer educational campaigns highlighting desirable traditional qualities of Label Rouge poultry (Born and Stevenson 2002).

Some crucial questions remain. Will increasing supermarket power shift the distribution of economic margin in the Label Rouge sector in directions that work to the permanent disadvantage of the farmer-based quality groups and their partners? Furthermore, will current patterns continue, resulting in greater differentiation in economic vi-

Table 10. Structural and financial characteristics of farms in three poultry systems

System	Scale (birds/year)	Physical capital investment	Median annual net income	Maximum annual net income	Net income/bird
Large, industrial[a]	400,000	$450,000	$22,000	$30,000–$60,000	$0.035–$0.05
Midsize, alternative value chain (Label Rouge)[b]	40,000	$15,000–$125,000	$14,000	$20,700	$0.35
Small, direct marketing[c]	600–4,000 (smaller growers), 14,500 (average for larger growers)	$2,000–$5,000 (varies according to processing equipment purchases)	$10,000–$12,000	$30,885 (average for larger growers)	$2.00–$3.00

[a]See Farmers' Legal Action Group 2001.

[b]Estimates of farmer income from Label Rouge are taken from research conducted by scientists at UREQUA, a French national research center that focuses on issues of quality in the food system; interview with UREQUA scientists on May 24, 2001.

[c]Figure for midwestern pastured poultry farmers who direct market their birds were generated by two research projects conducted under the auspices of the Center for Integrated Agricultural Systems, University of Wisconsin–Madison. See CIAS Research Briefs. Nos. 46, 57, and 63. http://www.wisc.educ/cias/research/livestoc.html#poultry.

ability between the larger nationally marketed brands like Loué and the smaller regionally marketed *filières*? Much will depend on how French consumers respond to more detailed farm gate price information that the smaller brands highlight regarding the excessive margins being claimed by the larger supermarkets.

Lessons for Constructing a "Middle" in the U.S. Poultry Industry

The pivotal lesson from the Label Rouge model is the importance of combining poultry production systems that turn out high-quality, premium-commanding products with farmer-centered business organizational structures that are large enough to achieve processing and distribution efficiencies and to penetrate conventional food retail outlets. Large industrial poultry enterprises cannot accomplish the first, and small direct marketing enterprises cannot achieve the second.

Currently, several more farmer-centered strategies are being explored in the United States in order to create what may someday constitute a middle sector in poultry similar to what has developed in France.[17] The first strategy seeks through farmer cooperatives to scale up pastured poultry enterprises beyond direct marketing. The second seeks to reorganize traditional contract poultry producers into midscale enterprises that will produce higher quality products and earn farmers significant increases in net income. A brief case study of each farmer-centered strategy is presented below with observations related to the lessons learned from the Label Rouge model. Providing context is table 10 that compares structural and financial characteristics of three systems of broiler poultry production.

Wholesome Harvest

Begun in 2001, Wholesome Harvest is a farmer-centered limited liability company (LLC) in the U.S. Midwest seeking to scale up organic pastured poultry sales beyond the level of direct marketing.[18] In addition to pastured poultry (chickens, ducks, and turkeys), Wholesome Harvest produces and sells organic pasture-raised beef and lamb. As of summer 2003, nearly forty farmer-owners in Iowa, Illinois, Minnesota, and Wisconsin participated in the enterprise. Poultry producers currently grow broiler chickens in batches of 200 to 2,000.

In addition to organic certification, Wholesome Harvest emphasizes pasture rearing and farmer ownership in order to differentiate its products from organic poultry raised within the industrial model. Initial marketing strategies have targeted white tablecloth restaurants and upscale supermarkets in university towns in the Midwest. Advertising has been accomplished inexpensively, primarily through a Web site (www.wholesomeharvest.com—last accessed April 22, 2006) and free feature stories in regional newspapers and television programs. Their poultry product line emphasizes high-quality products ranging from whole birds through cutups to processed products. Initially all poultry products sold are frozen to ensure longer storage and shelf life.

Enterprise capitalization has been kept deliberately low by contracting out trucking, processing, cold storage, and legal and accounting services. As of summer 2003, Wholesome Harvest's paid staff consisted only of a sales manager, operations manager, administrator, and bookkeeper. The enterprise's board of directors is comprised of farmer-owners.

In general terms the enterprise's goal is to expand beyond its midwestern base to become a more nationally recognized brand. Furthermore, Wholesome Harvest seeks to construct a business model that will enable poultry producers to net $1 to $2 per bird. Finally, the end goal is to support a range of diversified meat farms raising poultry, beef, and/or lamb that annually will earn an economically sustainable income.

Bay Friendly Chicken

Still in the planning stages, Bay Friendly Chicken seeks to reorganize traditional poultry production in the Chesapeake Bay region.[19] This business aims to provide a strong contrast to the industrial enterprises that dominate poultry production on the Delmarva Peninsula. Planned eventually to operate at the significant scale of 6.2 million broilers annually—supplied by twenty-five to thirty growers who each provide 250,000 to 300,000 birds per year—Bay Friendly Chicken would still account for only 1 percent of the broilers raised in the Chesapeake Bay region.

Under the slogan "Better Taste, Better Health, Better Bay," Bay Friendly Chicken seeks to strongly differentiate itself from the area's industrial brands like Tyson and Purdue and from the area's natural brands like Pennsylvania-based Bell and Evans. Key dimensions of this differentiation are high-quality poultry products (from air-chilled processing and no antibiotics), high environmental standards (from composting manure and efficient use of water and energy), fair treatment of growers (through doubling farmers' net incomes), high labor standards (offering livable wages and decent benefits for the entire workforce), stakeholder decision making (by investors, growers, workers, and community representatives), and local and regional control (local ownership and permanent wealth production for the Chesapeake Bay bioregion).

Consistent with its regional image, the company will focus its sales on the large consumer market within two to three hours trucking drive from the enterprise's proposed processing center in Maryland; thus it will start first with Maryland, Virginia, and Washington DC markets and gradually move to the entire Chesapeake Bay bioregion including Delaware, Pennsylvania, and New Jersey. Within this market the business will focus on small retail outlets like health food and gourmet stores, caterers, and white tablecloth restaurants, with the possibility of expanding later to supermarkets. It will target a range of consumers, including activists associated with labor unions, churches, or environmental groups that support Bay Friendly Chicken's social and environmental agendas. Initially, advertising will be accomplished through a range of low-cost techniques including an interactive Web site, strategic collaborations with community organizations and labor unions, and joint marketing ventures with nonprofit groups.

In contrast to Wholesome Harvest, Bay Friendly Chicken plans to own significant capital infrastructure over time. The enterprise's business plan calls for eventual ownership of its own hatchery, feed mill, air-chilled processing plant, and delivery trucks. Current plans involve gradually working into these levels of capitalization by first contracting out functions such as chick hatching, feed milling, product distribution, and manure management. Consistent with this phased-in approach are plans to construct a moderately scaled, conventional processing plant that later can be expanded to include the more expensive air-chilling technology.

Overall enterprise economics are projected to be profitable by the third and fourth years of operation. The grower economics pivot around the enterprise's commitment to double the net income of poultry farmers associated with Bay Friendly Chicken. Current thinking regarding ways to operationalize this commitment is to guarantee growers a premium of an additional $10,000 per year based on reports published in the *Baltimore Sun* that the average conventional Delmarva chicken grower nets between $8,000 and $10,000 per year.

Final Observations from the Label Rouge Experience

Like Label Rouge, both Wholesome Harvest and Bay Friendly Chicken base their enterprises on the sale of consistently high-quality poultry products to discerning consumers. Challenges for the two U.S. enterprises will be to construct a rigorous series of HACCP-based quality control tests along the supply chain. This task will be easier for Bay Friendly Chicken if it owns the major components of production, particularly the air-chilled processing plant. On the other hand, Wholesome Harvest will benefit from third-party organic certification. Third-party certification of quality with government oversight is important to consumers in both France and the United States.[20]

Both enterprises legitimately tout their farmer-centered organization but otherwise focus on different product attributes intended to resonate with consumers. On the production side, Wholesome Harvest's commitment to pasturing enables strong claims regarding animal welfare and product taste.[21] Beyond air chilling, Bay Friendly Chicken will be challenged to differentiate itself from conventional poultry on production criteria. However, its social and environmental standards are extremely high for the industry and should resonate strongly with targeted activist consumers. Bay Friendly Chicken will benefit from its regional identity as it positions itself as a strong alternative to the conventional practices that have significantly degraded the Chesapeake bioregion. The value of a regional identity is less clear for Wholesome Harvest.

Beyond these considerations specific aspects of supply chain organization differ between Wholesome Harvest and Bay Friendly Chicken. Wholesome Harvest plans to outsource such functions as chick rearing, feed milling, and bird processing regularly. Bay Friendly Chicken plans to own the capacity for performing these functions within the enterprise over time.

The Label Rouge *filière* model emphasizes strategic alliances up and down the supply chain and ownership of expensive capital assets only when quality alternatives are not available, as with the Loué and Landes farmer groups' significant capital contribution for the development of poultry genetics. The development of long-term, equitable strategic alliances with dependable partners throughout the value chain will be important for Wholesome Harvest and for Bay Friendly Chicken should it modify its long-term plan to own major capital infrastructure. Combinations of the two models are also options, as with Bay Friendly Chicken owning its own processing plant but outsourcing other functions. In addition to bird rearing, poultry genetics, poultry processing, brand development, and product distribution need to be of high quality, whether through strategic alliances or within enterprise capacity.

Both enterprises plan to be solidly profitable by the third and fourth years of operation. As for economic returns to farmers, the two enterprises operate differently. Wholesome Harvest farmers will likely compare their situation with that of pastured poultry producers in the Midwest who direct market their broilers. As table 10 indicates, such direct marketers can expect to net between $2–$3 per bird. Wholesome Harvest's goal of returning to farmers a net income of $1–$2 per bird would appear attractive to poultry producers who are not keen on overseeing the processing, distribution, and marketing of their birds. As one enterprise in a diversified meat farm, the rearing of 10,000 chickens (four batches of 2,500) could generate between $10,000 and $20,000 of net income.

The context for poultry growers on the eastern shore of Maryland is quite different. They most likely will compare their situation to contract production with one of the large conventional integrators in the region. Data from the business plan of Bay Friendly Chicken and from research on poultry contracts nationwide indicate that farmers in these situations net somewhere between 3.5 and 5 cents per bird (see table 10). The $10,000 premium built into Bay Friendly Chicken's business plan would raise net income to about 7.5 cents per bird. With 250,000 broilers per year (2 houses x 25,000 x 5 batches), this plan would result in an annual net income roughly equal to that in the scenario for Wholesome Harvest's farmers.

In conclusion farmer-centered enterprises like Wholesome Harvest and Bay Friendly Chicken are still in the early stages of constructing viable poultry enterprises in the middle between direct marketers and the contract producers. At the farm level it seems clear that alternative poultry enterprises like these are best positioned as *one* of several profitable enterprises on diversified farms. Viewed through the lens of the Label Rouge model in France, the key challenges for U.S. poultry farmers include establishing creative, effective organizations that will ensure the raising of high-quality, certified poultry products; constructing efficient and equitable processing and distribution systems; and developing significant market power through the successful engagement of a growing sector of discerning U.S. consumers. Although the challenges are significant, once clearly recognized, they also may represent strategic opportunities.

However, unlike the Label Rouge quality groups in France, enterprises like Wholesome Harvest and Bay Friendly Chicken experience a very different governmental context for their efforts. Thus far enterprises developing models that might renew an agriculture-of-the-middle in the United States have had to forge their way with minimal formal support from the public sector. This difference between the French experience with Label Rouge and the current situation in the United States may be the most significant. Longer-term implications for the prospects of an agriculture-of-the-middle in the United States therefore remain to be seen.

Notes

1. While there is no comprehensive list of all small specialty poultry producers in the United States, the membership roster of the American Pastured Poultry Association includes nearly five hundred producers in forty-five states.

2. Interview with Agnes Laszczyk-Legrendre, executive director of Syndicat national des labels avicoles de France, France's national federation of poultry labels, Paris, May 17, 2001.

3. Interview with Laszyzyk-Legrendre.

4. Interview with Marie-Agnes Gatinois, Commission nationale des labels et des certifications, Paris, May 18, 2001.

5. Our inspection of the fresh poultry section of a French supermarket in Paris in the spring of 2000 revealed that the price of whole chickens raised under industrial conditions was 17 FF/kilogram, while the prices of whole birds sourced from three different Label Rouge quality groups ranged from 32 FF/kilogram to 39 FF/kilogram.

6. Interview with Bertile Sylvander, Unité de recherche sur l'économie des qualifications agro-alimentaires (UREQUA), the French national research institute on food quality issues, Le Mans, May 21, 2001.

7. Interview with Laszczk-Legrendre.

8. Interview with Sylvander.

9. Interview with Bernard Lassault, UREQUA. Le Mans, May 21, 2001.

10. Interview with Jean Baptiste D. Dorval, export manager for Sasso Genetics, Sabres, May 24, 2002.

11. Interview with Lassault. A visit to a Paris supermarket in the spring of 2001 revealed that the retail price for Loué poultry products was 2 FF/kilogram higher than for the Gers brand, which in turn was 4 FF/kilogram higher than for the Janze red label brand.

12. Interview with Laszczyk-Legrendre.

13. Interview with Lassault.

14. Interview with UREQUA researchers, Le Mans, May 24, 2001.

15. Interview with Lassault.

16. In 2001 the Label Rouge sector sold 113 million birds, a reduction from peak sales of 130 million birds in 2000. See http://www.synalaf.com/english/facts_and_figures/index.htm (last accessed April 9, 2006).

17. We focus on U.S. poultry enterprises that are farmer owned or farmer centered, as is the case with the Label Rouge quality groups. Other types of poultry enterprises currently exist in the United States that are midscale compared to companies like Tyson and Perdue but that either raise their own birds or interact with farmers through contracts very similar to the largest integrators. Examples in the western United States include Petaluma Poultry in California (http://www.petalumapoultry.com—last accessed April 22, 2006), which produces roughly 11 million industrially raised organic birds per year, and the Nebraska-based MBA Partners, which also sells more than 11 million chickens per year under its Smart Chicken label that emphasizes air chilling (http://www.smartchicken.com—last accessed April 22, 2006). Firms in the eastern United States include the Pennsylvania-based Farmers' Pride Company, selling more than 25 million birds per year designated as "natural" under its Bell and Evans label (http://www.bellandevans.com—last accessed April 22, 2006).

18. We obtained information related to Wholesome Harvest from interviews with company officials, restricted use of the enterprise's business plan, and the company's Web site. Additional information on Wholesome Harvest can be obtained through the Web site (http://www.wholesomeharvest.com—last accessed May 6, 2006) or by writing Wende Elliot, Wholesome Harvest, Colo IA 50056.

19. We obtained information related to Bay Friendly Chicken from interviews with company officials and use of the enterprise's several business plans. Additional information on Bay Friendly can be obtained by writing Michael Schuman, 3713 Warren Street, N.W., Washington DC 20016.

20. This point is reinforced in "Attracting Consumers with Locally Grown Products" (2001), a survey of consumers in four midwestern sites prepared for the North Central Initiative for Small Farm Profitability (http://www.farmprofitability.org/local.htm—last accessed April 22, 2006).

21. For evidence that taste is a key consideration influencing consumer purchases of pastured poultry, see page 116, Periera (2000).

References

Born, H., and G. W. Stevenson. 2002. "Label Rouge: An Ecolabel with the Potential to Increase Small Farm Viability?" Paper presented at the Conference on Ecolabels and the Greening of the Food Market, Boston.

Cochrane, W. 2003. *The Curse of American Agricultural Abundance: A Sustainable Solution*. Lincoln: University of Nebraska Press.

Farmers' Legal Action Group. 2001. *Assessing the Impact of Integrator Practices on Contract Poultry Growers*. http://www.flaginc.org/pubs/poultry.htm.

Goldschmidt, W. 1978. *As You Sow*. Montclair NJ: Allanheld, Osmun.

Heffernan, W., and M. Hendrickson. 2002. "Multi-National Concentrated Food Processing and Marketing Systems and the Farm Crisis." Paper presented at the annual meeting of the American Association for the Advancement of Science, Symposium: Science and Sustainability, Boston. http://www.foodcircles.missouri.edu/paper.pdf (last accessed April 22, 2006).

Newton, D., and R. Hoppe. 2002. "Typology of America's Small Farms." Paper presented at the Third National Small Farm Conference. Albuquerque NM.

Paybou, F. 2000. Technical and economic feasibility study of adopting French label rouge poultry systems to Illinois. Master's thesis, University of Illinois.

Periera, K. 2000. Pastured poultry: Its potential as a sustainable agricultural system in Wisconsin." Master's thesis, University of Wisconsin–Madison.

Strange, M. 1988. *Family Farming: A New Economic Vision*. Lincoln: University of Nebraska Press.

Welsh, R., and T. Lyson. 2001. "Anti-Corporate Farming Laws, the Goldschmidt Hypothesis and Rural Community Welfare." Paper presented at the annual meeting of the Rural Sociological Society, Albuquerque NM.

Westgren, R. 1999. "Delivering food safety, food quality, and sustainable production practices: The label rouge poultry system in France." *American Journal of Agricultural Economics* 81:1107–11.

8. Eating Right Here

The Role of Dietary Guidance in Remaking Community-Based Food Systems

Jennifer Wilkins

Many forces shape the food system and not the least is what every one of us chooses for breakfast, lunch, and dinner, day after day. Most people are concerned to some extent with eating right, that is, choosing foods that are healthful and that together form a diet that reduces one's risk of chronic disease such as heart disease, diabetes, and cancer. Fewer of us regularly endeavor to make food choices that in addition to being healthful are supportive of local community-based food systems or in some way contribute to the sustainability of the food system. Further, because federal dietary guidance in the United States is based almost exclusively on the diet-health relationship, food system considerations traditionally are lacking from these tools of nutrition education. Federal guidelines are found in the U.S. Department of Agriculture (USDA) 1992 *Food Guide Pyramid* and more recently in the USDA 2005 Web-based *MyPyramid* as well as the USDA and U.S. Department of Health and Human Services (DHHS) 2000 and 2005 *Dietary Guidelines for Americans*.

This chapter proposes that there is a role for dietary guidance in rebuilding community-based food systems. Implications and issues that arise from developing regionally specific dietary guidance are explored. One regional food guide, designed for the northeastern United States, provides the basis for this exploration.

Dietary Guidance in the United States

A food guide is a nutrition education tool that translates recommendations on nutrient intake into recommendations for food intake (Welsh, Davis, and Shaw 1992). Davis, Britten, and Myers write that "[t]he his-

Table 11. Comparison of the 2000 and 2005 *Dietary Guidelines for Americans*

2000 dietary guidelines	2005 dietary guidelines
A. Aim for fitness Aim for a healthy weight. Be physically active everyday.	Adequate nutrients within calorie needs • Consume a variety of nutrient-dense foods and beverages within and among the basic food groups while choosing foods that limit the intake of saturated and trans fats, cholesterol, added sugars, salt, and alcohol. • Meet recommended intakes within energy needs by adopting a balanced eating pattern, such as the USDA Food Guide or the DASH Eating Plan.
B. Build a healthy base Let the Pyramid guide your food choices. Choose a variety of grains daily, especially whole grains. Choose a variety of fruits and vegetables daily. Keep food safe to eat.	Weight management • To maintain body weight in a healthy range, balance calories from foods and beverages with calories expended. • To prevent gradual weight gain over time, make small decreases in food and beverage calories and increase physical activity. Physical activity • Engage in regular physical activity and reduce sedentary activities to promote health, psychological well-being, and a healthy body weight. • To reduce the risk of chronic disease in adulthood, engage in at least 30 minutes of moderate-intensity physical activity above usual activity at work or home on most days of the week. • To help manage body weight and prevent gradual, unhealthy body weight gain in adulthood, engage in approximately 60 minutes of moderate- to vigorous intensity activity on most days of the week while not exceeding caloric intake requirements.
C. Choose sensibly Choose a diet that is low in saturated fat and cholesterol and moderate in total fat. Choose beverages and foods to moderate your intake of sugars. Choose and prepare foods with less salt If you drink beverages, do so in moderation.	Food groups to encourage • Consume a sufficient amount of fruits and vegetables while staying within energy needs. Two cups of fruit and 2½ cups of vegetables per day are recommended for a reference 2,000-calorie intake, with higher or lower amounts depending on the calorie level. • Choose a variety of fruits and vegetables each day. In particular, select from all five vegetable subgroups (dark green, orange, legumes, starchy vegetables, and other vegetables) several times a week. • Consume 3 or more ounce-equivalents of whole grain products per day, with the rest of the recommended grains coming from enriched or whole grain products. In general, at least half the grains should come from whole grains. • Consume 3 cups per day of fat-free or low-fat milk or equivalent milk products.

Fats

- Consume less than 10 percent of calories from saturated fatty acids and less than 300 mg/day of cholesterol, and keep trans fatty acid consumption as low as possible.
- Keep total fat intake between 20 and 35 percent of calories, with most fats coming from sources of polyunsaturated and monounsaturated fatty acids, such as fish, nuts, and vegetable oils.
- When selecting and preparing meat, poultry, dry beans, and milk or milk products, make choices that are lean, low fat, or fat free.
- Limit intake of fats and oils high in saturated and/or trans fatty acids, and choose products low in such fats and oils.

Carbohydrates

- Choose fiber-rich fruits, vegetables, and whole grains often.
- Choose and prepare foods and beverages with little added sugars or caloric sweeteners, such as amounts suggested by the USDA Food Guide and the DASH Eating Plan.
- Reduce the incidence of dental caries by practicing good oral hygiene and consuming sugar- and starch containing foods and beverages less frequently.

Sodium and potassium

- Consume less than 2,300 mg (approximately 1 tsp of salt) of sodium per day.
- Choose and prepare foods with little salt. At the same time, consume potassium-rich foods, such as fruits and vegetables.

Alcoholic beverages

- Those who choose to drink alcoholic beverages should do so sensibly and in moderation—defined as the consumption of up to one drink per day for women and up to two drinks per day for men.
- Alcoholic beverages should not be consumed by those who cannot restrict their alcohol intake, women of childbearing age who may become pregnant, pregnant and lactating women, and children and adolesents.

Food safety

- Know and follow proper food handling practices.

torical objective of food guides has been to interpret dietary standards and recommendations into simple nutrition education tools that are useful to consumers" (2001, 881). The *Food Guide Pyramid* (USDA 1992) was designed to convey the themes of variety (multiple food groups), proportionality (appropriate numbers of servings from each group), and moderation (restrictions of fat and sugar). In 2005 the USDA replaced the 1992 *Food Guide Pyramid* with the *MyPyramid* food guide system. In addition to a simplified graphic symbol, *MyPyramid* (USDA 2005) provides Web-based resources allowing for individualized recommendations for daily food intake based on gender, age, and level of physical activity. The *Dietary Guidelines for Americans* (USDA/DHHS 2000, 2005) that forms the basis of food and nutrition education programs in the United States is meant to govern the design of the food guidance system and to provide more general overarching principles of a sound diet (see table 11).

The 2005 *MyPyramid* and the *Dietary Guidelines for Americans* are designed to respond to recent criticisms that the earlier versions failed to communicate effectively the current state of knowledge about the diet and health relationship. Central to the criticism of the 1992 *Food Guide Pyramid* was that it failed to differentiate between good and bad fats or between complex and refined carbohydrates. The 2000 *Dietary Guidelines* were criticized for being confusing and bowing to food industry pressure by carefully avoiding wording that suggests eating *less* of anything (Nestle 2002).[1]

Looked at from a local community-based food system perspective, however, the nutritional limitations of our federal dietary recommendations and guidelines are only part of the problem. In a nation with vastly different food-producing environments, conventional food guides and dietary guidelines are completely without geographic context. If these tools are to encourage diets that support local food systems, they need to take account of regional differences in agricultural production. One such example of contextualized dietary guidance the *Northeast Regional Food Guide* (Wilkins and Bokaer-Smith 1996) (and accompanying dietary guidelines) will be used here to examine some of the issues that arise when dietary guidance is embedded in a geographic context.[2] Because the *Northeast Regional Food Guide* was based on the 1992 *Food Guide Pyra-*

mid and 2000 *Dietary Guidelines*, these (as opposed to the 2005 revisions of both) will serve as the appropriate documents for comparison in this chapter.

Expanding the Purpose of Dietary Guidance

Developers of the 1992 *Food Guide Pyramid* and the 2000 *Dietary Guidelines for Americans* utilized data on the nutrition status of Americans, nutritional standards, food consumption practices, food availability, food composition, and food cost (Davis, Britten, and Myers 2001) to make dietary recommendations. It is the limited scope of the health focus of dietary guidance that a growing number of nutritionists and leaders in sustainable agriculture have criticized as insufficient for informing food choices in a time of resource constraints and environmental stresses as well as consolidation, centralization and specialization in food production, processing, and retailing. As Gussow and Clancy put it, "[i]n our time, educated consumers need to make food choices that not only enhance their own health but also contribute to the protection of our natural resources. Therefore, the content of nutrition education needs to be broadened and enriched not solely by medical knowledge, but also by information arising from disciplines such as economics, agriculture, and environmental science" (1986, 270). They proposed the term "sustainable diets" to describe food choices that might contribute not only to the eater's health but also to the sustainability of the U.S. agricultural system. Sustainable diets, they suggest, would be seasonally varied in accordance with the local agricultural harvest and rich in foods grown and processed within their region (Wilkins 1995).

Regional Food Guides

The standard food guides and dietary guidelines in the United States are not particular to the agriculture and food system of specific places. By contrast, regional dietary guidance provides tools in the form of food guides and dietary guidelines to help consumers choose healthful, seasonally varied diets from foods available within a given geographic region. The goals of regionally specific dietary guidance are two-fold: (1) to promote health and decrease chronic disease risk and (2) to support

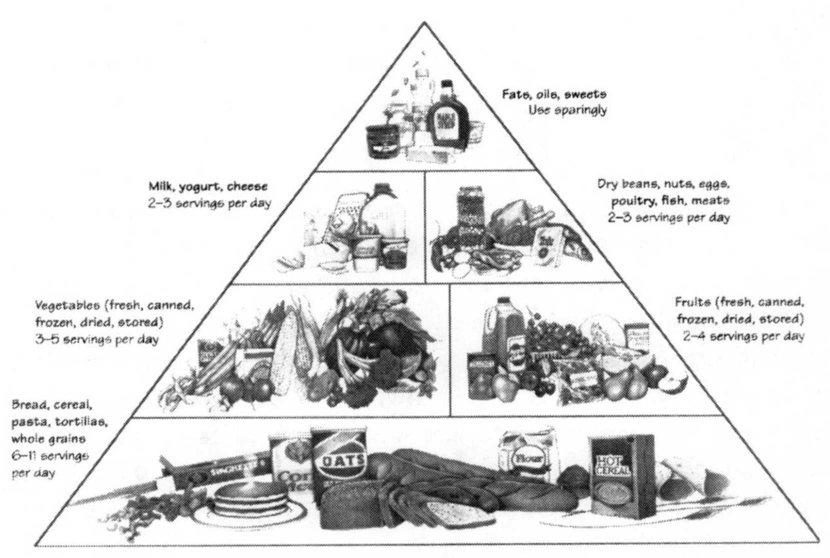

Figure 1. *Northeast Regional Food Guide*. Source: Wilkins and Bokaer-Smith. 1996. Note: Number of servings is based on the *Food Guide Pyramid*, USDA, 1992.

the development and reinvigoration of sustainable community-based food systems. One of the earliest examples of regional food guides is the *Northeast Regional Food Guide* (NERFG) that was developed for the northeastern United States.

The NERFG (see figure 1) is the first food guide developed for a multi-state region of the United States. This food guide is graphically similar to the national food guide, the USDA *Food Guide Pyramid*. The NERFG is pyramidal in shape, has six food groups, suggests numbers of servings from each group, and encourages variety, proportionality, and moderation. Like the USDA *Food Guide Pyramid*, this regional food guide is based on the U.S. *Dietary Guidelines for Americans*.

With these features, however, the similarities end. The *Northeast Regional Food Guide* and accompanying dietary guidelines differ in important ways from the federal food guide. At the top of the *Food Guide Pyramid* are symbols for fats and sugars, while at the top of the NERFG actual foods that are essentially sources of fats and sugars (jams, jellies, honey, maple syrup, butter, oil, and so on) are pictured. These foods, of which consumers are well-advised to eat less, represent important niche markets for regional producers and processors. The NERFG acknowledges that most consumers are likely to continue eating at least some of these top of the Pyramid foods. Therefore the NERFG's recommending regional food products like local maple syrup can help strengthen the agricultural economy of northeast communities.

Another graphic difference between the NERFG and the *Food Guide Pyramid* is that the regional food guide pictures a wider variety of specific foods within each group, representing those that are currently, have been in the past, or could be (given soil types and climate) available from the regional food system. Still in keeping with the regional nature of the NERFG, fruits such as oranges and bananas, prominent on the USDA *Food Guide Pyramid*, are absent on the NERFG.

Another graphic difference is that on the *Food Guide Pyramid* the meat group is titled, "meat, poultry, fish, beans, eggs, and nuts." On the NERFG the word order for this high protein food group is changed to "dry beans, nuts, eggs, fish, poultry, and meat" to be more consistent with the dietary guidelines emphasis on plant foods in the diet.

While the produce featured on the *Food Guide Pyramid* is shown in the fresh form only, the NERFG pictures fruits and vegetables in a variety of forms (fresh, canned, frozen, and stored). Eating a diet based on the availability of locally grown foods means that the form of the foods eaten likely will vary throughout the year. For example, instead of fresh tomatoes, canned, sun-dried, or otherwise processed tomato products would be consumed in the winter.

Finally, seasonal lists of fruit and vegetable availability accompany the pyramid graphic of the NERFG. In addition to conveying what foods are available in the region, it is important in helping consumers be aware of when local foods are likely to be available from fresh harvest or from stored sources only. State departments of agriculture typically provide seasonal produce harvest calendars as a public service, and this information should be an integral part of a regional food guide.

The dietary guidelines that accompany the NERFG are the same as those in *Dietary Guidelines for Americans*. However, in addition to the A, B, and C sections of this publication, the NERFG dietary guidelines include a "D" for "Develop the Local Food System." Under this head-

Table 12. *Northeast Regional Food Guide* dietary guidelines

A. Aim for fitness

B. Build a healthy base

C. Choose sensibly

D. Develop the local food system
Choose a diet with plenty of foods produced locally.
Choose a diet with plenty of foods processed locally.
Choose a variety of fresh fruits and vegetables when they are available from local farmers.
Choose a variety of root vegetables during the fall, winter, and early spring.
Minimize your total food mile intake.

ing are further guidelines: choose a diet with plenty of foods produced locally; choose a diet with plenty of foods processed locally; choose a variety of fresh fruits and vegetables when they are available from local farmers; choose a variety of root vegetables during the fall, winter, and early spring; and minimize your total food mile intake (see table 12).[3]

Issues and Questions Concerning Seasonally Varied, Locally Based Diets

The development and implementation of regional dietary guidance suggests an expectation that consumers adopt eating patterns that are reflective of agricultural production, which is constrained by local soils and climate, and of processing, which may or may not exist locally. This expectation raises several issues and questions, some thornier than others. At the consumer level is the question of nutritional adequacy and acceptability of local and seasonally varied diets. Understanding of what foods are grown locally, when they are available, and how to prepare them are also consumer issues. If regional dietary guidance becomes more accepted, another issue is how such food guides are interpreted by consumers and how the information they contain is translated into making food choices. Finally, regional dietary guidance raises issues related to the adequacy of local agricultural production and the availability of processing and other market channels. The remainder of this chapter is devoted to exploring these issues through research related to the *Northeast Regional Food Guide* or through the New York and northeastern context.

Nutritional Adequacy

The first and perhaps most basic issue is nutritional adequacy. Whatever its current production, can a given area produce diverse crops in adequate quantity to meet the nutritional needs of the population living there? There are actually two issues here. One relates to quality: Are the soils and climate in a region capable of producing the diversity of crops and livestock that together yield the nutrients required to meet human nutritional needs? The other relates to quantity: can a region produce enough volume to meet the nutritional needs of the population that lives there?

To explore the first of these questions in the context of the northeastern U.S., we conducted a nutrient analysis of the NERFG (Wilkins and Gussow 1997). The study involved constructing two two-day menus (one ovo-lacto and one including animal flesh) for each season using selections of foods from the NERFG that could be grown in sufficient quantities in the Northeast in that season if demand existed. Using the Nutritionist IV computer program, the menus were analyzed for energy; the macronutrients; vitamins A, C, and E; beta-carotene; thiamin; riboflavin; niacin; folate; iron; calcium; and zinc. Results showed that the calorie distribution for the two diets was consistent with established recommendations, and with few exceptions the values for all nutrients measured were above the U.S. Recommended Dietary Allowances.

These results indicate that it is possible to obtain a nutritionally adequate diet from the diversity of foods that can be grown in the northeastern region. Assuming adequate production levels, one would still need to assume adequate processing and storage capacity in the local food system. Given our current reliance on (and preference for) imports of food from other regions and countries, there is little perceived need for such capacities. The current relatively limited supply of storage and processing capacity would become apparent in the face of significant transition to more regional food self-reliance.

Adequacy of Regional Food Supplies

To address the second interpretation of nutritional adequacy—the quantity aspect—comparisons between current consumption, production, and recommended intakes are needed. Such comparisons have been undertaken for consumption, dietary recommendations, and production of vegetables and fruits in the state of New York (Peters et al. 2002, 2003). These studies show that, while New York State produces a wide diversity of both fruits and vegetables, it produces enough of only a few commodities to meet current and recommended levels of intake.

Based on northeastern regional data from the national Food Commodity Intake Database (FCID),[4] the average New Yorker consumes about 180 pounds of fruit per year (compared to the national average of 140 pounds).[5] In 1999 New Yorkers consumed 3.2 billion pounds of fruit all together. Nearly 2 billion pounds of this fruit was consumed as juice. According to agricultural statistics from 1994 to 1998, about 1.5 billion pounds of fruit were harvested each year from farms in New York—a little less than half of the pounds consumed.

But this gross comparison of production volume and consumption data does not mean that nearly half of the fruit consumed by New Yorkers is produced in the state. Indeed, much of New York State's production is sold into national markets, and grocery stores also buy from these national markets. Almost all of the state's fruit harvest, 94 percent, was in apples and grapes. The rest was mostly pears, peaches, cherries (sweet and sour), and strawberries. Apples, which account for 73 percent of New York fruit production and 56 percent of the farmland planted to fruit, are the only fruit produced in excess of consumption levels. Production of all other fruit falls well below current consumption, and of course, many of the most popular fruits—bananas, oranges (and their juice), and all other tropical fruits—are not produced at all in the state because of climate constraints.[6] Considering fruit alone, eating seasonally varied and locally based diets in New York would mean eating more of the fruits that the state can produce (fresh during harvest and as juice, canned, frozen, or dried at other times of the year) and less of those that it lacks the natural endowments to produce.[7]

When considering how vegetable consumption and dietary recommendations compare with production in New York, gaps are similar to those existing for fruit. While New Yorkers consume about 160 pounds per capita of vegetables per year, they fall short of the dietary recommendations and most of the vegetables consumed are from out-of-state sources. Relative to consumers in other regions, New Yorkers consumed more spinach, winter squash, eggplant, kale, turnips, and artichokes, all of which are produced in New York, indicating some correspondence between vegetable consumption patterns and in-state production.

Like fruit, much of New York's vegetable production is sold into national markets, and many of the vegetables available in local grocery stores come through national distribution channels. While New York currently produces more beets, cabbage, onions, pumpkins, snap beans, and sweet corn than New Yorkers consume, a relatively small proportion of total vegetable consumption comes from the state.

Meeting the recommendation for vegetable intake (three to five servings daily) throughout the year from instate production in New York would require significant shifts and expansion in production and a rebuilding of the processing industry. New York farms could meet current demand for many vegetables if farmland now producing "excess" vegetables or nonvegetables were planted with needed crops. Of the vegetables most consumed in New York (nearly 80 total), only nine cannot be grown in the state, and these account for a mere 0.4 percent of consumption.

Another consideration is the form (that is, fresh, frozen, stored, or canned) and the specific vegetables that are consumed throughout the year. If people varied the kinds and forms of vegetables they consumed month to month, emphasizing roots and processed vegetables in the winter and fresh vegetables in the summer and fall, eating patterns would better reflect the state harvest calendar. This scenario assumes the state possessing adequate processing capacity, which is not the case for many New York grown vegetables. It is also important to recognize that New Yorkers on average consume far fewer vegetable servings than are recommended on the *Food Guide Pyramid*. If New Yorkers suddenly started consuming in accordance with these guidelines, state vegetable production levels would account for an even smaller proportion of demand.

At the national level, Young and Kantor (1999) estimated adjustments in crop acreage that would need to occur in order to meet changes in food demand if the American diet became more consistent with *Food Guide Pyramid* recommendations. They estimated that an increase of 5.6 million additional acres would need to be put into production to meet recommendations for intake of fruits, vegetables, and dairy products. This change is small overall (about 2 percent of the average area of U.S. cropland planted in 1991–1995). However, Young and Kantor also noted that significant decreases in acreage could be anticipated for single commodity groups like sweeteners, fats and oils, fruits, and some vegetables. They also stressed that, because of land and climatic differences across the country, regional specialization in production logically would occur.

While absolute self-sufficiency is an unrealistic and, some would argue, an ill-advised goal, Duxbury and Welch (1999) have explored the capacity for food self-sufficiency in the U.S. Northeast. They claim that, although current production in the region is not at all oriented toward self-sufficiency, the Northeast and other areas with long winters could meet the food needs of the population if production was refocused in that direction.

The issue of regional and national nutritional adequacy also raises another question: can every region have a local food system? Relative regional food independence may present a dilemma for regions with climates and soil-types ill-suited for diverse agricultural production—the arid Southwest, for example. How should such areas be considered with respect to regional dietary guidance and local food systems? In these cases where diverse agricultural production is possible only with significant input of fossil fuel and already over-committed ground water supplies, regional dietary guidance may be neither feasible nor practical despite the best efforts of individual local eaters (Nabhan 2002). In the long run, of course, and increasingly in the short run as well, significant food production may not be possible in such regions. Scarce water resources for food production and various domestic uses (from watering lawns to washing cars to taking showers to doing laundry) have thus far not seemed to curb accelerating population growth. Whether growth should be restricted in areas with limited water supplies and a dependence on food imports is a question beyond the scope of this chapter.

Consumer Acceptance

Another issue related to regional dietary guidance is the acceptability of diets composed of local foods and varying by season. Are consumers willing (and indeed able) to vary the foods they eat and the forms in which they are consumed throughout the year in ways more consistent with local production and processing? We explored this question using a telephone survey of 500 consumers in the northeastern region (Wilkins, Bokaer-Smith, and Hilchey 1996).

When asked whether nutritional needs could be met with local crops, level of concern seems to depend on the time of year being considered. Of the consumers surveyed in the northeastern region, 72 percent agreed with the statement, "in winter, it is necessary to import fruits and vegetables to get enough variety for a healthy diet." However, nearly 80 percent of the respondents disagreed with the statement, "even in summer, there is not enough variety in local fruits and vegetables to maintain a healthy diet." These results indicate that consumers perceive that for part of the year, the northeastern food system produces enough quantity and variety to support a nutritionally adequate diet. But they do not believe that the region's food system would provide sufficiency during the winter months. Given the short growing season of the Northeast relative to other regions, eating locally year-round would require adequate storage, value-added processing, and effective distribution networks.

Further, because of now well-established importation of fresh fruits and vegetables when they are out of season at home, consumers have grown accustomed to year-round availability. Survey results indicate that it is important to most consumers to have certain fruits and vegetables, particularly tomatoes, lettuce, broccoli, and strawberries, available *fresh* throughout the year. So not only are consumers concerned that local production is inadequate to meet nutritional needs, they also may be unwilling to forego year-round consumption of certain produce items.

However, the survey of northeastern consumers did reveal some distinct seasonal differences in the *form* in which fruits and vegetables are consumed. Seventy-eight percent of the respondents reported that there were some fruits they eat fresh only in the summer and early fall but not in the winter (for example, melons, peaches, watermelons, and berries). In the case of fresh vegetables, two-thirds of the respondents indicated that there were some vegetables that they ate fresh only in the summer and early fall but not in the winter (particularly sweet corn, tomatoes, string beans, squashes, and asparagus). There seems also to be seasonal variability in consumption of processed (that is, canned or frozen) fruits and vegetables. Half of the respondents indicated that they ate peaches, pears, fruit salad, pineapple, and strawberries in canned or frozen form more often in the winter and early spring (before some of these are available fresh from local harvests) than they did in the summer and fall. An even greater proportion of survey respondents (67 percent) indicated greater consumption of canned and frozen vegetables (particularly sweet corn, peas, string beans, broccoli, and mixed vegetables) in the winter and early spring than when they are available fresh from local sources. These results suggest the presence of seasonally varied eating patterns that correspond at least in part to harvest calendars in the Northeast.

Usability

Food guides are created to provide a bridge between nutrition science and the general public. To be effective in shaping consumer food choices, a food guide needs to be easily understood, memorable, and appropriate for the intended audience(s). Indeed, the strength of a food guide lies in its "usability"—how readily consumers can apply the concepts communicated in the guide in making actual food choices in the marketplace.

While usability, it would seem, is a fundamental criterion for the development and dissemination of dietary guidelines and food guides, little research has been done to explore how consumers apply food guide information, such as from the *Food Guide Pyramid*. *Perceived* usability has been studied in the development of food guides, as has change in knowledge, attitudes, and behavior. Usability testing looks at an intermediate step between increasing knowledge and changing behavior. It looks at how food guides can be used to change behavior.

In one of the only studies of actual usability, Hunt, Gatenby, and Rayner (1995) evaluated the ability of consumers to apply the underlying concepts of *Britain's National Food Guide* when making food choices. In this evaluation of usability, subjects were asked to perform various tasks in order to assess the impact of receiving information about the guide on people's understanding and recall of the guide's concepts. Evaluation tasks examined four capabilities: substitution (of foods with others of equivalent nutritional value), comparison (of the healthfulness of sample meals), sorting (foods into appropriate groups), and ingredient selection (for appropriate proportion in a composite food).

We evaluated the usability of the *Northeast Regional Food Guide* using tasks modified from those used by Hunt, Gatenby, and Rayner (1995). We sought to assess the ability of consumers to apply three concepts underlying both the USDA *Food Guide Pyramid* and the NERFG (balance, proportionality, and variety) and two (seasonality and localness) that were particular to the regional food guide (Ryan and Wilkins 2001). Subjects, twenty-seven participants in the Women, Infants, and Children (WIC) program, were randomly assigned to either a control group, which received instruction on the *Food Guide Pyramid*) or an experimental group, which received instruction on the NERFG).

In the first task, "Ranking Menus for Healthfulness," subjects ranked four different menus in order of their healthfulness, that is, how closely they reflected the U.S. *Dietary Guidelines*. The second task, "Local/Seasonal 5-a-Day," required that subjects choose five fruits and vegetables from a list of ten that were available from local sources at the time of the study (September and October). In the third task, "Shopping Basket Substitution," subjects selected an alternate food for four different items identified by the researcher with the goal of maintaining nutritional equivalence in the basket. In the final task, "Seasonal Shopping Basket," subjects identified the season represented by the foods in four different shopping baskets containing foods from all food groups in the approximate ratio of the *Food Guide Pyramid*.

From these tasks meant to simulate food decisions that consumers actually face in the marketplace, we compared the usability of the NERFG with that of the *Food Guide Pyramid*. We also assessed understanding of the seasonal and local concepts. Differences in scores on the fourth task ("Seasonal Shopping Basket") revealed that individuals who received information about the NERFG were better able to apply seasonal and local concepts when making food choices than were those who received information on the *Food Guide Pyramid* alone (Ryan and Wilkins 2001).

Local and Seasonal Food Knowledge and Skills

Because the *Northeast Regional Food Guide* incorporates concepts not common to dietary guidance, understanding of terms such as seasonal and local becomes important. When advised to eat locally based and seasonally varied diets, how do consumers understand and apply these guidelines? In one study shoppers were asked to describe the concepts of local foods and seasonal foods and to name foods that were local, not local, seasonal and not seasonal (Wilkins, Bowdish, and Sobal 2002).

Results suggest a complex and multidimensional conceptualization for the term seasonal when applied to food. Conceptualization of the term local was less complex but still revealed several dimensions. Many foods named as local were fruits and vegetables, and many named as not local foods were tropical fruits. Many foods named as seasonal were fruits and vegetables, while many named as not seasonal were staple foods like meat, breads, and cereals. Locality and seasonality have specific agricultural meanings. Clarifying and reinforcing these meanings are important if consumers are to apply dietary guidance effectively in the context of a local food system.

In general, eating more seasonally and locally may require expansion, or at least shifts, in food knowledge and skills. As consumers incorporate new and possibly more whole foods into their diets, they may need to acquire new and different food preparation skills or reclaim forgotten ones to make optimal use of less familiar or infrequently used food items. Since most supermarkets obscure rather than reveal the character of the local food supply, consumers will need to gain greater awareness of just what foods are grown locally, when they are available, and in what form, in order to adopt eating patterns that are more aligned with seasonal variations in local agricultural production.

Food System Gaps

In order for consumers to shift their dietary patterns in ways that support local agriculture, current gaps in the food system need to be filled. As described in this volume's chapter by Tom Lyson, consolidation and centralization in the food system have resulted in the disappearance not only of previously diversified crop production in a given area but also of a web of food processors, market outlets, and local distribution channels. In several areas, including the Northeast, such links and infrastructure must be rebuilt to make possible the consumption of local foods out of season in a preserved form. In the current system, crops grown in one area usually are transported to another for processing and then shipped back to where they were produced to be consumed.

Indeed, regeneration of local processing and distribution networks may be essential to the development of more sustainable local food systems in the Northeast, as in other regions of the country. For as is widely recognized, today's consumers often lack the skills and/or the desire to prepare for the table not only foods that are presently unfamiliar, but any raw foods at all. This inability or unwillingness to cook from scratch is widely attributed to a perceived lack of time in households where adults are often at work during most daylight hours. Unless locally produced convenience products are available, the percent of the population that will opt for local, seasonal foods may remain small, and supporting durable local food systems may remain a marginal venture. Some new, inventive labeling strategies, such as those described in the chapters in this volume by G. W. Stevenson and Holly Born and by Elizabeth Barham, may be necessary in order to help consumers identify local value-added products when they are available. Such food products not only could save cooks time in the kitchen but also facilitate broader participation in support of the local food system.

Evaluation of the nutritional adequacy of the *Northeast Regional Food Guide* suggests that it is possible to produce enough variety to meet human nutrient requirements. However, from consumption data, we know that current production of several crops falls far short of need on a population basis in the Northeast and most likely in several other regions. In regions lacking sufficient rainfall and fertile soils, regional diets would be limited in both variety and quantity.

Given current distortions in productive capabilities of some areas through energy intensive agriculture, extensive irrigation, and long-distance transport of massive amounts of food, several regions now are populated beyond their agricultural capacity to regain more than a minimum level of food self-reliance. In other regions, such as the Northeast and Midwest, which are relatively rich in water resources and have fertile soils and adequate growing seasons, entertaining the idea of greater self-reliance in food, even at current population levels, makes a certain amount of sense. In these contexts regional dietary guidance offers a strategy for developing consumer demand that can help remake community-based food systems.

Acknowledgments

Development of the Northeast Regional Food Guide was supported with a grant from Cooperative State Research, Education, and Extension Service.

Notes

1. In the interval of time between the writing of this chapter and publication of this book, the USDA released *MyPyramid* (2005). However, since the *Food Guide Pyramid* (USDA 1992) is more graphically similar to the *Northeast Regional Food Guide* and readily conveys the concepts of proportionality, balance, and moderation in a manner consistent with the latter, the older USDA version has been retained for this chapter. Likewise, the 2005 *Dietary Guidelines for Americans* supersedes the 2000 version upon which the dietary guidelines for the *Northwest Regional Food Guide* were modeled. Table 11 provides a comparison of the 2000 and 2005 *Dietary Guidelines for Americans*.

2. For a full-sized poster of the *Northwest Regional Food Guide*, contact the author at jhw15@cornell.edu.

3. The concept of food miles comes from a study of the weighted average distance food travels in a local versus a conventional food system as reported in Pirog et al. (2001).

4. For this study national survey data from the Food Commodity Intake Database (FCID) were used to estimate per capita consumption of fruit in New York. The FCID is constructed using the 1994-96 Continuing Survey of Food Intake by Individuals (CFSII) plus a supplementary survey of children (ages <10) conducted in 1998 by the EPA and ARS. For more information, see: http://www.ars.USDA.gov/services/docs.htm?docid=7828 (last accessed May 6, 2006)

5. Much of the difference can be accounted for in fruit juice consumption. New Yorkers drink an average of 65 pounds of orange juice per year, compared to an average of 48 pounds nationwide (Peters et al. 2003).

6. Bananas are the most popular fresh fruit, for example, and New Yorkers eat an average of 14 pounds of bananas per person each year. In comparison, they eat 12 pounds of apples. For juices each New Yorker drinks 65 pounds of orange juice per year but only 19 pounds of apple juice (Peters et al. 2003).

7. New York fruits include apples, grapes, peaches, plums, pears, cherries, nectarines, apricots, currants, blueberries, raspberries, and blackberries as well as cantaloupe, honeydew, and watermelon (Peters et al. 2003).

References

Davis, C. A., P. Britten, and B. Myers. 2001. "Past, present, and future of the Food Guide Pyramid." *Journal of the American Dietetic Association* 101 (8): 881–85.

DHHS/USDA. 2003. "Announcement of Establishment of the 2005 Dietary Guidelines Advisory Committee and Solicitation of Nominations for Membership." *Federal Register* 68 (94): 26280.

Duxbury, J. M., and R. M. Welch. 1999. "Agriculture and dietary guidelines." *Food Policy* 24:197–209.

Gussow J. D., and K. Clancy. 1986. "Dietary guidelines for sustainability." *Journal of Nutrition Education* 18 (1): 270–75.

Hunt P., S. Gatenby, and M. Rayner. 1995. "The format of the National Food Guide: Performance and preference studies." *Journal of Human Nutrition and Dietetics* 8:335–51.

Nabhan, G. P. 2002. *Coming Home to Eat: The Pleasures and Politics of Local Foods*. New York: Norton.

Nestle, M. 2002. *Food Politics: How the Food Industry Influences Nutrition and Health*. Berkeley CA: University of California Press.

Peters, C. J., N. Bills, J. L. Wilkins, and R. D. Smith. 2002. *Vegetable Consumption, Dietary Guidelines and Agricultural Production in New York State: Implications for Local Food Economies*. Research Bulletin 2002–07. Ithaca NY: Department of Applied Economics and Management, Cornell University. http://aem.cornell.edu/research/researchpdf/rb0207.pdf (last accessed May 4, 2006).

———. 2003. *Fruit Consumption, Dietary Guidelines, and Agricultural Production in New York State: Implications for Local Food Economies*. Research Bulletin 2003–02. Ithaca NY: Department of Applied Economics and Management, Cornell University. http://aem.cornell.edu/research/researchpdf/rb0302.pdf (last accessed May 4, 2006).

Pirog, R., T. Van Pelt, K. Enshayan, and E. Cook. 2001. *Food, Fuel, and Freeways: An Iowa Perspective on How Far Food Travels, Fuel Usage, and Greenhouse Gas Emissions*. A report for the Leopold Center for Sustainable Agriculture. Ames IA: Iowa State University. http://www.leopold.iastate.edu/pubs/staff/ppp/index.htm (last accessed April 9, 2006).

Ryan, C., and J. L. Wilkins. 2001. "Using tasks to measure consumers' ability to apply food guide recommendations: Lessons learned." *Journal of Nutrition Education* 33(5): 293–96.

USDA. 1992 (revised slightly 1996). *The Food Guide Pyramid*. Home and Garden Bulletin. No. 252. Center for Nutrition Policy and Promotion.

———. 2005. *MyPyramid*. http://www.mypyramid.gov (last accessed April 10, 2007).

USDA and DHHS. 2000. *Nutrition and Your Health: Dietary Guidelines for Americans*, 5th edition. Home and Garden Bulletin. No. 232. Washington DC: U.S. Government Printing Office.

———. 2005. *Nutrition and Your Health: Dietary Guidelines for Americans*, 6th edition. Washington DC: U.S. Government Printing Office.

Welsh, S., C. Davis, C., and A. Shaw. 1992. "Development of the Food Guide Pyramid." *Nutrition Today* 27 (6): 12–23.

Wilkins, J. L. 1995. "Seasonal and local diets: Consumers' role in achieving a sustainable food system." In *Research in Rural Sociology and Development*, edited by H. K. Schwarzweller and T. A. Lyson, 149–66. Greenwich CT: JAI Press.

Wilkins, J. L., and J. C. Bokaer-Smith. 1996. *The Northeast Regional Food Guide*. Ithaca NY: Cornell Cooperative Extension, Cornell University. http://www.nutrition.cornell.edu/foodguide/archive/index.html (last accessed May 15, 2006).

Wilkins, J. L., J. C. Bokaer-Smith, and D. Hilchey. 1996. *Local Foods and Local Agriculture: A Survey of Attitudes Among Northeastern Consumers*. Survey Report. Ithaca NY: Division of Nutritional Sciences, Cornell Cooperative Extension.

Wilkins, J. L., E. Bowdish, and J. Sobal. 2002. "Consumer perceptions of seasonal and local foods: A study in a U.S. community." *Ecology of Food and Nutrition* 41:415–39.

Wilkins, J. L., and J. D. Gussow. 1997. "Regional dietary guidance: Is the Northeast nutritionally complete?" In *Agricultural Production and Nutrition. Proceedings of an International Conference*, edited by W. Lockeretz, 23–33. Medford MA: Tufts University School of Nutrition Science and Policy.

Young, C. E., and L. S. Kantor. 1999. "Moving toward the food guide pyramid: Implications for U.S. agriculture." In *America's Eating Habits: Changes and Consequences* (AIB-750), edited by E. Frazao, 403–23. Washington DC: USDA Economic Research Service.

9. Community-Initiated Dialogue

Strengthening the Community through the Local Food System

Joan S. Thomson, Audrey N. Maretzki, and Alison H. Harmon

Never doubt that a small group of committed citizens can change the world; indeed, it's the only thing that ever has.

Margaret Mead

Complex, crosscutting issues affect local food systems. They include urban sprawl, rural ambiance, growth management, open-space preservation, and local agriculture's economic vitality. Another issue is hunger, which may involve food security, food access, the existence of food deserts, local food self-sufficiency, and healthy eating. Environmental quality issues ranging from water consumption limits caused by drought to such pollution as odor, noise, and manure associated with farming operations often must be addressed.

Any of these issues can be a concern in our communities, with the major worries of a community depending on its sociopolitical and economic situation that includes the vitality of its local food system. This food system is "the process by which food is produced (grown, raised, harvested, or caught), transformed by processing, made available for purchase, consumed, and eventually discarded" (Harmon, Harmon, and Maretzki 1999, 14). To sustain a locally vibrant and economically viable food system requires that food and agriculture be included in the vocabulary a community uses to describe itself.

Most individuals in any community express limited concern about such issues. Those who actively engage in the community decision-making process represent only a small segment of the population (Dietz and

Rycroft 1987). Often these individuals coalesce around a particular issue. To move from the concerns of a few to consensus within the larger community requires expanding the dialogue across the community. Through this process of expansion a community or region has the opportunity to ask itself if it wants a locally viable food and agriculture system as it defines its desired future. For such a vision to be articulated and then realized mandates direct local involvement, both in its definition and its subsequent translation in the local context (Pfeffer and Lapping 1995).

Framing the Issue

Most of those not involved with an issue become aware of it through mass media—newspapers, radio, and television. Media serve as both windows and gatekeepers. As windows, media introduce, expose, and acquaint residents with issues beyond just the individuals' own concerns. As gatekeepers, media determine what is or is not reported as well as how the topic is covered. In the reporting process information is organized; facts and their local interpretation can be highlighted, ignored, or downplayed, thus influencing how readers, listeners, and viewers think about the issue. Community-based media play a critical role in interpreting how issues affect residents' everyday lives (Lauterer 1995). All news is local. The frame within which information is organized then becomes the window through which residents of a community become aware of an issue and perhaps motivated to become actively involved with it.

Although each media outlet in a community targets a distinct segment of the community and should not be overlooked, print media usually offer more options for exposure. In fact, most newspapers are local. Among the 1,457 daily newspapers in the United States in 2002, 85 percent had circulations under 50,000. The American Society of Newspaper Editors classifies these newspapers as small. Among the 1,238 newspapers with circulations under 50,000, 84 percent had circulations under 25,000 (*Editor and Publisher* 2003). Circulation size and coverage area help to define a newspaper's audience and therefore what it publishes and the local sources on which it relies. In smaller circulation markets such sources are vital to news content. Media reflect what a community is and shape what it will become.

Becoming a source for news depends on many interrelated factors but one, source credibility, is consistently cited (Powers and Fico 1994). Providing timely, unbiased, accurate information, explained without being judgmental, enhances one's probability of being a preferred news source (Stringer and Thomson 2000).

Media exposure is crucial in making the local food system a part of a community's agenda because such awareness requires moving the discussion of issues from small groups sharing common interests to community-based conversations. Building this salience means engaging the larger community. Food is a common human experience. Communication that connects the issue to everyday life, putting a face on the issue, is perceived to be most effective (FoodRoutes Network 2002). Public dialogue should reflect the diversity of the local community. Not everyone, however, will share the same level of awareness, understanding, or knowledge of the issues or even be concerned about the same issues. As this educational process unfolds, information is shared; motives, interests, and values, both individual and collective, are revealed to clarify, further define, resolve conflicts, and develop common understandings of concerns identified. As different constituencies within the community participate, competing interests must be resolved before a collective vision can evolve to enable a community to pursue its preferred future.

Communities are not monolithic. Public conversations on whether a local food system is part of a community's economic and social structure and is contributing positively to its quality of life will not be value free. Nor will they be conflict free. Words used to describe farmland can reflect the competing interests and understandings among community residents. Suburbanites often view farmland as open space, a public amenity. To farmers this same land is their economic foundation on which their financial stability is based. These descriptions convey different values and thus different uses for the same resource and reflect the underlying tension and sometimes overt hostility that can erupt as the population of a community changes. Such differences have the potential to erode community trust that is crucial in developing community consensus. For example, in market interviews with shoppers, Thomson and Kelvin (1996) found that, among those concerned about retaining farms in southeastern Pennsylvania, the most frequently voiced concern was not wanting more housing development (33 percent of 1569). Most respondents enjoyed seeing open farmland (98 percent of 1202).

To address differing local realities, strategies to structure the dialogue are available to establish boundaries regarding the content and direction of the public conversation, ensuring that the focus remains on the issue. Involving local media in this dialogue increases the likelihood that the issue will be covered from a community perspective. Media involvement is essential to frame the issue in order to build awareness among the larger public about the issue. This involvement expands the number of residents who initially are interested, concerned, and willing to become active participants in a community-directed strategy to retain and/or develop a sustainable local food system.

Through this process a community can redefine itself through language that resonates with the larger community, that is, through the words communities use to describe themselves, through government policies and regulations, and through citizen advocacy and funding (Abel 2000; Bridger 1996, 1997). The words people choose to communicate about an issue reflect how they understand and interpret the topic. The language used also shapes their future understanding of the issue. However, competing interests do not necessarily share a similar understanding, even if they use the same words. Only through interaction can a shared sense of understanding and community develop. Finding common ground is essential.

Community Strategies to Build Common Ground

In his chapter in this volume Thomas Lyson defines six characteristics of civic agriculture. This chapter focuses on one: strategies to ensure that food and agriculture are defined as an integral part of the economic and social fabric of the community in order to engage those who eat as well as those who derive their livelihoods from the food system. Identifying strategies for public participation appropriate for a community provides the framework for such discussions. Through such discourse, shared values and common ground can be identified, creating awareness, developing a common language, and providing participants with

Table 13. Public dialogue strategies: strengths and weaknesses

Strategy	Objective	Strengths	Weaknesses
Community forums	To move from individual to collective action on issues of common concern	• Are issue focused, providing a venue for diverse stakeholders to express their views. • Are structured frameworks that facilitate dialogue. • Engage a diverse mix of stakeholders. • Attract media participation that expands the audience. • Facilitate rapid learning about local food system issues. • Show how food connects people as consumers and as professionals.	• Involve an extensive collaborative planning process. • Require strong leaders/organizers to ensure diversity of participants, quality dialogue, and action plans leading to outcome activities. • Require follow-up with media to ensure increased coverage of food system issues. • Require ongoing coordination/follow-up to maintain momentum.
Study circles	To become informed about an issue through small group study	• Provide an opportunity to educate small groups about an issue. • Offer an in-depth learning opportunity. • Focus on individual actions.	• Offer limited follow-up. • Are time intensive for participants • Sustainability often depends on facilitator's enthusiasm. • Offer individual, as opposed to organizational, representation in groups making it difficult to generate commitments to action on a community or regional level.
Focus groups	To understand, not determine action on, a topic	• Provide opportunities to discuss issue in depth from multiple perspectives. • Are useful in obtaining information on a defined topic with an identified group.	• Follow-up is responsibility of those who conduct group. • Are time and resource intensive. • Involve small numbers. • Need to synthesize ideas from multiple focus groups.

common knowledge to explore, influence, shape, and redefine issues to build community consensus and action strategies. This consensus, based on broad local participation, will enable an agenda for action to be defined and implemented by the community. Through such strategies, public policy evolves.

Multiple strategies—community forums, study circles, and focus groups accompanied by supporting educational resources—are available for communities to use in order to facilitate local dialogue about their food system. As table 13 presents, multiple factors determine the strategy that is most appropriate and viable for a community. Focus groups can help a community identify key issues and understand the diversity of perspectives community residents hold on these issues, providing input for a more inclusive, in-depth community discussion. Those who want to acquire a more comprehensive understanding of a particular topic might join a study circle. First, and most importantly, strategies should be selected that are appropriate for what those who seek to foster dialogue on the local food system want to accomplish.

According to Thomson, Abel, and Maretzki (2001a), public dialogue can enable citizens to

- Become aware of others who share common interests about food system issues;
- Learn specifics about their community food system (such as the extent of hunger and farmland loss) and how they can put their knowledge to work;
- Become part of creating a solution that addresses community concerns (such as food safety, sprawl, keeping local farms in business, or keeping supermarkets from closing in a neighborhood);
- Be prepared to contribute in meaningful ways to public decisions on how to deal with community concerns, both in the short-term and via longer term initiatives;
- Build partnerships and coalitions among and across diverse constituencies—interest groups, civic organizations, government agencies and organizations, elected officials, businesses (including the media), and concerned citizens.

This process forces people to consider the ideas and opinions of others as they create a vision for the future of their community food system (Smith and Maretzki 2000). Simultaneously, the process can facilitate the development of trust and respect among those involved. Focusing on the community's assets rather than problems can reaffirm the commitment to action. Ensuring quality discourse is critical to developing a shared vision and common ground on which to act.

Community Forums

Edible Connections (Nunnery, Thomson, and Maretzki 2000) is a food communications forum that offers local communities a way to expand their conversations on local food systems issues by involving a cross section of residents. Two distinct elements distinguish *Edible Connections* from other citizen dialogue models. These features are its focus on the inclusion of media representatives as key forum participants and its structured format so that the preidentified issue is the center of the dialogue. Participants should therefore span the food system from growers to consumer advocates, including media personnel. Bringing media representatives into the mix is an important step in communicating concerns about a sustainable food system to an audience broader than just forum participants.

Edible Connections is structured around six elements:

- *Setting the table* defines the goals of the forum and the topic or issue that will be addressed within the local context.
- *Food as lifestyle* focuses on how participants interact with the identified food system issue as consumers and as members of families, cultures, and organizations.
- *Food as livelihood* explores how the food system issue represents a point of common economic connection for many different occupations in a community; often these occupations are not perceived as agricultural in nature.
- *Food as connection* explores how the issue connects consumers to the local environment, the local food system, and to each other.
- *Town meeting* allows forum participants via facilitated discussion to clarify questions, explore elements of the food system that are

locally desired, and to identify actions that can be taken to address the issue discussed in order to strengthen the local food system.
- *Celebration of local foods*, featuring locally grown and processed foods, allows participants to continue discussions begun during the forum as well as to initiate networking that can lead to future action.

The planning process is central to the forum's outcome. The forum itself can encourage people to support local agriculture, educate others about the local food system, and commit to meeting further to address issues identified at the forum (Thomson, Abel, and Maretzki 2001b). Forum outcomes have been as varied as the topics addressed. Forums focusing on community awareness and understanding tend to result initially in participants taking individual action such as buying more local produce or talking to others about local food system issues (Thomson, Abel, and Maretzki 2001b). If those on the committee organizing the forum have previously worked together, then an organizational or community initiative will more likely be the outcome. Such initiatives may range from establishing farmers' markets to creating food policy councils to articulating agriculture's role in a region's future comprehensive plan.

Our Food—Our Future: Enhancing Community Food Security through Local Action offers communities another model around which to structure their dialogue on the food system (Wilkins et al. 2002). This tool is particularly useful for communities interested in planning and conducting activities and initiating projects to increase food security and to promote sustainable local agriculture. A twenty-seven-minute, made-for-television video is the centerpiece of *Our Food—Our Future*; the video highlights four successful and highly innovative community food projects in the states of New Jersey, New York, Pennsylvania, and West Virginia.

The video visualizes, for example, how adding value to a local commodity such as apples can enhance the economic viability of a community, thereby increasing local food security, a key to community mobilization that can lead to food system change. Moving from such images on television to concerted local action, however, requires an effective educational strategy. To facilitate this process, a program guide supplements the video. This guide provides detailed information on how to conduct workshops in which the video is used to stimulate discussion about the concept of community food security and strategies to address it. Participants should then be able to link resources available in the community so as to make their community more food secure.

Both *Edible Connections* and *Our Food—Our Future* structure the forum's dialogue in order to establish boundaries regarding the content and direction of the dialogue, allowing the focus of the forum to be the issue. Involving local media in this dialogue helps ensure that the issue will be covered from a broad community perspective and framed in such a way that awareness among the larger public about the issue can grow. This involvement expands the number of residents who initially are interested, concerned, and willing to become active participants in a community-directed strategy to retain and/or develop a more sustainable local food system.

Holding Future Search conferences is another community forum strategy, and such conferences have been used to explore various topics of common interest even beyond those associated with the local food system among a defined constituency. This approach has two goals: to help diverse, often large, groups discover values, purposes, and projects they have in common and to enable people to create a preferred future together and to begin implementing it right away (Weisbord and Janoff 2000).

A Future Search conference usually takes place over several days and consists of five tasks (Weisbord and Janoff 2000):

- Looking back;
- Identifying trends affecting the current state of affairs and identifying how forum participants currently are addressing local issues;
- Highlighting what forum participants presently are doing that makes them proud and that generates regrets;
- Looking ahead to develop a common vision for the future;
- Formulating action plans.

Six rural counties in northern New York used Future Search to generate community awareness, planning, and action related to community food security in the region (Pelletier et al. 1999). Each county conducted its own conference, yet each produced similar ideas regarding how citizens wanted to improve food and agriculture in their communities. Pelletier et al. note that "[t]he action agendas reflected a strong interest in re-localizing many food system activities, strengthening the economic viability of local agriculture, improving access to healthful local foods, strengthening anti-hunger efforts, and strengthening education about larger food system issues in addition to consumer nutrition education" (1999, 414).

As the name Future Search implies, the purpose of these conferences is to generate an agenda for action based on the vision defined during the event. Both *Edible Connections* and *Our Food—Our Future* represent strategies that are more fluid. No specific expectations are predefined regarding the outcomes. Using the approach of either enables local communities to structure forums to meet their specific local objectives. These objectives might include informing community members, mobilizing citizen action, or encouraging the media to increase coverage of food system issues (see table 13).

For any one of the community forum strategies, the need to develop ownership for follow-up activities must be recognized, or these efforts can become lost in the group process. Individuals and organizations need to be identified in order to take leadership roles and sustain those roles over time. If multiple tasks are to be carried out, coordination across committees and individuals needs to occur. Groups also need to address how they will maintain continuing commitment among those involved as well as addressing attrition that is likely to occur over time.

Study Circles

Study circles are small peer-led discussions that provide community members with the opportunity to learn about and act upon important social and political issues (Study Circles Resource Center 2005). These circles allow participants to increase their understanding about issues facing their communities and to brainstorm alternative actions that they can take as individuals (see table 13). According to Smith and Maretzki, "[t]hey bring the wisdom of ordinary individuals to bear on difficult issues" (2000, 7). Study circles usually involve no more than twelve people who meet three to six times to discuss an issue of importance to them, to their communities, and to society. Often educational resources are provided to help inform and frame the discussions.

A series of study circles was used to engage people not directly involved in agriculture in discussions regarding their concerns about the food system (Wagoner and Thomson 1995). In their final meeting participants discussed action steps that they might take, agreeing that as informed consumers they could support local farmers and processors by choosing locally produced foods. They also suggested that convincing people to change their purchasing and consumption habits needs to begin through "fun" events such as garden tours, food fairs, and whole foods potlucks.

Study circles can be one outcome of a community forum. The strength of the study circle approach is that small groups of citizens are encouraged to explore an issue in depth and formulate strategies for changes that they can make as individuals within their communities (see table 13). A weakness of this approach is that it involves only small numbers of individuals, and thus changes at the community level usually are not articulated.

Focus Groups

Focus groups primarily have been used as a research and marketing tool in order to help those sponsoring the focus group to gain understanding of the topic for which the group was gathered together. In a community setting such in-depth discussions can help to identify citizen awareness, understanding, and interest in the local food system as well as give voice to those who in a larger forum might be less likely to speak up. The insights of the latter can provide critical input to strengthen other community initiatives.

A focus group, usually involving no more than ten to twelve individuals, is a structured discussion in a "permissive, non-threatening environment" lasting from one to three hours in order to obtain the participants' perspectives on a defined topic (Krueger 1994). These individuals often represent a mix of constituencies and thus may have had limited previous association with one another.

Focus group discussions can ensure that diverse concerns about the community's food system are not overlooked. For these conversations to be useful, however, participants need to be familiar with the subject being considered. Focus groups often are carried out to obtain information as input into another initiative, rather than to develop action plans (see table 13). Those who organize the focus groups are responsible for any follow—up that occurs.

Supporting Community Initiatives

Educational resources can be valuable supplements in support of citizen dialogue strategies, as they allow individuals to gain background information on issues to better inform community conversations. Also, such resources can be used in multiple settings, not only for those focusing on the local food system.

To understand the food and agricultural system in one's community and also to expand those involved in defining the issue, community-initiated and implemented needs assessments can provide considerable insight. Many approaches are available that can be tailored to address a community's particular concerns within its available resources (Siedenburg and Pothukuchi 2002). Local people interviewing other local residents can provide further insight into the local context. By collecting, analyzing, and interpreting locally generated information, local residents have a vested interest in interpreting and disseminating this information to the larger community.

Undertaking a broader analysis of a region's food system has resulted in several guides that not only are useful to residents but that can serve as models for food system analysis in other settings (Hora and Tick 2001; Northeast Sustainable Agriculture Working Group 2002). To strengthen the local food system, local findings often can be further clarified and explained on the basis of previous surveys and related research (Green and Hilchey 2002; Wildfeuer 2002).

Food system mapping is another tool that can be carried out informally as part of a local educational program or workshop and used to answer the seemingly simple question of "Where does our food come from?" (Harmon, Harmon, and Maretzki 1999). On the community level, food system mapping using Geographic Information Systems (GIS) can address a variety of questions and bring about better understanding of a particular issue, for example, the degree to which low-income neighborhoods have access to such food resources as farmers' markets and grocery stores (Food Trust 2001). Regional food guides like the *Northeast Regional Food Guide* described in Jennifer Wilkins' chapter in this volume also can increase awareness regarding locally available foods and how to use them.[1]

Grassroots involvement lends legitimacy to the concerns of citizens and provides energy to sustain action (Pelletier et al. 1999). Sharing this information through supporting resources such as a county food system atlas (Hinrichs et al. 2002) and special reports, like one that addresses the need for more supermarkets in Philadelphia (Food Trust 2001), provides an even larger cross section of the community with the opportunity to learn more about these issues from a local perspective. Ideally, this exposure results in better-informed public discourse.

Each of the community-based strategies discussed offers some supporting resources with which to acquaint users on how to implement the process as well as on suggested content for the community initiative being carried out. The more directly related these resources are to the interests of the community, the more likely organizers of such initiatives will find these resources useful.

Access to information does not ensure attention to an issue. However, such information is essential if the public is to become aware of an issue and wants to understand and address it within the community. Although no community exists in isolation, the quality of individual and community life is significantly influenced by individual and public actions at the local level. If a community wants a vibrant, economically viable local food system with a strong, sustainable agricultural base, then the residents of the community must move beyond individual conversations to more public dialogue in which the community's food issues can be explored by a variety of stakeholders. In order to define their common needs, local citizens must initiate an activity of civic process that involves consumers as well as producers and others who derive their livelihoods from the food system. Carrying out appropriate community-based strategies can enable a community to build stronger consensus regarding what it currently is and what it desires to be as it develops a shared vision for a sustainable local food system.

Notes

1. The *Northeast Regional Food Guide* (1996) was prepared by J. L. Wilkins and J. C. Bokaer-Smith, Cornell Cooperative Extension, Cornell University. A downloadable version is available at http://www.nutrition.cornell.edu/foodguide/archive/index.html (last accessed May 15, 2006). For a full-sized poster, contact the author at jhw15@cornell.edu.

References

Abel, J. L. 2000. "Assessing the involvement of Pennsylvania professional planners in food system activities." Master's thesis, Pennsylvania State University.

Bridger, J. C. 1996. "Community imagery and the built environment." *Sociological Quarterly* 37 (3): 353–74.

———. 1997. "Community stories and their relevance to planning." *Applied Behavioral Science Review* 5 (1):67–80.

Dietz, T. M., and R. W. Rycroft. 1987. *The Risk Professional*. New York: Russell Sage Foundation.

Editor and Publisher International Yearbook: The Encyclopedia of the Newspaper Industry. 2003. 83rd ed. Vols. 1–3. New York: Editor and Publisher.

FoodRoutes Network. 2002. *Where Does Your Food Come From? Recipes for Communicating Effectively about American Agriculture*. Millheim PA: FoodRoutes Network.

Food Trust. 2001. *The Need for More Supermarkets in Philadelphia: Special Report*. Philadelphia: Food Trust. http://www.thefoodtrust.org (last accessed April 7, 2007).

Green, J., and D. Hilchey. 2002. *Growing Home: A Guide to Reconnecting Agriculture, Food, and Communities*. Ithaca NY: Community, Food, and Agriculture Program, Cornell University.

Harmon, A. H., R. S. Harmon, and A. N. Maretzki. 1999. *The Food System: Building Youth Awareness through Involvement. A guidebook for educators, parents, and community leaders*. A Keystone 21: PA Food System Professions Education Project. University Park PA: College of Agricultural Sciences, Pennsylvania State University.

Hinrichs, C., S. Gradwell, M. Russell, and W. VanDyk. 2002. *Audubon County Food Systems Atlas: Exploring Community Food Systems*. Ames IA: Department of Sociology, Iowa State University. http://www.leopold.iastate.edu/pubs/other/files/auduboncountyfoodsystematlas.pdf (last accessed April 7, 2007).

Hora, M., and J. Tick. 2001. *From Farm to Table: Making the Connection in the Mid-Atlantic Food System*. Washington DC: Capital Area Food Bank.

Krueger, R. A. 1994. *Focus Groups: A Practical Guide for Applied Research*. Thousand Oaks CA: Sage.

Lauterer, J. 1995. *Community Journalism: The Personal Approach*. Ames IA: Iowa State University Press.

Northeast Sustainable Agriculture Working Group. 2002. *Northeast Farms to Food: Understanding Our Region's Food System*. Belchertown MA: NESAWG/New England Small Farm Institute.

Nunnery, S., J. S. Thomson, and A. N. Maretzki. 2000. *Edible Connections: Changing the Way We Talk about Food, Farm, and Community: A Planning Guide for Conducting a Food Communications Forum*. With supporting video. University Park PA: College of Agricultural Sciences Publication Distribution Center, Pennsylvania State University.

Pelletier, D., V. Kraak, C. McCullum, U. Uusitalo, and R. Rich. 1999. "Community food security: Salience and participation at community level." *Agriculture and Human Values* 16:401–19.

Pfeffer, M. J., and M. B. Lapping. 1995. "Prospects for a Sustainable Agriculture in the Northeast's Rural/Urban Fringe." In *Research in Rural Sociology and Development*, edited by H. K. Schwarzweller and T. A. Lyson, 67–93. Greenwich CT: JAI Press.

Powers, A., and F. Fico. 1994. "Influences on use of sources at large U.S. newspapers." *Newspaper Research Journal* 15 (4): 87–97.

Siedenburg, K., and K. Pothukuchi. 2002. *What's Cooking in Your Food System? A Guide to Community Food Assessment*. Venice CA: Community Food Security Coalition.

Smith, J., and A. N. Maretzki. 2000. *Citizen Dialogue: A Guidebook*. University Park PA: Pennsylvania State University.

Stringer, S. B., and J. S. Thomson. 2000. "Sources of Agricultural News: An Evaluation of Pennsylvania Media." Paper presented at the annual meeting of the Agricultural Communicators in Education, Washington DC.

Study Circles Resource Center. 2005. "What is a study circle program?" http://www.studycircles.org/en/page.whatisastudycircle.aspx (last accessed May 7, 2006).

Thomson, J. S., J. L. Abel, and A. N. Maretzki. 2001a. "Edible Connections: A model for citizen dialogue used to discuss local food, farm, and community issues." *Journal of Applied Communications* 85 (1): 25–42.

———. 2001b. "Edible Connections: A model to facilitate citizen dialogue and build community collaboration." *Journal of Extension* 39 (2) (April). http://www.joe.org/joe/2001april/a5.html (last accessed April 9, 2006).

Thomson, J. S., and R. E. Kelvin. 1996. "Suburbanites' perceptions about agriculture: The challenge for media." *Journal of Applied Communications* 80 (3): 11–21.

Wagoner, P., and J. S. Thomson. 1995. *Final Report of the Food System Study Circles Conducted in Southeastern Pennsylvania, April-June 1995*. Unpublished.

Weisbord, M. R., and S. Janoff. 2000. *Future Search: An Action Guide to Finding Common Ground in Organizations and Communities*. San Francisco: Berrett-Koehler.

Wildfeuer, M. 2002. *Farmers, Food, and the Modern Consumer: A Review of Consumer Surveys, Local Food Initiatives, and Eco-Label Programs.* Kutztown PA: Rodale Institute.

Wilkins, J. L., A. N. Maretzki, M. W. Hamm, J. D. Paddock, K. Asher, and E. Tuckermanty. 2002. *Our Food—Our Future: Enhancing Community Food Security through Local Action; A Community Food Security Program Guide.* Ithaca NY: Cornell University.

Part III
The Importance of Place and Region in Remaking the Food System

10. Retail Concentration, Food Deserts, and Food-Disadvantaged Communities in Rural America

Troy C. Blanchard and Todd L. Matthews

For many residents of the United States purchasing groceries is a minor inconvenience rather than a major obstacle. In 1995 a standard shopping trip for the average U.S. family involved a six-mile drive lasting no more than 12.5 minutes (U.S. Department of Transportation Federal Highway Administration 2001). For some Americans, however, especially those in rural areas, the time and distance traveled to purchase groceries is significantly longer. A report by the Economic Research Service of the U.S. Department of Agriculture indicates that residents of rural communities in the Lower Mississippi Delta endure a far different journey to the shopping center (Kaufman 1998). For example, rural counties in the Delta average one supermarket per 190.5 square miles. Additionally, over 70 percent of the low-income population in the Delta traveled thirty or more miles to purchase groceries at supermarkets in an effort to avoid high-priced smaller grocers and inadequate quality food sold at convenience stores and gas stations.

Researchers studying similar conditions in parts of the United Kingdom have described areas with limited access to food as "food deserts" (Furey, Strugnell, and McIlveen 2001). In the case of Northern Ireland declining neighborhoods in large cities lost all grocery stores and markets, leaving the population without access to any type of food retailer. In contrast, the notion of food deserts has been largely overlooked by both researchers and policy makers in the United States.

One possible explanation for this omission is the important distinction between quantity and quality of food retailers. In the UK the *absence* of food retailers was the central issue driving the recognition of food desert populations. In the United States the proliferation of convenience stores and gas stations ensure that some type of food is accessible to almost all residents. However, the quality and pricing of food products available in U.S. convenience stores and supermarkets varies dramatically. Consumers purchasing food at a convenience store pay a premium for access to food products. Additionally, consumers choose from a smaller variety of food products that may not be suitable for the maintenance of a healthy diet. Thus the application of the food desert concept in the United States elucidates a great divide between those with and without access to low-cost, high-quality foods.

In this study we apply the concept of food deserts to U.S. nonmetropolitan areas in an effort to understand inequalities in food access for rural residents. We develop a measure of food access for U.S. nonmetropolitan counties in order to examine how the restructuring of nonmetropolitan retailing has created food deserts. We also explore the characteristics of individuals who reside in food deserts in order to understand who is affected by them.

The Emergence of Food Deserts in the United States

The retail distribution of food is a central concern for U.S. nonmetropolitan areas. Simply put, if U.S. nonmetropolitan retail food sales activity among supermarkets and supercenter stores (hybrid stores offering groceries and discount merchandise) becomes concentrated within limited geographic areas, such as one or two cities or towns within a county, persons outside of these retail centers become isolated from convenient access to low-cost, quality food. For these residents the remaining choices, such as small convenience stores, gas stations, and restaurants, offer few prospects for the maintenance of a quality diet. Populations facing these conditions reside in food deserts and must undertake lengthy commutes in order to access food sold in supermarkets and supercenters.

The creation of food deserts in the United States has occurred gradually during the past thirty years. The impetus for the shift from a large number of widely dispersed small-scale local grocers to a concentration of supermarkets and supercenters located in a limited geographic area has been fueled by the globalization of food production and distribution that has resulted in a handful of corporations controlling the majority of sales, as detailed in this volume's chapter by Thomas Lyson (see also Lyson and Raymer 2000). Globalization allows supermarket and supercenter chains to purchase large quantities of food from suppliers in order to sell at lower prices. The buying power possessed by such large retail chains as Wal-Mart, Target, Sam's Club, Albertson's, and others provides these corporations a distinct advantage over smaller chains and mom and pop grocers. Especially in the South, Midwest, and West, the entrance of a large retailer into a nonmetropolitan community can have a substantial negative impact on the level of retail activity and result in the loss of small retail establishments and decline in the size of the retail labor force (Blanchard et al. 2003).

Studies of globalization in the retailing industry have focused on discount merchandise superstores and the concentration of food sales into large chain supermarkets. Combined, these two types of retail outlets account for 89.8 percent of all grocery and "food for off premise consumption" sales in the U.S. (U.S. Bureau of the Census 2001a). Studies of both types of retailers follow a common theme: the buying power of large chains reduces the viability of smaller establishments. While discount merchandisers accounted for only 12.1 percent of all grocery sales in 1997, their share of grocery sales grew by 9.3 percent from 1992 to 1997, and they represent the fastest growing segment of firms selling groceries. In 1999 the dominant firm in this category, Wal-Mart supercenters, ranked fifth in total grocery sales in the United States (Kaufman et al. 2000). Additionally, studies of discount merchandisers focus exclusively on Wal-Mart because of its unique effect on competing small establishments.

Studies of the impact of Wal-Mart indicate that the entrance of a Wal-Mart store restructures local retail markets. In a study of the effects of the presence of Wal-Mart in fourteen Missouri counties, Keon, Robb, and Franz (1989) found that the number of retail stores in counties with a Wal-Mart declined during the 1980s. Stone (1995) also documented the concentration of retail activity within thirty-four Iowa towns. Towns without a Wal-Mart experienced declines in sales, number of retail establishments, and sales tax, while those with a Wal-Mart experienced substantial growth in sales revenue and employment. Prior studies also suggest that local retailers in rural communities report an "environment of hostility" when large retailers enter local markets (Shils 1997). Research contrasting the location of Wal-Mart stores to other large retailers attributes the success of Wal-Mart to its strategy that targeted small southern towns in which competition was negligible (Barnes et al. 1996; Shils 1997; Graff 1998).

The food stores industry (supermarkets, grocery stores, and convenience stores) has followed a similar trend over the past fifty years. Until the mid-1980s the consumer market for food stores serving metropolitan populations could be characterized as a single metropolitan area; the consumer market for food stores serving rural communities could be characterized as a town with as few as 1,000 residents (Kaufman, Newton, and Handy 1993; Stone 1995). During this era the viability of small mom and pop grocers varied. In 1982 the percentage of sales accounted for by the four largest food store firms operating in a metropolitan area ranged from 90.6 percent in Iowa City, Iowa, to 27 percent in the Appleton-Oshkosh, Wisconsin Standard Metropolitan Statistical Area (SMSA).

However, changes in the technology of food distribution and corporate mergers led to a major restructuring in food retailing beginning in the mid-1980s. For rural areas in Iowa, Stone (1995) suggests that many small towns (1,000 or fewer residents) lost local grocery stores, forcing residents to travel to larger towns for groceries. In the 1990s supermarkets in rural and urban markets also competed with discount chains for sales. From 1992 to 1997 the percentage of total grocery sales in the United States accounted for by food stores declined by 8.9 percent, while the discount merchandisers gained 9.3 percent of grocery sales (U.S. Bureau of the Census 2001a). Additionally, the number of supermarket and grocery stores in the United States declined from 73,357 in 1992 to 69,461 in 1997, representing a 5 percent decline in the number of stores. This pattern follows the general one of decline in the number of food stores over the past forty years (Kaufman, Newton, and Handy 1993).

Implications for Food Desert Populations

The increasing concentration of food retailing activity has clear implications for food access among nonmetropolitan residents. Studies documenting price differentials in food costs between urban and rural areas find that rural residents, especially the rural poor, pay more for groceries because of lack of access to large supermarkets that offer more competitive prices than smaller grocers (Kaufman et al. 1997). Thus nonmetropolitan residents experiencing physical or economic resource limitations are at a distinct disadvantage to those without these limitations. For example, persons experiencing physical disabilities may be less able to travel long distances to secure low-priced, high-quality food. This problem is compounded by the lack of public transit systems available to nonmetropolitan residents.

The poor in food deserts also experience a severe disadvantage. Studies of food pantry clients indicate that the vast majority of persons using food pantries to meet food needs do not own vehicles (Daponte et al. 1998; Molnar et al. 2001). Additionally, the food stamp program limits total assets held by program participants to no more than $6,550, constraining the ability of a family to own reliable transportation and receive food stamp benefits (Molnar et al. 2001). If a low-income family on a tight budget owns an unreliable vehicle, the family must redirect money away from food expenditures into car maintenance.

Thus the changes in the food retailing industry have "distanced out" many disadvantaged nonmetropolitan populations from supermarkets and superstores. These residents may be forced to rely on convenience stores or on small grocery establishments that may not offer foods promoting a healthy diet. For the poorest of the poor a food pantry may be the only option (Daponte et al. 1998).

Measuring Food Deserts

To our knowledge prior research on food access has not attempted to measure the concept of food deserts in the United States. In the UK researchers have measured food deserts through site visits and direct observation of neighborhoods in order to determine the absence or presence of food retailers. Although direct observation of a given neighborhood or community is an ideal methodological approach, the objective of this study is to measure the level of food access in the 2,275 nonmetropolitan counties in the continental United States. Because of the scope of our study, we rely on secondary data on food retailers and the distribution of the U.S. nonmetropolitan population derived from the 1999 Zip Code Business Patterns (ZCBP; U.S. Bureau of the Census 2002b) file and the 2000 Census of Population and Housing Summary File 1 (U.S. Bureau of the Census 2001b). ZCBP data provide information on the number of businesses by type of business for all zip codes in the United States.

We classify food desert population as those residents of a county residing ten or more miles from a supermarket or supercenter. Our choice of a ten-mile radius assumes a point-to-point drive time of approximately twenty minutes, traveling at an average rate of speed of thirty miles per hour. To estimate the percentage of a county's population residing ten or more miles from a supermarket or supercenter, we used ARCVIEW Geographic Information System (GIS) mapping software, which provides a means to estimate the distance between county residents and food retailers.

We selected zip codes that contained at least one supermarket or supercenter/wholesale club in 1999 and obtained latitude and longitude coordinates in order to create centroids (the center of the zip code area) for these zip codes from the 1999 U.S. Bureau of the Census Zip Code File (U.S. Bureau of the Census 2002b). We then estimated the total population residing outside of a ten-mile radius of the zip code centroid using population counts tabulated at the block group level (a very small subdivision of a county). All block groups falling outside of the ten-mile radius of a zip code containing a supermarket, supercenter, or wholesale club are designated as food desert populations. We then tabulated the total number of persons residing in a food desert and divided by the total county population to obtain the proportion of the county population residing in a food desert. In the process of estimating food desert populations, we also adjusted for the type of roads available to the local population. A more detailed technical description of this methodology for measuring food access can be found in Blanchard and Lyson (2002).

We classify nonmetropolitan counties as food deserts if the proportion of the county's population in a food desert is greater than the median proportion for the region of the United States in which the county is located. For example, in the western region of the United States the median proportion of the population residing in a food desert is .63.

Thus all counties in western states with a proportion of the population in food deserts greater than .63 are classified as food desert counties. We define region using the Bureau of the Census designation. We classify counties based on the regional median for two reasons. First, the distribution of the proportion of the population in a food desert among counties is highly skewed. Second, we chose regional medians rather than the national median because the size of a county (square miles) varies across regions. Thus a large county may have a higher proportion of the population in a food desert because of our use of a ten-mile radius to capture food desert populations. Classifying food desert counties regionally rather than nationally avoids overclassification of western counties as food deserts and underclassification of southern and northeastern counties.

Analytical Strategy

In our descriptive analysis we employ both maps and tables to describe food desert counties. We present a national map of all food desert counties and a map of severe food desert counties using GIS software. Severe food desert counties are defined as counties in which the total county population resides in a food desert.

Our tabular data address two issues. First, we provide information on other types of food retailers present in food desert counties; these retailers include small grocers, convenience stores, gas stations, fruit and vegetable markets, fast food restaurants, and full-service restaurants. This information will identify the alternatives available to consumers without convenient access to a supermarket, supercenter, or wholesale club. Our information on other types of food retailers comes from the 1999 County Businesses Patterns data from the Bureau of the Census (U.S. Bureau of the Census 2001c). These data report the number of businesses by type of business for all U.S. counties.

Second, we provide sociodemographic characteristics of food desert residents. The data reported come from the 2000 Census of Population and Housing Summary File 3 (U.S. Bureau of the Census 2002a). The sociodemographic characteristics of food desert populations provide information on the type of persons residing in food deserts and point to the specific policy needs of food deserts.

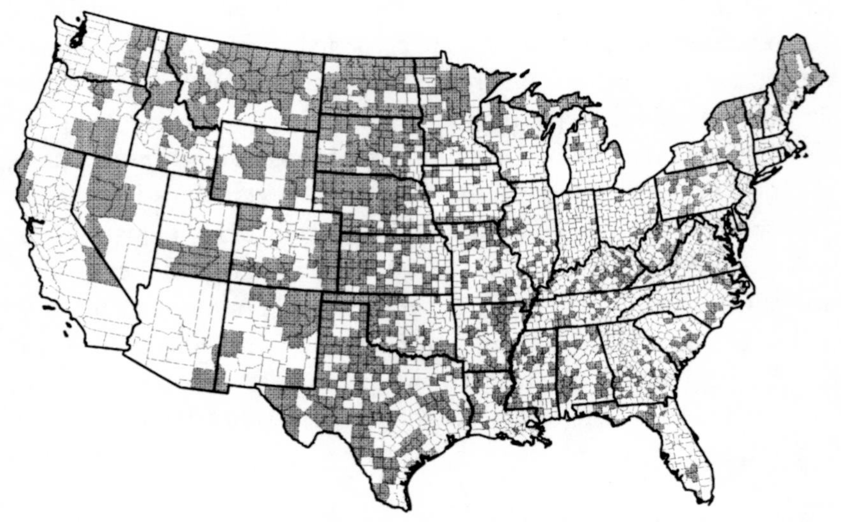

Map 1. U.S. nonmetropolitan food desert counties, 2000

Descriptive Results

Map 1 shows nonmetropolitan counties in the United States that we classify as food desert counties. Only Massachusetts, Rhode Island, Connecticut, and New Jersey have no food desert counties. Several states, primarily but not exclusively in the Midwest and Mountain West, have a majority of their land area composed of food desert counties. These states include Maine, Texas, Kansas, Nebraska, North Dakota, South Dakota, and Montana.

Three key trends emerge from an analysis of map 1. First, food desert counties tend to cluster together, both within and between state boundaries. Second, there is a high concentration of food desert counties stretching from the Rocky Mountains east into the western part of the Great Plains and from the Canadian border to the Mexican border. Virtually all of the nonmetropolitan counties in Montana, eastern Wyoming, eastern Colorado, northeastern New Mexico, North Dakota, South Dakota, Nebraska, western Kansas, western Oklahoma, Texas, and western Minnesota qualify as food desert counties. Finally, in addition to the Great Plains, there are concentrated areas of food desert counties within the southeastern United States. These areas are those traditionally linked to conditions of hardship and deprivation and in-

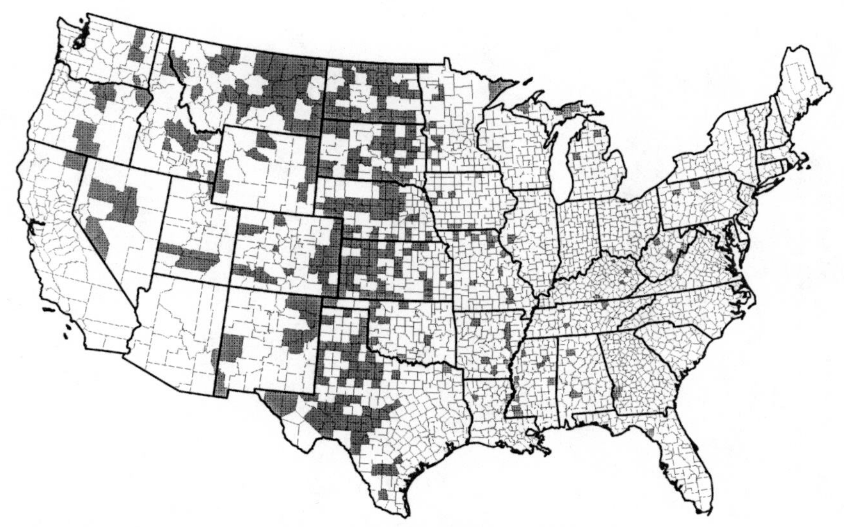

Map 2. U.S. nonmetropolitan severe food desert counties, 2000

clude the Mississippi Delta and Black Belt counties; the Appalachian Mountain region of Kentucky, West Virginia, and northwestern Virginia; and eastern sections of North Carolina and South Carolina.

In map 2 the nonmetropolitan counties in the United States that we classify as severe food desert counties are shown. Compared to map 1, far fewer counties qualify for this status, though again, some interesting observations can be drawn. Clustering of severe food desert counties still is apparent in the western portion of the Great Plains states, including Montana, eastern Wyoming, eastern Colorado, northeastern New Mexico, North Dakota, South Dakota, central Nebraska, western Kansas, and western Texas. There are also counties in Oklahoma and Minnesota that qualify as severe food desert counties. Regarding the southeastern United States, the only clusters of severe food desert counties are very small clusters in the Mississippi Delta region and the West Virginia-northwestern Virginia Appalachian border country.

Table 14 reports the number of food retailers per 10,000 persons by region for food desert and nonfood desert counties in nonmetropolitan areas. In each region there are more supercenters and supermarkets in nonfood desert counties. These differences are particularly pronounced in the South, Midwest, and West. It is interesting to note that in the

Table 14. Number of food retailers per 10,000 by region for nonmetropolitan food desert and nonfood desert counties

	Supercenters	Supermarkets	Small grocers	Convenience stores	Gas station convenience stores	Fruit and vegetable markets	Fast food restaurants	Full-service restaurants
Northeast								
food desert	0.02	0.44	3.63	1.58	5.02	0.12	6.33	10.42
nonfood desert	0.05	0.76	2.50	1.53	3.87	0.19	7.79	10.26
South								
food desert	0.03	0.19	4.12	1.36	6.08	0.04	4.94	5.76
nonfood desert	0.08	0.78	2.55	1.25	5.87	0.10	6.48	5.97
Midwest								
food desert	0.01	0.19	5.45	0.80	5.46	0.02	5.91	10.26
nonfood desert	0.04	0.86	2.25	0.79	5.03	0.07	6.95	8.16
West								
food desert	0.00	0.18	5.84	1.04	5.45	0.01	7.32	14.47
nonfood desert	0.05	0.84	2.50	1.13	4.30	0.06	8.13	11.94

nonmetropolitan food desert counties of the West there are virtually no supercenters such as Wal-Mart or Target.

In terms of small grocers there are more small grocery stores in food desert counties across the country. The difference between county types is most extreme in the midwestern and western states. It also should be noted that across regions and county type there are substantially more small grocery stores than supermarkets or superstores in nonmetropolitan areas. There are also few differences between county types in the number of convenience stores; however, in the Northeast and the West, there are more gas station convenience stores in food desert counties. And, there are more gas station convenience stores in nonmetropolitan areas than general convenience stores. Also noteworthy is that in each region there are more fruit and vegetable markets in nonfood desert counties. Fast food restaurants are somewhat more likely to be found in nonfood desert counties than food desert counties, particularly in the Northeast and the South. Full-service restaurants are more likely to be located in food desert counties in the Midwest and the West. With the exception of the South, where gas station convenience stores are found as frequently as full-service restaurants, this category of food retailer appears most commonly in nonmetropolitan food desert and nonfood desert counties.

In table 15 regional differences in demographic and socioeconomic characteristics between food desert and nonfood desert counties are compared. Across regions food desert counties have a greater percentage of rural residents than nonfood desert counties. This large rural population is particularly evident in the Midwest, where almost 86 percent of food desert counties are rural, as opposed to 59 percent of nonfood desert counties. Racial-ethnic disparities are slight, except for the percentage of Hispanics in food desert counties in the South (10 percent) versus nonfood desert counties (5 percent) and the percentage of Native Americans in food desert counties in the Midwest (almost 4 percent) versus nonfood desert counties (1 percent).

In each region the percentage of the population living below the poverty line is higher in food desert counties as opposed to nonfood desert counties. In the South the percentage in poverty is higher than the other regions in both food desert and nonfood desert counties. Correspond-

Table 15. Sociodemographic characteristics of nonmetropolitan food desert and nonfood desert counties by region

	Northeast		South		Midwest		West	
	Food desert	Nonfood desert	Food desert	Nonfood desert	Food desert	Nonfood desert	Food desert	Nonfood desert
Demographic characteristics								
Population	48,013	63,730	17,090	31,825	11,098	29,247	10,786	32,062
% Rural	74.84	61.89	77.79	62.52	85.72	59.22	78.00	52.15
% Hispanic	1.30	1.51	10.30	5.34	1.94	2.41	11.68	12.66
% Non-Hispanic white	95.50	94.97	71.10	74.75	92.44	93.81	82.58	79.32
% Non-Hispanic black	1.33	1.58	16.56	17.03	0.88	1.25	0.45	0.65
% Non-Hispanic Native Americans	0.53	0.30	0.85	1.33	3.50	1.10	3.35	4.79
% Foreign born	0.81	0.77	0.40	0.44	0.36	0.41	0.57	0.77
Socioeconomic characteristics								
% in Poverty	12.66	10.39	20.31	18.10	13.28	10.92	15.47	14.84
Median family income	$40,210	$45,422	$33,830	$36,412	$38,533	$42,941	$38,125	$41,641
% Unemployed	6.37	5.54	6.83	6.59	4.70	5.15	6.15	7.48
% Less than HS diploma	19.39	17.93	32.49	30.25	19.19	18.73	18.05	17.48
% HS diploma	41.73	39.56	34.70	34.42	37.70	39.42	33.09	29.55
% Some college	22.69	22.88	21.13	22.18	28.65	27.07	31.43	32.86
% BA or better	16.19	19.62	11.67	13.15	14.45	14.71	17.43	19.62
% Households with no vehicle	7.52	7.56	9.30	9.19	5.83	6.47	5.67	5.92
Number of disabilities per 1,000 persons	343.93	313.46	442.13	431.22	312.59	312.90	320.03	319.54

ingly, within each region there is a gap of several thousand dollars in median family income between nonfood desert counties and food desert counties. Again in the South, median family income is lower for both food desert and nonfood desert counties than in any other region of the country. In fact, residents of food desert counties in the other regions have a higher median family income than residents of nonfood desert counties in the South.

Educational differences between food desert and nonfood desert counties also are found, with residents of food desert counties being more likely to have received less than a high school education and less likely to have received a bachelor's degree. Residents of the South are much more likely to have less than a high school education than residents of other regions, regardless of whether or not they live in a food desert county. This pattern also holds for the percentage of households with no vehicle and the number of disabilities per 1,000 persons, both of which are higher in the South than other regions, regardless of food desert status.

Our study points to a central finding regarding the prevalence and severity of food deserts in U.S. nonmetropolitan areas. Residents living in food desert areas will pay higher prices for groceries or incur a greater travel cost to access the large food retailer that may offset the savings available at these stores. Our findings suggest that small grocers and gas and convenience stores are the likely alternatives in the absence of access to supermarkets and supercenters. More importantly, healthy alternatives, such as fruit and vegetable markets, are less prevalent in food desert areas. This absence is especially troubling for vulnerable segments of the population such as low-income individuals and the disabled who comprise a greater share of the population in food deserts. For these persons it may not be feasible to shop at a large food retailer because of travel cost and time considerations. This issue is especially problematic in the South where the percentage of households without a vehicle is greatest.

The key implication of the food desert dynamic is that populations such as the poor and the less educated already experiencing high risk of poor dietary intake and nutrition-related illness may experience even greater risks as a result of living in a food desert. Food deserts may compound ongoing and severe nutritional problems and further exacerbate the socioeconomic gradient in health status. More specifically, food deserts may limit the capacity of populations to meet recommended servings of fruits and vegetables because fresh produce is rarely available in convenience and gas station food retailers. A recent report summarizing the link between fruit and vegetable consumption and such major health problems as heart disease, stroke, some forms of cancer, and pregnancy complications underscores the health risk of poor nutrition (Hyson 2002). If food deserts do indeed influence nutritional intake, the social and economic costs of food deserts are substantial. Increased public health care expenditures through Medicaid and lost productivity due to poor health may hamper economic development and limit the viability of nonmetropolitan communities.

Acknowledgments

Funding for this study was provided by the Food Assistance Research Small-Grants Program of the Southern Rural Development Center and Economic Research Service and by the Research Initiation Program at Mississippi State University.

References

Barnes, N., A. Connell, L. Hermenegildo, and L. Mattson. 1996. "Regional differences in the economic impact of Wal Mart." *Business Horizons* 39:21–26.

Blanchard, T., M. Irwin, C. Tolbert, T. Lyson, and A. Nucci. 2003. "Suburban Sprawl, Regional Diffusion, and the Fate of Small Retailers in a Large Retail Environment, 1977–1996." *Sociological Focus* 36 (4): 313–31.

Blanchard, T., and T. Lyson. 2002. "Access to Low Cost Groceries in Nonmetropolitan Counties: Large Retailers and the Creation of Food Deserts." Paper presented at the Measuring Rural Diversity Conference, Washington DC. http://srdc.msstate.edu/measuring/blanchard.pdf (last accessed May 9, 2006).

Daponte, B, G. Lewis, S. Sanders, and L. Taylor. 1998. "Food pantry use among low-income households in Allegheny County, Pennsylvania." *Journal of Nutrition Education* 30 (1): 57–63.

Furey, S., C. Strugnell, and H. McIlveen. 2001. "An investigation of the potential existence of 'food deserts' in rural and urban areas of northern Ireland." *Agriculture and Human Values* 18 (4):447–57.

Graff, T. 1998. "The Locations of Wal-Mart and Kmart Supercenters: Contrasting Corporate Strategies." *Professional Geographer* 50:46–57.

Hyson, D. 2002. *The Health Benefits of Fruits and Vegetables: A Scientific Overview for Health Professionals*. Wilmington DE: Produce for Better Health Foundation.

Kaufman, P. 1998. "Rural poor have less access to supermarkets, grocery stores." *Rural Development Perspectives* 13 (3):19–26.

Kaufman, P., C. Handy, E. McLaughlin, K. Park, and G. Green. 2000. *Understanding the Dynamics of Produce Markets: Consumption and Consolidation Grow*. AIB758. Washington DC: USDA-ERS.

Kaufman, P., J. MacDonald, S. Lutz, and D. Smallwood. 1997. *Do the Poor Pay More for Food? Item Selection and Price Differences Affect Low-Income Household Food Costs*. AER-759. Washington DC: USDA-ERS.

Kaufman, P., D. Newton, and C. Handy. 1993. *Grocery Retailing Concentration in Metropolitan Areas, 1954–1982*. ERSTB1817. Washington DC: USDA-ERS.

Keon, T., E. Robb, and L. Franz. 1989. *Effect of Wal-Mart Stores on the Economic Environment of Rural Communities*. Columbia MO: Business and Public Administration Research Center and College of Business and Public Administration, University of Missouri.

Lyson, T., and A. Raymer. 2000. "Stalking the wily multinational: Power and control in the U.S. food system." *Agriculture and Human Values* 17:199–208.

Molnar, J., P. Duffy, L. Clayton, and C. Bailey. 2001. "Private food assistance in a small metropolitan area: Urban resources and rural needs." *Journal of Sociology and Social Welfare* 28 (3):187–209.

Shils, E. 1997. *The Shils Report: Measuring the Economic and Sociological Impact of the Mega-Retail Discount Chains on American Small Business*. Philadelphia: Wharton School, University of Pennsylvania. http://www.lawmall.com/rpa/rpashils.htm (last accessed April 30, 2006).

Stone, K. 1995. "Impact of Wal-Mart stores on rural communities." *Economic Development Review* 13:60–69.

U.S. Bureau of the Census. 2001a. *1997 Economic Census, Retail Trade Subject Series, Summary*. Washington DC: U.S. Government Printing Office.

———. 2001b. *Census 2000 Summary File 1, United States*. Washington DC: U.S. Department of Commerce, Economics, and Statistics Administration.

———. 2001c. *County Business Patterns, 1999*. Washington DC: U.S. Department of Commerce, Economics, and Statistics Administration.

———. 2002a. *Census 2000 Summary File 3, United States*. Washington DC: U.S. Department of Commerce, Economics, and Statistics Administration.

———. 2002b. *Zip Code Business Patterns, 1999*. Washington DC: U.S. Department of Commerce, Economics, and Statistics Administration.

U.S. Department of Transportation Federal Highway Administration. 2001. 1995 npts *Databook*. Oak Ridge TN: Oak Ridge National Laboratory.

11. Localization in a Global Context

Invigorating Local Communities in Michigan through the Food System

Michael W. Hamm

It would not be unreasonable for United States residents to ask themselves the question: what type of food system do I want? How about a food system in which we know where a significant percentage of our food comes from? How about one in which production, processing, distribution, and waste handling are consistently done in an environmentally sensitive manner? How about one in which the democratic principles set forth at this country's founding are made stronger and not weakened through economic consolidation and monopolization? How about one in which the farmers who grow our food are honored as heroes and not marginalized as commodity producers? How about one in which every consumer and person working in the food system has the opportunity to reach their potential and is not limited by less than living wage jobs, poor nutrition, and substandard education? How about one in which food is a right and working honestly is a responsibility?

This framing argues for a food system operating within a context of economic optimization, environmental harmonization, and ethical actualization (Bawden 2003), in which these three components are simultaneously consulted to improve outcomes in the course of development. It argues for more community-based food systems in which relationships among people and with natural resources are primary—what Thomas Lyson has termed civic agriculture in an earlier chapter in this volume (see also DeLind 2002; Lyson 2004).

Strong tendencies in our present food system would, at first blush, work against this possibility. Globalization has proceeded at a rapid pace with increases in food imports, food exports, and the worldwide distribution of foods. Global trade battles continue over importation of GMO crops as well as the ability of governments to support their indigenous farmers. World trade agreements have consistently expanded the reach and regulation of global food actors while attempting to inhibit efforts at local food system development. But simultaneously there has been growth of interest in and activity around community-based, more localized food systems throughout the United States. This activity is reflected in the emergence and development of the Community Food Security Coalition and its instrumental role in the 1996 U.S. Farm Bill's provisions for $2.5 million, now $5 million, in annual funding for Community Food Projects, described in further detail by Audrey Maretzki and Elizabeth Tuckermanty in this volume.

Wilkins and Gussow (1997) surveyed consumers in the Northeast and found 98 percent felt that maintaining farm viability is important, with a majority seeing one helpful strategy as buying local produce. Wimberley et al. (2003) reported 71 percent of respondents willing to pay more for food grown locally, 71 percent willing to pay more if methods to protect the environment were used, 77 percent saying government policy should help protect family farms, and 59 percent saying family farms should be supported even if this means higher food prices. These studies are at odds with earlier work that indicated less interest in the source of produce (see, for example, Brumfield, Adelaja, and Kimberly 1993). This change may indicate an emerging shift in consumer attitudes and priorities, reinforced perhaps by the spread of farmers' markets and community supported agriculture (CSA) over the last decade. The number of farmers' markets in the United States increased by 111 percent between 1994 and 2004, with more than 3,700 markets in existence at the time of this writing (USDA-AMS, "Farmers Market Facts"). The first CSA in the United States began in Massachusetts in 1986, and within thirteen years more than 1,000 existed across the country (Henderson and Van En 1999). This heightened public interest in community-based food supplies, or "food with the farmer's face," flows from several beliefs:

- That local food is healthier;

- That local food helps build community;
- That the ecosystem services of local farms should be supported;
- That farmland should be retained and supported;
- That connection to an agrarian way of life is desirable.

This interest in purchasing food locally, as well as directly from the farmer, has expanded beyond fresh fruits and vegetables to include locally processed and value-added products. Many farmers' markets now include fresh and frozen meats, local bakery products, and dairy products as well as flowers. One truly can eat according to the USDA's *Food Guide Pyramid* and also include a beautiful floral centerpiece from the items at sale at many farmers' markets. Still, even with all of this interest, the total dollar value of locally produced, locally marketed farm food products is only a very small percentage of total farm gate receipts. For example, Michigan farmers had total farm gate receipts of $3.77 billion in 2002 (USDA-NASS, 2002) with only $37.3 million (or one percent) going to direct market sales for human consumption. We can hope, however, that it is probable that people's desire for locally produced food is not sated by the relatively narrow framework of farmers' markets, farm stands, and CSAs that currently exist. In fact, the above-mentioned studies indicate there is an opportunity to significantly expand marketing and engage broader market outlets for locally produced and processed food.

More specifically, we can ask: what would it take for a population to consume 10, 20, or 50 percent of its food from local sources? In order to understand potential opportunities and barriers, it is useful first to understand consumption patterns. Using Michigan as a case study is instructive for several reasons. First, it has a population of nearly 10 million people on 57,000 square miles of land (density of 175 people per square mile) (U.S. Bureau of Census 2000, 2001). In terms of human land use, Michigan ranks ninth in the country but has a population density near the mean for the fifty states (Michigan has 474 per square mile as opposed to the fifty states' average of 416 per square mile) (Demographia 2000). Second, Michigan has a very diverse agriculture, which leads the nation in production of ten crops and which ranks in the top five in thirty-two others (Michigan Dept. of Agriculture 2001–2006);

this ranking puts the state near the top among the fifty U.S. states in overall agricultural diversity. Thus it is a moderately densely populated state with large total farmland acreage and considerable agricultural production diversity. This combination provides a strong base for conceptualizing a more community-based food system.

But Michigan also epitomizes a threatened agriculture. It has been estimated that at current projected rates 71 percent of the farms between 50 and 500 acres in size will be lost between 2000 and 2040 (Levy 2001), a loss that represents nearly 17,000 "farms in the middle." While much attention within Michigan focuses on the issue of preserving the farmland base for agriculture, it is clear that attention needs to be focused equally on preserving the other three F's of agriculture—farming, farmers, and farms (Hamm 2001).

The Consumption Side

Current consumption patterns in the United States provide an opportunity for enhancing local consumption within existing dietary habits as well as to enhance general dietary quality. While the rate of obesity in the United States continues to escalate (Centers for Disease Control and Prevention 2006), a concerted effort to increase the percentage of food from local sources can address this latest threat to our nation's health. In fact, current consumption patterns are far from the target established by the USDA and NIH. According to the 1994-96 Continuing Survey of Food Intake of Individuals data (Kantor 1998), the Midwest population consumed on average less than amounts recommended for each section of the *Food Guide Pyramid*, except at the peak of the pyramid—added fats and oils (see table 16). Current consumption equals approximately 977 pounds of food from the five main portions of the *Food Guide Pyramid*. When added fats and oils along with soda and fruit drinks are included, total consumption rises to 1,320 pounds per year; adding coffee, tea, and alcohol increases total consumption to over 1,700 pounds per year. On a poundage basis, a person would need to consume 98, 195, and 489 pounds of food from state agricultural production to get 10, 20, or 50 percent of these food totals. Given that average consumption is below that recommended, a healthier diet would increase food consumption from the key parts of the pyramid by 27 percent.

Table 16. Food consumption in the Midwest

	gm/day[a]	lbs/day[a]	Yearly pounds[a]	% of recommended intake[b]	lbs/day recommended	Yearly pounds
Dairy	300	0.66		68	0.97	
Meat	248	0.55		87	0.63	
Fruit	163	0.36		60	0.60	
Vegetables	195	0.43		94	0.46	
Grain	311	0.68		90	0.76	
Subtotal A	1,217	2.68	977		3.42	1,245
Fats/Oils	14	0.03				
Sugar/Sweetener	25	0.06				
Subtotal B		2.77	1,009			
Fruit drinks	102	0.22				
Soda	342	0.75				
Subtotal C		3.74	1,365			
Coffee	382	0.84				
Alcohol	97	0.21				
Total		4.79	1,750			

[a]From USDA-ARS 1997
[b]From Kantor 1998

Table 17. Michigan population consumption and food expenditures

	Population or households in Michigan[a]	Basis[b]	10% from Michigan[b]	20% from Michigan[b]	50% from Michigan[b]
Consumption relative to current consumption patterns (pounds-individual)	9,938,444	9,712,354,337	971,235,434	1,942,470,867	4,856,177,169
Consumption relative to recommended consumption patterns (pounds-individual)	9,938,444	12,377,090,561	1,237,709,056	2,475,418,112	6,188,545,281
Difference between current and recommended	—	2,664,736,224	266,473,622	532,947,245	1,332,368,112
Consumption relative to dollar expenditures (food at home-family)	3,785,661	$7,336,050,740	$733,605,074	$1,467,210,148	$3,668,025,370
Consumption relative to dollar expenditures (ideal $) (food at home-family)	3,785,661	$9,231,805,856	$923,180,585	$1,846,361,171	$4,615,902,927
Difference between current and recommended	—	$1,895,755,116	$189,575,512	$379,151,023	$947,877,558

[a] From U.S. Bureau of the Census 2001

[b] Poundage based on table 16 current poundage intake; dollar expenditures based on Bureau of Labor Statistics (U.S. Dept. of Labor 2002).

In 2001 consumers in the United States spent almost $739 billion on food (USDA-ERS 2005). Of this amount approximately $440 billion was spent on food consumed at home and $298 billion on food consumed away from home. Nationally, the average household spends about 13.6 percent or $5,174.12 of its $38,045 in disposable income on food, while the average Midwestern family spends 13.4 percent or $5,254.54 of its $39,213 (U.S. Dept. of Labor 2002). In these data food consumed away from home is limited to a total value descriptor with no categorization. Thus this analysis assumes relative distribution within the *Food Guide Pyramid* away from home was equal to that at home. Table 17 outlines expenditures for the at-home component of the 977 pounds of *Food Guide Pyramid* expenditures at 10, 20, and 50 percent intake from Michigan sources.

The population of Michigan spends approximately $7.3 billion on home-consumed food within the dairy, meat, fruit, vegetable, and grain groups annually—double the annual farm gate receipts in Michigan. Including food consumed away from home adds approximately $5.6 billion. In other words, there is enormous economic potential for local, community-based sourcing. Relating expenditure to recommended consumption is illustrative. While the average midwestern family spends $1,938 on *Food Guide Pyramid* items consumed in the home, consuming according to recommended intakes would yield $2,564 in expenditures. Across the entire Michigan population this differential totals nearly $1.9 billion in expenditures or 53 percent of total Michigan farm gate receipts.

Consumer expenditures provide a number of possibilities for action that include:

- Shifting 10 percent more of current at-home expenditures toward Michigan-grown products would increase local sales by $734 million.
- Shifting 10 percent plus modifying the diet to increase intake toward recommended levels would increase local sales by $923 million.
- Simply shifting the intake toward recommended levels with 10 percent of this food coming from local sales would increase sales by $190 million.

Raising the percentage of sales of Michigan products from 10 to 20 or 50 percent increases the total sales volume that much more. At the highest levels, assuming that all families in Michigan consumed 50 percent of their at-home food purchases from Michigan farms, there would be approximately $4.6 billion in sales representing about 129 percent of total current farm gate receipts. Outside-the-home expenditures add tremendously to the potential.

The Production Side

The production side of the food system in Michigan demonstrates the impressive agricultural diversity in the state. In an attempt to develop an understanding of what Michigan produces relative to its population and to determine the land base requirements, plant and animal production were investigated. Agricultural production in the state was analyzed utilizing recent agricultural production data.[1] Grain servings produced were calculated based on the assumption of approximately sixteen servings per pound with no factor for losses from field to table. Fruit servings were calculated utilizing fruit production data from 2001 and 2002. These figures were corrected for losses at the farm, nonedible components, and cooking and retail losses (Kantor 1998; Peters et al. 2002). Vegetable servings were calculated from 2000 production data and corrected in a like manner as for fruit. The annual survey of vegetable production does not include a number of minor vegetables and thus underestimates total production. An analysis of the 1997 Census of Agriculture demonstrates that vegetables not included in the annual survey increase the total vegetable acreage by only 2 to 4 percent and thus were not considered a major source of error. Meat servings include all egg, chicken, turkey, beef, pork, and dry/edible bean production within Michigan and were calculated from established *Food Guide Pyramid* serving parameters (Shaw et al. 2003). Meat servings were corrected for primary losses to consumer (eggs) or cooking (meats) as well as retail loss. Cow carcass weights were calculated at 745 pounds as the average carcass weight per slaughtered animal (USDA Market News 2002). Pig carcass weight was calculated at 74 percent of live weight.[2]

Michigan currently produces a broad range of the nutrient needs of its population (see table 18). On average, in good years the state pro-

Table 18. *Food Guide Pyramid* servings required, consumed, and produced in Michigan

	Population	Grain	Fruit	Vegetables	Meat (oz.)	Dairy
2000 Population servings required	9,938,443	28,534,451,773	9,932,117,857	13,314,367,727	19,556,987,182	8,542,748,350
Servings production (High)[a]	—	34,406,400,000	7,603,906,867	11,029,000,653	21,233,930,325	1,259,469,697
% servings recommended[b]	—	121	77	83	109	132
% servings consumed[c]	—	134	128	88	125	194
Servings production (low)[d]	—	31,516,800,000	3,590,070,221	6,261,009,745	15,393,244,160	11,259,469,697
% servings recommended[b]	—	110	36	47	79	132
% servings consumed[c]	—	123	60	50	90	194

[a]Servings production (high) utilizes 2001 grain production, 2001 fruit production, 2000 vegetable production with potatoes, and 2002 bean production within meat.

[b]Percentage of servings recommended is servings production/servings required if all ate according to *Food Guide Pyramid*.

[c]Percentage of servings consumed is serving production/servings consumed utilizing CSFII, 1994–96 USDA 1999).

[d]Serving production (low) utilizes 2002 grain production, 2002 fruit production, 2000 vegetable production without potatoes, and 2001 bean production within meat group.

Table 19. Michigan acreage utilized for production of *Food Guide Pyramid* components

Crop category	Acreage utilized for production	Acreage required for production of current consumption	Acreage required for recommended consumption
Dairy	1,085,406	559,991	823,516
Meat, beans (utilizing existing meat group sources)	1,577,466	1,263,995	1,452,868
Fruits	115,500	96,375	150,865
Vegetables	252,415	280,596	304,719
Grains	560,000	417,985	464,428
Total acreage	3,590,767	2,618,942	3,196,395
% of Michigan farmland	36.4	26.5	32.4

duces about 103 percent of its population's overall food needs. In lesser years it produces about 77 percent. A large drop in fruit production occurred in 2002 as opposed to 2001 due to a late frost that destroyed more than 90 percent of the cherry crop and cut the apple crop by 46 percent, the pear crop by 70 percent, the plum crop by 95 percent, and the peach crop by 67 percent.

To identify the total Michigan land mass required to feed the population, an analysis was conducted identifying acreage for fruits, vegetables, grains, and feed for livestock. Livestock feed requirements utilized Michigan production standards and were used to calculate acreage requirements for corn, soybeans, corn silage and haylage, hay, and wheatlage as well as pasture.[3] Yields were utilized from data sources described above. Acreages required for production were summed within each category of the *Food Guide Pyramid* in order to determine total utilization within Michigan (see table 19). These acreages then were corrected for production requirements relative to current food group consumption as well as for recommended consumption.

Currently, Michigan utilizes about 36 percent of its agricultural land base to produce the food delineated in this study (see table 19). Some-

Table 20. Michigan acreage requirements for animal production within current midwestern consumption patterns

	% of total	Servings required	Servings consumed	Acres for servings consumed	Acres for servings required
Beef	24.1	4,706,660,816	4,094,794,910	1,571,452	1,806,267
Pork	19.9	3,895,167,572	3,388,795,787	164,699	189,309
Chicken	22.8	4,463,212,842	3,882,995,173	216,675	249,052
Other poultry	16.2	3,164,823,652	2,753,396,577	122,238	140,503
Eggs	7.1	1,379,538,515	1,200,198,508	80,422	92,439
Legumes	10.0	1,947,583,786	1,694,397,894	38,548	44,308
Total	100	19,556,987,182	17,014,578,848	2,194,034	2,521,878

Note: Percentage in each food category from USDA-ARS (1997), with grams consumption of frankfurters, sausages, luncheon meats; fish/shellfish; and mixtures—mainly meat, poultry, fish—being equally distributed among beef, pork, chicken, and other poultry. Other poultry utilized as Michigan turkey production.

what less than 27 percent of the agricultural land base is required to produce for Michigan residents' current consumption patterns and about 33 percent to produce recommended intakes.

However, there are several caveats to this analysis of farmland use. First, a broad grouping of foods into the food groups ignores microconsumption patterns within a given food group. For example, consumers eat and drink a variety of citrus fruits, but none are grown in Michigan. Second, dry beans were fully allocated to the meat group and account for 50 percent of the total meat servings produced within Michigan.

While 50 percent meat as dry beans is not a realistic assessment of the meat group intake patterns of Michigan residents, it does provide a sense of land needs with a modest meat intake while consuming sufficient foods from this group. In this scenario 67 percent of agricultural land still would be available (1) for trade, (2) for providing a buffer to production variations, and (3) for nonfood production agricultural land uses. If current consumption patterns within the meat group are taken into account, the land requirements moderately increase to about 36 percent of acreage for current consumption and about 43 percent for recommended consumption.

Table 20 gives a breakdown of the land requirements approximating the intake patterns of foods from the meat and bean group. The total land requirements for production of foodstuffs in this group increase with greater meat consumption from 1.26 to 2.19 million acres for current consumption and from 1.45 to 2.52 million acres for recommended consumption levels. Significant amounts of this acreage are due to requirements for cow-calf combinations in beef cattle production.

Consumer-Producer Linkages

Conceptualizing the above within a framework that requires minimal activity by individuals to produce maximal effect is useful in considering options for an action plan. With average family food expenditures at about $5,200 per year, spending $10 per month (or about 2 percent of total food expenditures) on locally produced agricultural products would mean total economic activity of $454 million per year in sales.

What does this amount mean with respect to community-based food systems? Within Michigan the above data illustrate both great potential

Table 21. Population, farmers' markets, and CSAs for selected states

Rank (by farmers' markets/ 100,000)	Rank (by CSAs/ 1,000,000	Area	Population census[a]	Farmers' markets[b]	Farmers' markets/ 100,000	CSAs[c]	CSAs per 100,000
1	1	Iowa	2,926,324	146	4.989	34	1.162
2	3	Wisconsin	5,363,675	149	2.778	54	1.007
3	10	Kansas	2,688,418	67	2.492	6	0.223
4	2	Oregon	3,421,399	58	1.695	36	1.052
5	8	New York	18,976,457	269	1.418	80	0.422
6	4	Massachusetts	6,349,097	87	1.370	55	0.866
7	5	Washington	5,894,121	78	1.323	51	0.865
8	7	Pennsylvania	2,281,054	156	1.270	59	0.480
9	6	Minnesota	4,919,479	55	1.118	31	0.630
10	14	Indiana	6,080,485	67	1.102	11	0.181
11	12	California	33,871,648	365	1.078	74	0.218
12	16	Illinois	12,419,293	129	1.039	11	0.089
13	13	Michigan	9,938,444	66	0.664	21	0.211
14	11	Ohio	11,353,140	72	0.634	25	0.220
15	16	New Jersey	8,414,350	46	0.547	11	0.131
16	17	Texas	20,851,820	99	0.475	15	0.072
17	18	Florida	15,982,378	63	0.394	9	0.056
18	9	Missouri	5,595,211	None listed	—	17	0.304

[a]From U.S. Burea of Census 2001

[b]From USDA-AMS 2002

[c]From Wilson College 2006

and great challenges. People tend to think of eating locally as consuming fresh, in-season fruits and vegetables. This certainly appears to resonate, as seen by the growth of farmers' markets and CSAs; while both have become increasingly diverse with respect to provision of other food items, they remain strongly fruit and vegetable focused. Yet even here there is much opportunity. Table 21 presents figures on farmers' markets and CSAs per 100,000 population for selected states in 2002. These data do not include roadside stands at farms which, depending upon the state, can contribute significantly to direct sales. Overall the numbers for farmers' markets and CSAs are probably underestimated, but these numbers still enable comparisons of relative density across states. A considerable spread is evident in the number of farmers' markets per 100,000 people, with a mean of 1.09 and a range of from 4.99 for Iowa to 0.39 for Florida. CSAs demonstrate an even larger range per 100,000 people, with a mean of 0.32 and a range of from 1.62 in Iowa to 0.06 in Florida. Of the states sampled, Michigan is near the bottom with figures of 0.66 for farmers' markets and 0.21 for CSAs.

Which begs the question: what would it look like if everyone targeted Iowa farmers' market and CSA densities? For Michigan such a goal would mean in excess of 495 farmers' markets with potential retail sales of over $79 million (Payne 2002). Payne identified average sales per customer for the North Central region of $265 and average sales per market of $171,000, amounts that would give a total customer pool of approximately 320,000 people with 495 farmers' markets. This consumer number is less than 5 percent of Michigan's population (and probably a number of customers at some Michigan farmers' markets are vacationers).

It is somewhat more difficult to analyze the impact of moving CSAs to a broader level. If the CSA/population ratio were similar to Iowa, there would be at least 115 CSAs in Michigan. Cooley and Lass (1998) analyzed three CSAs in the Massachusetts area. These had an average of 230 members and charged on average $380 per share. For Michigan, increasing from 21 to 115 CSAs with average membership and share price (adjusted for annual increases of 5 percent) would yield nearly $10.5 million in local food sales and consumption. This figure may be low, with Cooley and Lass' data interpreted as CSA share prices being significantly underpriced. In total, increasing these two direct marketing vehicles to a per capita level equal to Iowa's would increase direct market sales in Michigan by $89 million, which would be 2.5 times the direct marketing of agricultural products for human consumption reported in the 2002 Census of Agriculture (USDA 2002). Furthermore, this total does not include the impact of farm stands throughout the state.

Another way to consider localization involves the type of production. Both organic agriculture and management-intensive rotational grazing offer marketing opportunities with a variety of benefits to communities. What if there were 300,000 acres of certified organic farmland in Michigan and 600,000 more acres of management-intensive rotational grazing land?[4] Currently there are about 25,000 acres of certified organic land, or about 0.25 percent of the total agricultural land.[5] A move to 3 percent would markedly reduce the marginalization of organic farming in the state and with an average farm size of about 215 acres in Michigan would mean about 1,400 organic farms.

Market studies find that almost 40 percent of the U.S. population now consumes organic food products, although there is variability between frequent and more occasional users (Organic Trade Association 2003). A twelve-fold increase would arguably provide greater opportunity for organic product cooperative and processing facility development with attendant market opportunities. More possibilities for consumers to purchase organic and/or pasture-fed local products would support movement toward 10 percent consumption from local sources as well as enhancing agricultural ecosystem services, opportunities for education, and possibly ecotourism.

This chapter has documented through a case study focused on Michigan both a methodology for understanding the production/consumption potential within a community, state, or region as well as opportunities to enhance relationships between those who grow our food and those who consume it. While Michigan has a much more diverse agriculture than many states today, historically most states had much greater agricultural diversity than currently evident. In other words, a state like Michigan has the immediate potential to build state and local infrastructures to more fully feed its population through civic engagement and support while maintaining space for ecologically and ethically based profitable export products. Some states may need to do more to rebuild their agricultural diversity, yet the capacity for their residents, too, to engage in a more sustainable civic form of food consumption is vast.

Notes

1. Information was accessed from monthly press releases (http://www.nass.usda.gov/mi/press/). Field crop production (Michigan Dept. of Agriculture 2003a), fruit production (Michigan Dept. of Agriculture 2003b), chickens/turkeys/egg production (Michigan Dept. of Agriculture 2003c), grain production (Michigan Dept. of Agriculture 2003d), beef cattle (USDA-NASS 2003a), pigs (USDA-NASS 2003a), vegetable production (USDA-NASS 2002) and potato production (Michigan Dept. of Agriculture 2002). "Milk Cows" uses 298,750 (2002 data) as the number of milking cows, relies on the 1997 Census of Agriculture number for milk cows, uses replacement heifers ratio as an index for determining the number of replacements (0.77 replacement heifers/milking cow), and assumes one-half of replacements in their year 1 and one-half in their year 2.

2. Personal communication with David Meisinger of the National Pork Board.

3. Chickens/turkeys data is from personal communication with Kevin Roberson, Dept. of Animal Sciences, Michigan State University; that for cow/calf combination is from personal communication with Dan Buskirk, Dept. of Animal Sciences, Michigan State University and is taken from the 2001 Integrated Resource Management program cow/calf producers; that for feedlot cattle is from personal communication with Steve Rust, Dept. of Animal Sciences, Michigan State University; that for pigs is from personal communication with Dr. Nathalie Trottier, Dept. of Animal Sciences, Michigan State University; and that for dairy cattle is from personal communication with Dr. David Beede, Dept. of Animal Sciences, Michigan State University.

4. Current estimates suggest that there are about 200,000 Michigan acres in use as some type of grazing for food production. Personal communication with Richard Leep, Dept. of Crop and Soil Science, Michigan State University.

5. From personal Communication with Dr. Jim Bingen, Dept. of Community, Agriculture, Recreation and Resource Studies, Michigan State University.

References

Bawden, R. 2003. "Systemic Discourse, Development and the Engaged Academy." Paper presented at the Forty-seventh Annual Meeting of the International Society for the Systems Sciences. Heronissos, Crete.

Brumfield, R. G., A. O. Adelaja, and L. Kimberly. 1993. "Consumer tastes, preferences, and behavior in purchasing fresh tomatoes." *Journal of the American Society of Horticultural Science* 118:433–38.

Centers for Disease Control and Prevention. 2006. *Overweight and Obesity Trends: 1991-2001 Prevalence of Obesity Among U.S. Adults by State.* http://www.cdc.gov/nccdphp/dnpa/obesity/trend/prev_reg.htm (last accessed May 8, 2006).

Cooley, J. P., and D. A. Lass. 1998. "Consumer benefits from community supported agriculture membership." *Review of Agricultural Economics* 20:227–37.

DeLind, L. 2002. "Place, work, and civic agriculture: Common fields for cultivation." *Agriculture and Human Values* 19:217–24.

Demographia. 2000. *Population Density per Human Use Land: Ranked by U.S. State: 1990.* Wendell Cox Consultancy. http://www.demographia.com/db-landstate-pophumandens.htm (last accessed April 8, 2006).

Hamm, M. W. 2001. "Farmland, farms, farming, and farmers: The four F's of food production." *Gastronomica* 1 (2):27–31.

Henderson, E., and R. Van En. 1999. *Sharing the Harvest: A Guide to Community-Supported Agriculture.* White River Junction VT: Chelsea Green.

Kantor, L. S. 1998. *A Dietary Assessment of the U.S. Food Supply: Comparing Per Capita Food Consumption with Food Guide Pyramid Serving Recommendations.* USDA. AER-772. Washington DC: U.S. Government Printing Office.

Levy, D., ed. 2001. *Michigan Land Resource Project.* Lansing MI: Public Sector Consultants.

Lyson, T. A. 2004. *Civic Agriculture: Reconnecting Farm, Food, and Community.* Medford MA: Tufts University Press.

Michigan Dept. of Agriculture. 2001–2006. *Facts About Michigan Agriculture.* http://www.michigan.gov/mda/0,1607,7-125-1572-7390—,00.html (last accessed April 8, 2006).

———. 2002. *Michigan Potato Production.* Agricultural Statistics Service. PR-02–03. http://www.nass.usda.gov/mi/press/pr0283.txt (last accessed April 8, 2006).

———. 2003a. *Michigan Annual Crop Summary.* Agricultural Statistics Service. PR-03–03. http://www.nass.usda.gov/mi/press/pr0303.txt (last accessed April 8, 2006).

———. 2003b. *Fruit Production 2002.* Agricultural Statistics Service. PR-03–08. http://www.nass.usda.gov/mi/press/pr0308.txt (last accessed April 8, 2006).

———. 2003c. *Poultry 2002 Summary.* Agricultural Statistics Service. PR-03–37. http://www.nass.usda.gov/mi/press/pr0337.txt (last accessed April 8, 2006).

———. 2003d. *Wheat Production Up.* Agricultural Statistics Service. PR-03–39. http://www.nass.usda.gov/mi/press/pr0339.txt (last accessed April 8, 2006).

Organic Trade Association. 2003. *Industry Statistics and Projected Growth.* http://www.ota.com/organic/mt/business.html (last accessed April 8, 2006).

Payne, T. 2002. *U.S. Farmers Markets—2000: A Study of Emerging Trends.* Washington DC: Agricultural Marketing Service, USDA. http://www.ams.usda.gov/directmarketing/farmmark.pdf (last accessed May 7, 2006).

Peters, C. J., N. Bills, J. L. Wilkins, and R. D. Smith. 2002. *Vegetable Consumption, Dietary Guidelines, and Agricultural Production in New York State: Implications for Local Food Economies.* Research Bulletin 2002-07. Ithaca NY: Department of Applied Economics and Management, Cornell University. http://hortmgt.aem.cornell.edu/pdf/smart_marketing/peters2-03.pdf (last accessed April 9, 2006).

Shaw, A., L. Fulton, C. Davis, and M. Hogbin. 2003. *Using the Food Guide Pyramid: A Resource for Nutrition Educators*. Washington DC: Food, Nutrition, and Consumer Services, Center for Nutrition Policy and Promotion, USDA. http://www.nalusda.gov/fnic/fpyr/guide.pdf (last accessed April 8, 2006).

U.S. Bureau of Census. 2000. *Census 2000: Land Area and Population by State*. http://www.governor.utah.gov/dea/rankings/states/statedensity.pdf (last accessed April 8, 2006).

———. 2001. Census 2000 PHC-T-2. Ranking Tables for States: 1990 and 2000. http://www.census.gov/population/cen2000/phc-t2/tab01.pdf (last accessed April 8, 2006).

USDA. 1999. Continuing Survey of Food Intake of Individuals: Data Tables: Food and Nutrient Intakes by Individuals in the United States, by Region, 1994-96 Table Set 13. http://www.ars.usda.gov/sp2userfiles/place/12355000/pdf/region.pdf (last accessed April 4, 2007).

———. 2002. *Market Value of Agricultural Products Sold Including Landlord's Share, Direct, and Organic: 2002 and 1997*. http://www.nass.usda.gov/census/census02/volume1/mi/st26_1_002_002.pdf (last accessed April 8, 2006).

USDA-AMS. 2002. Find a Farmers Market in Your State. http://www.ams.usda.gov/farmersmarkets/map.htm (last accessed April 8, 2006).

———. No date. *Farmers Market Facts*. http://www.ams.usda.gov/farmersmarkets/facts.htm (last accessed May 16, 2006).

USDA-ARS. 1997. Data Tables: Food and Nutrient Intake by Individuals in the United States, by Region, 1994–96: Table Set 13. http://www.ars.usda.gov/sp2userfiles/place/12355000/pdf/region.PDF (last accessed May 17, 2006).

USDA-ERS. 2005. Food CPI, Prices, and Expenditures: Food Expenditures by Families and Individuals as a Share of Disposable Personal Income. http://www.ers.usda.gov/briefing/cpifoodandexpenditures/data/table7.htm (last accessed April 8, 2006).

USDA Market News. 2002. *Estimated Weekly Meat Production under Federal Inspection*. http://www.ams.usda.gov/mnreports/wa_LS712.txt (last accessed April 8, 2006).

USDA-NASS. 2002. *2002 Census of Agriculture—Michigan*. http://www.nass.usda.gov/census_of_agriculture/census_by_state/michigan/index.asp (last accessed May 15, 2006).

———. 2003a. *Meat Animals Production, Disposition, and Income, 2002 Summary*. http://usda.mannlib.cornell.edu/reports/nassr/livestock/zma-bb/meat0403.pdf (last accessed April 8, 2006).

———. 2003b. *Poultry Production and Value: 2002 Summary*. http://usda.mannlib.cornell.edu/reports/nassr/poultry/pbh-bbp/plva0403.pdf (last accessed April 8, 2006).

U.S. Dept. of Labor. 2002. *Issues in Labor Statistics; Summary 02-04*. Bureau of Labor Statistics. http://www.bls.gov/opub/ils/pdf/opbils48.pdf (last accessed April 8, 2006).

Wilkins, J. L., and J. D. Gussow. 1997. "Regional Dietary Guidance: Is the Northeast Nutritionally Complete?" In *Agricultural Production and Nutrition. Proceedings of an International Conference*, edited by W. Lockeretz, 23–33. Medford MA: Tufts University School of Nutrition Science and Policy.

Wilson College. 2006. Robyn Van En Center for CSA Resources. http://www.wilson.edu/wilson/asp/content.asp?id=804 (last accessed May 16, 2006)

Wimberley, R. C., B. J. Vander Mey, B. L. Wells, G. D. Ejimakor, C. Bailey, L. L. Burmeister, C. K. Harris, M. A. Lee, E. L. McLean, J. J. Molnar, G. W. Ohlendorf, T. J. Tomazic, and G. Wheelock. 2003. "Food from our changing world: The globalization of food and how Americans feel about it." http://sasw.chass.ncsu.edu/global-food/foodglobal.html (last accessed April 8, 2006).

12. Assessing the Significance of Direct Farmer-Consumer Linkages as a Change Strategy in Washington State: Civic or Opportunistic?

Marcia Ruth Ostrom and Raymond A. Jussaume, Jr.

Washington State leaders commonly portray the state's agricultural industry as a sophisticated player in national and international markets. The Washington Wheat Commission heralds the fact that 85 percent of the wheat produced in the state is exported, primarily to Asian Pacific Rim Countries. The Washington Apple Commission proudly proclaims on its Web site that more than half of all apples grown in the United States for fresh eating come from Washington orchards. And the Washington State Potato Commission declares that just three hundred potato growers in the state produce 20 percent of the potatoes grown in the United States.[1]

These statements celebrate the conventional modern agricultural picture of technologically advanced, economically efficient farming operations producing a cornucopia of commodities. These commodities are either processed or shipped fresh to markets around the world. Deliberately crafted as part of an ambitious strategy to develop the state's natural resources, this image is perhaps exemplified by Columbia Basin agriculture involving large-scale, irrigated production in an arid desertlike environment (U.S. Dept. of the Interior 1964).

Certainly, this commonly portrayed view of Washington agriculture captures part of an important reality. In 2001 Washington State agricultural production was valued at $5.6 billion. But the history of Washington agriculture has long involved high capital investment (from the public and private sectors), sophisticated production and transportation technologies, and export-oriented marketing strategies. Irrigated orchards first were developed in areas east of the Cascade Mountains and west of the Columbia River in the late nineteenth century specifically with east coast and overseas markets in mind (Sonnenfeld, Schotzko, and Jussaume 1998). As one writer put it, "Our distant markets in the eastern states, Europe, Alaska and the Orient, are broadening every year" (Fletcher 1902, 14). Thus Washington agriculture corresponds to grand theorizing about an increasingly globalized capitalist system organized within universalized transnational practices and trade relationships (Sklair 1991).

However, as Sklair (1991), Whatmore (1994), and others point out, the global economic system is characterized by asymmetries. Individuals, communities, and regions are integrated to different degrees into that system with a highly uneven distribution of benefits. Indeed, throughout the world while some individuals, communities, and businesses see advantages to becoming more integrated into global trade structures, others struggle to resist or to create alternative spaces within, underneath, and outside the dominant economic system (Whatmore and Thorne 1997). The chapters in this volume by Thomas Lyson and by G. W. Stevenson and colleagues both explore the process and significance of such efforts in order to promote alternatives in the agrifood system.

In Washington State most farms are not large or capital intensive. They struggle to survive by repositioning themselves in different local, regional, and global markets. Since 1997 the state has lost over 4,000 farms, most of them small and midsized operations (USDA-NASS 2002). But the structure of the state's agriculture remains very complex. It is characterized by a great diversity of products, producers, climate zones, production practices, and varying ways that individuals respond to shifting political, environmental, and economic dynamics.

In this chapter we seek to advance understanding of this complexity, with an eye toward integrating global and local interpretations of change. We assume not only that global structures exist and matter in everyday lives but also that each locality is a place with distinctive ecological and social attributes and with people working to make their lives more meaningful and satisfying. We would agree with the assumption that globalization is necessarily the product of transnational practices and institutions and "the institutional, social, and cultural dimensions *of places*" (Lawson 1992, 2). We also would argue that locally based responses to global forces of change must necessarily be diverse because they reflect local conditions.

A significant challenge then is to assess the significance and efficacy of various local responses to changes in the agrifood system, from collective attempts to resist, reform, or transform the system to individual efforts by producers to sustain their livelihoods within a changing environment. From the standpoint of Washington agriculture, this challenge translates into some specific questions. What is the nature and extent of attempts by farmers to reposition themselves in relation to the mainstream global marketplace? Is their goal to resist, adapt, construct alternatives, or take advantage of global economic structures? Finally, to what extent is there congruence between the goals and interests of the state's farmers and consumers and hence potential for collective action to reshape local and regional agrifood relationships in order to better meet the needs of Washington citizens? Analysis of such questions can inform ongoing debates about whether it is possible to develop alternatives to the global agrifood system that perform better for farmers, communities, and the environment or whether such efforts can only achieve success at the margins. In other words, can efforts to create alternative food production and distribution systems in Washington establish new, lasting relationships among significant numbers of mainstream producers and consumers or will they remain fragile and vulnerable to co-optation or destruction by dominant market forces?

Until any transformation is complete, evidence can only hint at the possibility of significant long-term change. But theory and data can help investigate the nature and extent of changes thus far, particularly in light of dynamics and attributes of places and regions.

Our chapter focuses on the characteristics, prevalence, and significance of farmers' direct and local marketing activities in Washington State. We begin with a brief account of the argument over the benefits and limitations of establishing direct market connections between farmers and consumers at the local level as a way to reinvigorate family-scale farming and thus make food systems more responsive to local needs. We then present an analysis of data collected from a large sample of Washington agricultural producers and consumers in 2002. We describe current farm marketing strategies and explore factors most likely to explain farmer participation in direct or local marketing, including the ways producers frame their views on farm policy issues. We complement this analysis by examining consumer attitudes, policy views, and purchasing practices. Finally, we discuss the implications of this research for understanding the evolution and potential of locally based agrifood models to create lasting change within the context of increasing globalization.

Relocalization of Producer-Consumer Relationships as a Change Strategy

Recent academic debates turn on the question of which sources of human agency can address the deep-rooted problems of today's agrifood system. Where will resistance to the existing system be generated? Where will the foundational knowledge and motivational impetus to construct solutions come from? While the necessity of constructing an alternative, more sustainable and equitable agrifood system is widely proclaimed, there is little agreement about what constitutes the sustainable alternative, much less about effective and practical strategies for achieving this vision. Conceptualizations of change in the agrifood system based on polarized notions of local and global arenas for activity may limit our ability to apprehend more subtle aspects of change, including how they might manifest themselves in different settings.

Over the past decade theorists and activists have proposed that solutions to systemic problems of the market-driven global food economy lie in restoring and emphasizing local connections among agricultural production, consumption, and community development. Friedmann, best known for her work at a global food systems level, argues that the most promising solutions lie in "locality and seasonality" (1993, 228). Kneen (1989) argues that reversing the logic of today's agrifood system will require new adherence to the principle of proximity. The formula of Daly and Cobb (1994) for achieving global environmental stability similarly calls for the relocalization of food production and consumption. Mander and Goldsmith (1996) delineate the nature of threats posed to society under economic globalization and conclude that the primary solution is relocalization. Likewise, Korten (1995) argues for "localizing the global system" by managing interdependence in a way that supports and protects local efforts to determine people's "own rules of economic engagement." More recently, frameworks of civic agriculture (Lyson 2000) and food democracy (Hassanein 2003) have been developed to conceptualize how agrifood systems could become reembedded in civil society and become more responsive to local needs and decision makers.

But how workable is such a vision? Much theoretically informed research maintains that, subject to variations in forms, global political-economic forces are as significant in the agricultural sector as they are in any other industry (Bonanno et al. 1994; LeHeron 1993; McMichael 1997). Primary mechanisms for the expansion of the system are free-trade agreements and global pricing (Blank 1998). Thus farmers in the United States must adapt their production and marketing strategies in response to depressed global prices for agricultural commodities, while domestic costs steadily rise for land, labor, and other inputs. Furthermore, national and transnational agribusiness actors who dominate global markets have been adept at moving into newly emerging, high-value market niches and repositioning themselves as sustainable, organic, or authentic. Will they be equally adept at capturing local marketing systems?

Aside from the obvious problem of defining and operationalizing the term local, various limitations to the "go local" approach have been identified. Advocates for family farmers fear that local market development and other attempts to create direct linkages between farmers and consumers can only provide economic solutions for a small subset of farmers. This subset includes farmers located in close proximity to population centers and those who are small and flexible enough to adapt and market their production in accordance with local demand. From another standpoint Hinrichs (2003) points out that some connotations attached to the concept of local run directly counter to goals of social inclusiveness and equality by privileging a local or familiar social group over outside or less well-known social groups. Having analyzed the composition and tactical strategies of local, grassroots agrifood initiatives in California, Allen et al. conclude that, while many groups have succeeded at building small-scale alternative models for producing and distributing foods at a local level, movements that advance broader concerns of social justice "may be difficult to construct at a local scale" (2003, 61).

Hassanein (2003) highlights an important debate about whether the locally organized, decentralized grassroots agrifood initiatives currently emerging in the United States have any potential to mount a meaningful counter to powerful corporate and trade entities organized on a transnational scale. She references comments written by the editors of *Hungry for Profit* (Magdoff, Foster, and Buttel 2000) who note that "strategies such as farmers adding value to their produce by direct marketing and efforts to improve access to nutritious food by the poor can help people confront immediate problems in their everyday lives" (Hassanein, 2003, 77). Yet those writers also question whether implementing such tactics can ever provide anything beyond a "minor irritant" to corporate powers controlling today's food system. At issue here lies the question of whether grassroots citizens' agrifood initiatives can effectively introduce any measure of democratic control over economic systems that are essentially nondemocratic or whether meaningful agrifood system change can only be accomplished by first transforming the larger society as a whole.

In this chapter we explore whether direct marketing could be foundational in constructing alternatives to the global agrifood system or whether it could be nothing more than a niche marketing adaptation that at best benefits only a small minority of farmers and at worst opens up attractive new market opportunities for major corporate actors to take over. While we cannot predict a final outcome, our analysis provides a glimpse of the role that direct and local marketing currently play in the state of Washington.

Data and Methods

The first data set analyzed in this chapter comes from a large-scale mail survey of farmers from throughout Washington State. A sample stratified by county was drawn from the state list of farm operators maintained by the National Agricultural Statistics Service (NASS). Ten percent of farm households were sampled from each county. In addition, four counties that were thought to reflect distinct agricultural production systems in the state were selected for more intensive qualitative and quantitative data gathering on agricultural production as well as on food distribution and consumption. These four counties were (1) King County, a large urban county in which Seattle is located; (2) Skagit County, a rapidly urbanizing farming county in western Washington; (3) Chelan County, an eastern Washington county characterized by midsized tree-fruit orchards; and (4) Grant County, a large-scale irrigated agricultural county in the center of the state. In each of these four target counties three hundred farm households were sampled.

Questionnaires were sent out to 3,718 farm addresses in March 2002. Following Dillman's 2000 Tailored Design Method for mail surveys, a series of follow-ups was conducted with nonrespondents with a postcard reminder and two additional survey mailings. We received 1,201 completed surveys. Removing ineligibles and noncompleted returns from the original sample leaves a survey completion rate of 49 percent.[2] We consider this a reasonable completion rate for a farmer survey of this type.

A comparison of the farm characteristics of those who responded to our survey with the Agricultural Census of 1997 indicates that characteristics of our sample are similar to those of the state's agricultural producers (USDA-NASS 1997). Our respondents represent the state's diversity in terms of types of commodities produced, with 10 percent producing vegetables, 28 percent fresh fruits, 28 percent hay, and 35 percent cattle. We had slightly higher percentages of farmers reporting that they produced fruit and vegetables than the state averages reported in the Census. Our sample may also somewhat underrepresent small farms. Nearly 48 percent of the farmers in our sample had farm receipts of less than $25,000, as opposed to 55 percent of farmers in the Census. The lower percentage of smaller farms in our sample is not surprising, given that smaller farms with lower rates of participation in traditional farm organizations, commodity commissions, and government assistance programs are less likely to be included on state lists maintained by NASS.

The second data set we analyzed came from a telephone survey conducted over a two-month period in 2002 with consumers in the four target counties. The population for the survey consisted of all tele-

Table 22. Direct marketing methods used by farmers

	Vegetable growers (N = 118) %	Fruit growers (N = 332) %	All growers and farmers (N = 1,166) %
Roadside stands	46	21	12
Farmers' markets	35	15	8
U-Pick sales	23	13	7
CSA	8	2	2
Use any of direct methods above	59	30	20
Plan to increase direct marketing	49	23	25

phone households located within the four counties. As there is no universal list of all households in a particular region from which a random sample can be obtained, a random digit dialing approach was used to obtain the sample. The only households excluded by such an approach are those without telephones.

A random sample of 5,200 telephone numbers, with 1,300 in each target county, was selected. Of these 5,200 numbers 1,043 were determined to be business and/or nonworking numbers and were purged from the sample. This purge made the corrected sample 4,157. Interviewers asked to speak with the person living in the household, eighteen years of age or older, who was most involved with food buying for the household. A maximum of twelve call attempts was made to each number. Ultimately, 950 respondents, with at least 230 in each county, agreed to participate in the survey. The overall response rate was 23 percent, which currently is considered standard for a telephone survey.

The Extent and Character of Agricultural Direct Marketing

The results of our farm survey show widespread use of direct and local marketing strategies among Washington producers, including farms of different sizes, types, and locations. Farmer interest in increasing direct marketing in the future is also high, with more than a quarter of respondents stating that they plan to "do more direct marketing to consumers" within the next three years. We were especially interested in determining the extent to which farmers employed marketing strate-

Table 23. Direct marketing methods and total farm receipts

	Less than $250,000/year (N = 910) %	More than $250,000/year (N = 201) %	All growers and farmers (N = 1,111) %
Roadside stands	12	11	12
Farmers' markets	9	4	8
U-Pick sales	8	3	7
CSA	2	<1	2
Use any of direct methods above	20	12	20
Plan to increase direct marketing	26	23	25

gies that involved direct, personal contact between farmers and consumers, such as farmers' markets, roadside stands, u-pick sales, and community supported agriculture. As shown in table 22, a fifth of the farmers in our sample reported that they currently used at least one of these face-to-face marketing strategies.

Table 22 also shows that the percentage of farms that market directly to consumers is higher among farms that produce vegetables (59 percent) and fruits (30 percent). Livestock and grain farmers are far less likely to direct market. For example, less than 12 percent of cow-calf operators said they sold any products directly to consumers. Roadside stands are the most commonly employed form of direct marketing in Washington. Not only are they used the most frequently, but they are widely used by farms of different sizes, including large farms. In contrast, farmers' markets, U-pick, and community supported agriculture are most frequently utilized by smaller farms.

Table 23 illustrates the relationship between farm size, as measured by total farm receipts, and the use of various direct marketing strategies. Small and midsized farms (farm receipts of less than $250,000) are more likely to utilize direct marketing strategies and are slightly more likely to say that they plan to increase their use of direct marketing in the future. However, as shown below, when the effects of location and farm type are taken into account, farm size becomes a less significant factor.

In most cases direct marketing appears to be a supplemental rather than a primary marketing strategy. Just over 5 percent of survey respondents said that they sold all of what they produced via face-to-face direct marketing channels. However, among respondents in close proximity to Seattle, the percentage of farmers using direct marketing methods to sell all of their products more than doubles to 13 percent. Likewise, while 9 percent of all farmers statewide utilize direct marketing as their *primary* marketing strategy, near Seattle this percentage increases to 23 percent. Indeed, farmers in King County, the county surrounding Seattle, are eight times more likely to utilize direct marketing as their primary marketing strategy than their counterparts in counties on the eastern side of the Cascade Mountains.

Local sales of farm products are also important statewide. To avoid confusion over varying definitions of *local*, we specifically asked respondents to break down the percentage of their sales that went to consumers in (1) their county, (2) a neighboring county, (3) throughout Washington State, (4) in the United States, and (5) outside of the country. More than half of all respondents reported that they sell at least some of their crops to end users in their home county. Around 16 percent of farmers said that they sell all of what they produce to in-county consumers. However, these numbers are influenced by the large percentage of hay farmers. Over a third of all Washington farmers produce hay, and much of that hay is sold to neighboring farmers. Thus we should understand that *local* sales of agricultural products do not always mean sales of food to nearby residents.

To better understand and characterize structural and ideological factors associated with farmer participation in direct marketing, we analyzed the variation in use of direct marketing strategies. To do this analysis, we created a categorical variable for direct marketing that distinguished between those farmers who did not sell via any of the four face-to-face marketing channels (N = 934; 80.1 percent of the sample), those who sold less than half of what they produced via these channels (N = 113; 9.57 percent of the sample), and those who sold more than half in this manner (N = 108; 9.14 percent of the sample). We used this variable to explore the individual and structural conditions that were associated with a particular farmer's use of direct marketing. The multivariate technique we utilized was a maximum-likelihood ordinal logit estimation technique (see Maddala 1983 or McKelvey and Zavoina 1975). The value of analyzing these data in this manner is that it allows us to identify the variables that are most highly associated with the use of direct marketing after controlling for the influence of all the variables in the model. Table 24 shows the list of independent variables that were selected for analysis. Below we explain why we selected them.

An important constraint facing Washington farmers is what their farms are capable of producing. While some farmers have the ability to adapt what they produce to market conditions, others do not. For example, due to a lack of water and soil fertility, much of the land in Washington is unsuitable for anything other than cattle production. More than a third of all farms in the state are cow-calf operations. However, local and direct markets for cattle and other livestock products are severely constrained by farmers' lack of access to processing and inspection infrastructure because of the oligopolistic nature of the meat industry in the United States as well as from the necessity to adhere to health and food safety regulations that are associated with animal slaughter and meat storage.

Thus recognizing that different farmers face varying constraints depending on what they produce, we created three dummy variables to distinguish between farmers who produce (1) fresh produce, (2) grains, and (3) livestock (including dairy, poultry/eggs, sheep, goats, hogs, and cattle). After subtracting missing cases, 82.69 percent of all farmers (960 out of 1161) in our sample fall into one or more of these categories.

Another important consideration was the size of the farm. Some proponents argue that direct marketing offers special opportunities for small farmers to compete in an agrifood system that is dominated by industrial-scale, global, and national actors. Further, there is often a perception that direct marketing is not feasible on a larger scale. In order to test these assumptions, we selected two indicators for farm size. We believe that the number of acres managed by a farm operator is a poor indicator of size because acreage in Washington is so specific to particular commodities and ecological regions of the state. Cow-calf ranchers, for example, often own large tracts of poor-quality land that is suitable only for grazing and is priced accordingly. Thus we decided to use the presence of hired labor and farm gate receipts as indicators of size. One variable distinguished between farms that hired labor during the previous year and those that did not. Two additional variables were used to categorize farmers with low and high sales volume. Farms were categorized as having low receipts if they sold less than $25,000 in agricultural commodities and farm products in 2001, while high receipt farms were those which sold more than $250,000 worth.

In order to take account of ecological and demographic conditions, we differentiated between regions of the state. The region of Washington State west of the Cascade Mountain Range is characterized by high rainfall and a rapidly increasing population. Eighty percent of the state's population and nearly 40 percent of our farm respondents live in this western region. Farms in this region face intense land development pressure and complex environmental regulations but also have a milder climate, access to more water for irrigation, and direct access to large numbers of consumers. Conversely, eastern Washington agriculture has a long history of industrialized, export-oriented farming and is distant from major urban population centers (Selfa and Qazi 2005).

Beyond these factors we also sought to explore whether farmer ideology, farmer networks, and farmer access to information might play a role in direct marketing. To investigate whether direct marketing is more prevalent among farmers with particular ideological commitments to family farming, land stewardship, or free trade, three corresponding indicators were utilized: the degree to which a respondent agreed (1) that maintaining family farms is important to the future of his/her county, (2) that land should be farmed so as to protect its long-term productive capacity, and (3) that free-trade agreements will help his or her farm operation be profitable in the long term. We also looked at whether a farmer had any land that was being managed organically (either certified or uncertified) in order to see if there was any positive association between organic farming and direct marketing.

Next, in order to investigate whether connections to farm organizations, other farmers, or Cooperative Extension might be associated with use of direct marketing, we created three network variables. The first measured whether a farmer reported being "very involved" in a farm organization such as the Farm Bureau, the Grange, a commodity

Table 24. Individual and structural characteristics associated with farmers' use of direct marketing

	No. of farms	% of all farms	% of category that uses any form of direct marketing
Fresh produce farmers	398	34	33[b]
Grain farmers	205	18	7
Livestock farmers	508	43	15[b]
Use hired farm labor	615	52	19
Receipts < $250,000	530	47	22
Receipts > $250,000	203	18	12
West side county	457	39	29[b]
Has organic land	254	23	28[a]
Profamily farms	762	68	17
Pro-sustainability	663	59	19
Anti–free trade	501	45	20
Farm organization involvement	707	62	21[a]
Contact with other farmers	497	46	19
Contact with extension	427	37	22[a]

[a]Significant in the Logit Model at $P < 0.05$
[b]Significant in the Logit Model at $P < 0.01$

commission, or other growers' associations. A second distinguished between farmers who said that they obtained farm management information three or more times during the previous year from other farmers. The third variable measured whether the farmer had any direct, face-to-face contact with extension staff in the previous year, either on the farm or in the county office.

As mentioned earlier, we used a statistical procedure known as ordinal logit to assess which of these variables when controlling for the influence of the other variables was most important for understanding which types of farmers were most likely to utilize a direct marketing strategy. Those variables that were the most significant are identified in the far right-hand column of table 24. However, rather than report the coefficients for each variable from the logit model, we report the percentage of farmers in each variable category who utilize any form of direct marketing.

The characteristics that best explain whether a farm utilized direct farming are whether that farm produces fresh fruits and vegetables and whether it is in western Washington. The variable for livestock farmers is also highly significant, *but in a negative direction*. In other words, livestock farmers are significantly less likely to engage in direct marketing. In addition, while a greater percentage of farms that have less than $25,000 per year in sales market directly than do farms that have more than $250,000 a year in sales, these differences are not statistically significant. What this finding means is that once we control for the region where the farm is located and what a farmer produces, the relative size of the farm is not a good indicator of whether a farm sells directly to consumers. Put another way, a large percentage of the state's smallest farms that market directly to consumers are western Washington produce farms.

Three other variables are significantly associated with the likelihood of direct marketing, but at a lower level of statistical significance. These variables are whether the farm respondent managed any organic land, participated in a farm organization, or had contact with Extension. Thirty-seven percent of all respondents said that they had contact with Extension staff in the previous year. Of these respondents, nearly 22 percent also participated in direct markets. This finding should not be interpreted as evidence that having direct contact with Cooperative Extension or strong involvement in a farm organization *leads* to the use of direct marketing, but it does appear that farmers who practice direct marketing are slightly more likely to be active in farm-related information and social networks.

Our results do not suggest a very strong association between farmer ideological commitments to such ideals as agrarianism and sustainability and the use of direct marketing methods. Neither is size a particularly good indicator of whether a farm utilizes direct marketing channels. Instead, the most powerful explanatory variables reflect opportunity: those who can grow fresh produce and have access to consumer markets in major urban centers are most likely to market directly. However, while some farmers are able to change their marketing mix to take advantage of these emerging high-value markets, many cannot because of their farm's ecological conditions, geographical lo-

Table 25. Washington farmers' views on marketing and policy issues (percent agreeing or strongly agreeing)

	Fruit and vegetable producers	Cattle producers	Hay producers	Grain producers	Other livestock[a]	All growers and farmers[b]
"Grown in Washington" labels would help Washington farmers.	81	75	75	71	77	77
Free trade agreements help my farm's profitability.	26	16	20	37	14	23
Consumers should have more local food available.	56	62	57	41	67	57
Direct marketing helps keep local farms viable.	56	65	61	43	70	62
There is real demand for organic products in Washington.	36	44	39	20	57	39

[a]Includes producers of poultry, eggs, dairy products, hogs, sheep, goats, horses, and other animals.

[b]Some respondents are listed in more than one farm type.

cation, or preexisting infrastructure. These restrictions are particularly true for the many cow-calf producers, orchardists, and wheat growers in eastern Washington who currently appear to be gaining few advantages from the rapid expansion of direct marketing in the state (Qazi and Selfa 2005). Trying to find more profitable marketing options for these farmers remains a major challenge.

These variations in the opportunity structure for different types of farms are reflected in the respondents' views on farm and market policy. While most Washington farmers favor policies that encourage direct and local market development over international market development, table 25 shows that these opinions vary by farm type. For example, while most respondents said that direct marketing could help farms stay economically viable in their counties, grain producers were an exception. Similarly, reflecting the heavy reliance of grain producers on export markets in contrast to other types of farmers, these farmers are the most supportive of free-trade agreements and the least likely to state that consumers in their counties should have access to more locally grown foods. On the other hand, livestock producers feel most strongly that consumers should have access to more local foods and are the most likely to say that there is high consumer demand for organic items. Only one marketing topic elicits uniformly widespread agreement: over three-quarters of all respondents believe that a "Grown in Washington" labeling program would help Washington farmers. Overall, regardless of farm type, farmers appear to favor a state-level labeling program and other types of direct and local marketing efforts over an expansion of free-trade agreements.

Understanding Direct Food Purchasing by Consumers

Knowing the extent and nature of consumer demand for local farm products is critical for assessing the potential of direct and local market linkages needed to sustain the growing number of interested farmers. Our second data set measures the frequency of direct food purchasing by consumers and explores their primary motivations, constraints, and purchasing criteria. Further, we investigate whether there is any degree of consensus emerging among Washington consumers and farmers regarding policy changes that are needed in the agrifood system. Do pro-

Table 26. Consumer purchasing patterns in four Washington counties

Purchasing choice	% of respondents (N = 950)
Do not buy products directly from farmers	31
Buy directly from farmers once a month or less	43
Buy directly from farmers twice a month or more	26
Would pay 25 percent more for local	23
Interested in more direct purchases of:	
vegetables	82
fruits	81
eggs	52
dairy	44
beef	36
poultry	34

ducers and consumers have similar goals for participating in emerging direct markets, and if not, does a lack of consensus affect the potential for building a collective movement for agrifood system change?

Our research shows that in our four target counties, it is fairly common for Washington consumers to make at least some purchases directly from local farmers. More than a quarter of the consumers surveyed reported buying products directly from a local farmer twice a month or more during the growing season (see table 26). The majority of the respondents said that they would like to purchase more products directly from farmers. However, as shown in Table 26, interest levels varied substantially by product. While more than 80 percent of consumers were interested in buying more fruits and vegetables directly from local farmers, the demand for animal products like eggs, dairy, and meat was substantially lower. Nevertheless, for each type of farm product, at least one-third of consumers say they would like to increase their purchases from local farmers. Furthermore, nearly a quarter of respondents say they would be willing to pay 25 percent more for such local products.

In trying to understand what motivates consumers to seek out and buy local food products, we asked some general questions about their household food purchasing criteria. Table 27 sets out the food attributes, ranked in descending order of importance, considered most im-

Table 27. Consumer views on food purchasing criteria in four Washington counties (percent rating "very important")

	% of full sample (N = 950)	% with household income < $25,000 (N = 199)
Freshness	94	93
Taste	90	89
Nutritional value	77	69
Available where normally shop	74	79
Keeps local farms in business	70	75
Appearance	62	61
Price	59	73
Environmentally friendly	45	56
Grown in Washington	41	42
Grown locally	34	32
Organic	16	17

portant by the respondents. The top consumer values were freshness and taste, followed by nutritional value. For a majority of respondents price was not as important a factor as convenience. This finding corroborates information obtained from other questionnaire items. According to respondents, the main factors preventing them from buying food at farmers' markets were issues like "it isn't open at the right times" or "there isn't one in my neighborhood" rather than the cost. In comparison to other priorities, buying *local* food products was not an important consideration for most consumers. However, if the question is reworded to ask whether "helping to keep local farms in business" is important, the responses are far more favorable. Seventy percent of consumers felt that keeping local farms in business was very important.

Analyzed more closely, the data reveal interesting differences among different types of consumers. Table 27 also displays survey responses of lower-income consumers as contrasted with the full sample. One obvious difference is food price, which becomes far more important for households earning less than $25,000 annually. More than 73 percent of consumers in such households felt that price was a very important consideration in comparison to only 59 percent of the full sample. Similarly, food availability becomes a stronger concern for this group. Freshness and taste remained the top criteria for lower-income consumers, while this group was somewhat less likely to rate nutritional value as important compared to all respondents. Interestingly, lower-income consumers were somewhat more concerned than the full sample about environmental friendliness and keeping local farms in business.

Other differences among consumers emerge when the data are analyzed by county. The consumers from the most urban county, King County, were the least concerned about keeping local farms in business and the most concerned about purchasing organic products. Other than this stronger urban emphasis on organic, interest in the environmental attributes of food products did not appear to vary significantly by geographical region. Price, on the other hand, was far more important to consumers in eastern rather than western Washington. Overall the results of the consumer telephone survey suggest that consumer demand is far more regionally and culturally variegated than mass market theory would suggest. However, it is clear that for most consumers in Washington quality and convenience rather than ideological concerns about the environment or the globalization of the agrifood system constitute the most critical food purchasing criteria.

In further analyses we compared and contrasted consumer views as a whole with those of farmers on important food policy topics. When we posed the same survey questions to consumers and farmers, we found areas of both convergence and divergence. Consumers were more likely to think that free-trade agreements would help them than farmers were, while both sides overwhelmingly supported the idea of a Grown in Washington label. As reported in table 25, 77 percent of farmers were in favor of a Grown in Washington label. Curiously, while 94 percent of consumers thought that having a Washington label would help the state's producers, only 41 percent said that purchasing Washington food products was important to them (table 27). Perhaps some of these consumers thought that a Washington label would help producers succeed in out-of-state markets. Consumers also were very sympathetic toward the concept of the family farm. Even more consumers (90.1 percent) than farmers (86.6 percent) agreed with the statement that "maintaining family operated farms is important to the future of my county." A high percentage of farmers (74 percent) and an even higher percentage of consumers (82 percent) felt that "farmers should be paid for their participation in wildlife programs."

Not all views were so convergent, however. Consumers favored stronger policies regarding land use and were more concerned about genetically modified (GM) crops than farmers. Nonetheless, the survey results indicate important areas of agreement for policy building that supports the viability of family farms, provides positive environmental incentives, and creates regional labeling programs. Important questions remain, however, about the extent to which this stated pro–family farm sentiment and professed interest in purchasing local farm products can be leveraged to create actual changes in consumer shopping and political behaviors.

In this chapter we used data collected from large-scale surveys of Washington State producers and consumers to examine the potential and the limitations of direct marketing as a tool to revitalize local agriculture, create new kinds of community-based marketing circuits, and establish viable alternatives to agrifood globalization in the state. We explored and contrasted the attitudes and behaviors of producers and consumers with respect to the growth of alternative marketing relationships, the current policy landscape, and the degree to which local actors may be intentionally or unintentionally making changes that lead to more socially and environmentally responsive local food systems.

Our survey findings revealed that direct marketing is viewed as more legitimate and practiced more widely than originally expected. Moreover, both farmers and consumers expressed interest in further expanding direct market relationships, which already constitute more than a niche or a marginal activity. Contrary to the conventional wisdom that direct marketing is only practical for the smallest farms, we found that around 20 percent of farms statewide practice some form of direct marketing, including farms in the highest income categories. In addition, in most cases, direct marketing is used to supplement rather than replace wholesale marketing strategies. Rather than choosing *either* local or global markets, many Washington farmers appear to employ a mixture of marketing strategies as a means of reducing risk and gaining some control over uncertainties associated with volatile, globally linked wholesale markets. Direct markets have not developed for many of the commodities, such as grains and livestock products, produced and consumed in Washington. Proximity to urban markets and farm type explain farmer involvement in direct marketing far better than any type of ideological position on family farms, sustainable agriculture, or global trade agreements. Thus farmer participation in direct markets appears to be largely opportunistic rather than an intentional attempt to create a direct alternative to a highly industrialized food system or to build a more civic agriculture (Lyson 2000). Essentially direct marketing provides an opportunity for those with the right crops in the right places to improve their economic security.

A similar dynamic emerges among consumers. While significant numbers of consumers are making purchases directly from local farmers, their primary goal is not to improve the environment or the local economy. Instead, consumers appear to be most highly motivated by food quality, taste, nutrition, and convenience. Furthermore, while they are making some local purchases and say they want to increase them, consumers are most interested in locally obtaining specific components of their diet, that is, fruits and vegetables; they are not broadly committed to the principle of eating locally as a tool for social or environmental change.

Given that both farmer and consumer motivations for expanding direct market relationships appear to be far more practical than idealistic and more individualistic than collective, what is the significance of these emerging market linkages among Washington farmers and consumers? Are these alternative marketing circuits succeeding at creating and preserving new more democratic social spaces around the circumstances of food production, distribution, and consumption?

While not intentionally values-based or social movement-oriented such as movements for fair trade, ecolabels, or community supported agriculture, the expansion of direct marketing relationships in Washington shows some potential for furthering agrifood system change on two levels that warrant further investigation. First, direct marketing is succeeding at creating material changes for at least some Washington farmers and consumers, that is, providing new opportunities to generate farm income and new ways to acquire food. This is a fundamentally different kind of market exchange involving a unique social context and a particular physical space. Further research should investigate whether reembedding market transactions in such social relationships may unintentionally transform the participants. For example, do direct market consumers increase their knowledge and identification with local farmers as a result of their transactions? Do farmers gain opportunities to understand and to become more responsive to local market needs? More importantly, does direct market participation allow creation of new kinds of social networks that could become transformative at some stage? If so, the more commonplace these activities become in Washington, the greater their educational potential.

Second, to the extent that direct marketing keeps existing farms profitable and allows new farms to start up, it is developing and maintaining production capacity necessary for supplying more of the future local dietary needs. This function is especially critical in the rapidly urbanizing environment of western Washington. Given consumers' preferences for fresh and nutritious food products and their strong support for family farms and farmland protection, it seems entirely possible that vibrant citizen movements eventually could coalesce around agrifood system change. However opportunistic direct marketing may appear, its maintenance of the farming base could be essential for any future transition to a more locally controlled, civic-oriented agrifood system in Washington.

Finally, our analysis suggests that the process of remaking the food system necessarily will be a complex undertaking. A one-model-fits-all approach cannot work. This realization should not be surprising as any attempt to reembed food systems in local places necessarily will need to take into account the unique possibilities and challenges that are associated with local and regional histories, cultures, and ecological conditions.

In the case of Washington one of the biggest challenges for developing a more locally embedded food system is that many of the rural regions in the sparsely settled arid regions in the eastern two-thirds of the state primarily are used for producing cattle, hay, and grains.

These commodities are not as easy to market directly to consumers as fresh produce, and distance to urban markets remains an issue. In addition, in central Washington some items, like apples, are produced far in excess of any local market demand. Thus any attempt to create a successful, alternative agrifood system in Washington State will have to include approaches and networks that provide opportunities for cattle ranchers, orchardists, and grain farmers to capture higher values by diversifying at least some products out of conventional wholesale markets.

Acknowledgments

The authors wish to acknowledge the contributions of their survey research partner, Dr. Lucy Jarosz (University of Washington) and the funding provided by the W. K. Kellogg Foundation, the Washington Farming and the Environment Project, and the U.S. Department of Agriculture-National Research Initiative.

Notes

1. Further details are available on the Web sites of the Washington Wheat Commission ("Markets and Product Information," http://www.wawheat.com/markets.asp), the Washington Apple Commission ("Core Facts," http://www.bestapples.com/facts/index.html), and the Washington State Potato Commission ("Potato History," http://www.potatoes.com/growingpotatoes.cfm?section=growing-history.cfm#washington) (all sites last accessed April 15, 2006).

2. The farmer survey had 46 refusals and 1,047 ineligibles and returns. Ineligibles were defined as households that sold less than $1,000 in commodities in 2001 as well as those farm households that had moved, passed away, retired from farming, or received multiple surveys because they owned more than one agricultural property.

References

Allen, P., M. FitzSimmons, M. Goodman, and K. Warner. 2003. "Shifting plates in the agrifood landscape: The tectonics of alternative agrifood initiatives in California." *Journal of Rural Studies* 19:61–75.

Blank, S. C. 1998. *The End of Agriculture in the American Portfolio*. Westport CT: Greenwood.

Bonanno, A., L. Bush, W. H. Friedland, L. Gouveia, and E. Mingione, eds. 1994. *From Columbus to ConAgra: The Globalization of Agriculture and Food*. Lawrence KS: University Press of Kansas.

Daly, H., and J. Cobb Jr. 1994. *For the Common Good*. Boston: Beacon.
Dillman, D. A. 2000. *Mail and Internet Surveys: The Tailored Design Method*. New York: Wiley.
Fletcher, S. W. 1902. *Locating an Orchard in Washington*. Bulletin no. 51. Pullman WA: Agricultural Experiment Station, State College of Washington.
Friedmann, H. 1993. "After Midas' Feast: Alternative Food Regimes for the Future." In *Food for the Future: Conditions and Contradictions for Sustainability*, edited by P. Allen, 213–33. New York: Wiley.
Hassanein, N. 2003. "Practicing food democracy: A pragmatic politics of transformation." *Journal of Rural Studies* 19:77–86.
Hinrichs, C. 2003. "The practice and politics of food system localization." *Journal of Rural Studies* 19:33–45.
Kneen, B. 1989. *From Land to Mouth: Understanding the Food System*. Toronto: NC Press.
Korten, D.C. 1995. *When Corporations Rule the World*. San Francisco: Berrett-Koehler Publishers; Bloomfield CT: Kumarian Press.
Lawson, V. A. 1992. "Industrial subcontracting and employment forms in Latin America: A framework for contextual analysis." *Progress in Human Geography* 16:1–23.
LeHeron, R. 1993. *Globalized Agriculture*. Oxford UK: Pergamon Press.
Lyson, T. 2000. "Moving toward civic agriculture." *Choices* 15 (3): 42–45.
Maddala, G. S. 1983. *Limited-Dependent and Qualitative Variables in Econometrics*. Cambridge UK: Cambridge University Press.
Magdoff, F., J. B. Foster, and F. H. Buttel, eds. 2000. *Hungry for Profit: The Agribusiness Threat to Farmers, Food, and the Environment*. New York: Monthly Review.
Mander, J., and E. Goldsmith, eds. 1996. *The Case against the Global Economy: And for a Turn Toward the Local*. San Francisco: Sierra Club.
McKelvey, R. D., and W. Zavoina. 1975. "A statistical model for the analysis of ordinal level dependent variables." *Journal of Mathematical Sociology* 4:103–20.
McMichael, P. 1997. "Rethinking globalization: The agrarian question revisited." *Review of International Political Economy* 4:630–62.
Qazi, J., and T. Selfa. 2005. "The politics of building alternative agro-food networks in the belly of agro-industry." *Food, Culture, and Society* 8 (1): 45–72.
Selfa, T., and J. Qazi. 2005. "Place, taste, or face-to-face? Understanding producer-consumer networks in 'local' food systems in Washington State." *Agriculture and Human Values* 22 (4): 451–64.
Sklair, L. 1991. *Sociology of the Global System*. Baltimore: Johns Hopkins University Press.
Sonnenfeld, D., T. Schotzko, and R. A. Jussaume Jr. 1998. "The globalization of the Washington apple industry: Its evolution and impacts." *International Journal of the Sociology of Agriculture and Food* 7:151–80.
USDA-NASS. 1997. *1997 Census of Agriculture*. http://www.nass.usda.gov/census/index1997.htm (last accessed April 20, 2006).

———. 2002. Washington Quick Facts from the Census of Agriculture. http://www.nass.usda.gov/statistics_by_state/washington/publications/censo2brochure.pdf (last accessed April 9, 2006).

U.S. Dept. of the Interior. 1964. *The Story of the Columbia Basin Project*. Washington DC: U.S. Government Printing Office.

Whatmore, S. 1994. "Global Agro-food Complexes and the Refashioning of Rural Europe." In *Holding Down the Global*, edited by A. Amin and N. Thrift, 45-67. London: Sage.

Whatmore S., and L. Thorne. 1997. "Nourishing Networks: Alternative Geographies of Food." In *Globalizing Food: Agrarian Questions and Global Restructuring*, edited by D. Goodman and M. J. Watts, 287–304. London: Routledge.

13. Emerging Farmers' Markets and the Globalization of Food Retailing

A Perspective from Puerto Rico

Viviana Carro-Figueroa and Amy Guptill

The globalization of food retailing has enormous consequences for farmers, food processors, regional grocery chains, and other actors in the food system. A 2001 report to the National Farmers Union by Hendrickson and colleagues describes the fast pace of food retail consolidation in recent years (led by Wal-Mart) and discusses the implications of these trends for specific farming sectors and other smaller actors in the food system (Hendrickson et al. 2001). Among the most worrisome outcomes are new exclusive relationships between firms in highly concentrated processing sectors such as beef packing and new megaretailers that close out some of the best niche opportunities for small farmers to sell high-value perishables directly to local stores.

Guptill and Wilkins (2002) similarly found that, as local and regional chains are increasingly displaced by global giants, the possibilities for getting local foods into those stores become more remote. However, Guptill and Wilkins (2002) also note that some independent grocery stores are focusing on high quality perishable foods (produce, meats, and dairy) in order to distinguish themselves from cheaper stores, an effort that would benefit from close collaboration with local producers. Thus while overall trends in food retailing do not seem favorable for local farmers, the specific outcomes of these changes will vary by locality.

At the same time social scientists have noted that increasing corporate concentration in the food system has been accompanied by the proliferation of other kinds of food outlets that include farmers' markets, specialty stores, and subscription schemes. As supermarkets get larger and more centralized and some of their products become cheaper, it seems that producers and consumers alike forge new linkages in alternative outlets based on niche products and perhaps social and environmental values. Perhaps then marketing opportunities lost in the globalization of retailing simply shift to new kinds of systems.

Puerto Rico recently has seen these twin trends of concentration and innovation in food marketing: on the one hand, giant players in global food retailing like Wal-Mart are acquiring—in part or in whole—locally owned supermarket chains such as Amigo, leading to a competitive shake-up of local retailing. On the other hand, farmers, activists, and government agencies are establishing farmers' markets to pursue growing interest in locally and sustainably grown products and new, more favorable markets for farmers. What can explain these contrasting trends, and how will they affect consumers and local farmers?

To address these questions, we begin with some basic characteristics of the Puerto Rican food system. Though the Puerto Rico 2002 Census of Agriculture counts almost 18,000 farms on the island, three-quarters of them have an annual agricultural income of less than $10,000. The farming sector depends heavily on agricultural and social subsidies. Meanwhile, the great majority of the food consumed in Puerto Rico is imported, and close to 55 percent of Puerto Rican families use federal nutritional assistance programs to pay for those products (P.R. Departamento de Salud 2004; P.R. Junta de Planificación 2003). Most of the foods are distributed through a few large concentrated supermarket chains, some of which have been locally owned for decades (Carro-Figueroa 2002).

With this context in mind, we used a two-pronged approach to study the character and impact of recent food system trends in Puerto Rico. First, we gathered information on the global restructuring of wholesale and retail trade from secondary sources such as business magazines and newspaper accounts. In our discussion of this restructuring, we describe the history of food retailing and the particular shape of current trends in the supermarket sector.

Second, we conducted a basic study of five farmers' markets in Puerto Rico: a long-standing agricultural market in San Sebastian, two new farmers' markets in the mountainous central region of Puerto Rico

(Barranquitas and Aibonito), an organic farmers' market sponsored by a producer-consumer cooperative, and a WIC (Women, Infants, and Children) Itinerant Farmers' Market organized by the local Department of Agriculture and social service agencies to bring the WIC Farmers' Market Nutrition Program (FMNP) to Puerto Rico. We conducted interviews with the organizers of these markets in order to investigate their views on the market's goals, farmer participation, community reception, consumer sponsorship, general operational characteristics and perceived limitations and perspectives. All of the interviews were approximately two hours long, except the one with the current administrator of the San Sebastián market—conducted by phone—and those associated with the Barranquitas market, which were part of a broader study of that municipality's food system. We analyzed the information gathered in these two parts of the study in order to assess the potential future directions of agriculture and food in Puerto Rico.

Restructuring of Food Retailing in Puerto Rico: 1950 to the Present

One of the most striking visual signs of the modernization of Puerto Rico in the post–World War II years is the presence of large, modern grocery stores. As late as 1950 almost all Puerto Ricans purchased most of their nonperishables from their neighborhood *colmado* and perishables from specialized stores (like butchers) and *plazas del mercado*.[1] While retailing was small in scale, wholesaling and importing were highly concentrated, leading to high retail prices. Beginning around 1955, urbanization, population growth, and rising incomes spurred the growth of supermarkets, and the Puerto Rican government offered credit, technical assistance, and incentives to entrepreneurs opening modern grocery stores. The hope was that efficient, modern food retailing would lower the cost of food for Puerto Rican consumers and create an important amenity for North American companies considering opening branch plants in the island (Riley et al. 1970). The first big grocery stores to open in Puerto Rico, Grand Union and Pueblo, were established by North American retailers in the late 1950s. These stores offered lower prices by bypassing Puerto Rico's highly concentrated wholesale sector and importing directly from the United States. Other local supermarkets emerged to capture part of the market, but U.S.-

based chains dominated Puerto Rico's food distribution system until at least the early 1990s when local chains became more prominent (Carro-Figueroa 2002).

Government planners hoped that these modern grocery stores would create new marketing opportunities for Puerto Rican farmers, but the impact was mixed. Puerto Rico's dairy, egg, and chicken industries, the strongest sectors in the island's agricultural economy, did benefit from the new sector and continue to market products mostly through these stores. Traditional crop producers, however, were largely unable to accommodate the quality and packaging requirements of the big supermarket chains and had to continue selling their crops to truckers visiting their farms and to small *colmados*. Moreover, since the increase in international trade of the early 1960s, supermarkets tended to import produce from cheaper producers elsewhere in the Caribbean or North America. The plant foods that are most central to Puerto Rican cuisine—roots, tubers, plantains, rice, and beans—are those with the cheapest world prices.

Both traditional crop producers and small-scale retailers also lost ground in 1975 when the federal government extended the Food Stamp Program to Puerto Rico. While the amount of money people had to spend on food increased significantly, the new means for distribution of nutritional assistance benefited large-scale wholesalers, retailers, and food importers most (Ruiz and Choudhury 1978). In 1982 the coupons that recipients would redeem were replaced by a block grant that the Puerto Rican government then distributed to recipients in checks under a program called PAN.[2] Small *colmados* were the most hurt by the change since they usually lacked the cash flow needed to cash checks for their patrons.

Meanwhile, the supermarket sector grew due to the boosts in consumer spending on food and local investment in retail development. That investment reflected the millions of PAN funds that had been deposited in Puerto Rican banks, as well as deposits from U.S. corporations with branch plants on the island (Weisskoff 1985). Changes in federal and local tax regulations encouraged U.S. firms to space out the repatriation of their profits for several years (NACLA 1981). Developers used this capital to build suburban settlements with numerous shopping centers. As a result U.S. chain stores became even more of a presence on the island, forcing local retailers of all kinds to adapt to new competitive conditions (Gigante 1999). Many remaining *colmados* and some well-known supermarket chains like Grand Union disappeared during the economic restructuring that followed in the 1980s.

That restructuring continued into the 1990s with entrance of Wal-Mart. The first Wal-Mart opened in Puerto Rico in 1992. By the end of the decade eight more Wal-Marts and seven Sam's Clubs, a division of Wal-Mart, appeared on the island. More recently, Kmart and Wal-Mart both opened discount stores with full-line supermarkets inside (Gigante 2001a). At the same time, grocery stores faced competition from proliferating fast food chains as well as the expanded food sections of pharmacies, bakeries, and gas stations.

To survive this aggressive competition, conventional supermarkets had to undertake sizeable expansions, adding new products and departments in order to attract customers (Gigante 1999). Large supermarkets were hard pressed, but independent supermarkets and the remaining *colmados* were simply not surviving. From 1999 to 2000 industry sources reported that, although total bankruptcy cases decreased in Puerto Rico, bankruptcies of supermarkets and *colmados* increased 58 percent (Gigante 2001b).

As in the past this more recent restructuring is linked to changes in the distribution of food subsidies, particularly of PAN funds. In 2001 the PAN program introduced a debit card mechanism through which recipients are allowed to spend 75 percent of their allotment on food items and redeem a maximum of 25 percent in cash.[3] The "75/25 rule" further boosted the sales of food in general (Gigante 2001c; Rosa 2002a), but concentration in retailing grew apace.

Around this time, newspapers reported that Wal-Mart was negotiating with Amigo, Puerto Rico's largest locally owned supermarket chain, for the purchase of its thirty-three stores. A highly unusual coalition of worried retailers formed to denounce the "negative repercussions" of this deal for the local economy and lobby the government to block it. The coalition involved pharmacies, bakeries, a consumer cooperative, an association of retailers and wholesalers, the Wholesalers Chamber of Commerce, and the Farmers' Association. Some of these actors, usually the most ardent defenders of free enterprise, were now highly critical of the threat of monopoly. Two local groups, the Puerto Rico Products Association and the Puerto Rico Manufacturers Association, supported the deal, arguing for a free market and accepting Wal-Mart's promise to help local business to expand globally (Rosa 2002b).

The anti-Wal-Mart coalition's alarming allegations had some merit. Wal-Mart claimed that their stake in the whole Puerto Rican food market would be less than 19 percent (*Caribbean Business* 2002), but regional concentration was more acute. The U.S. Federal Trade Commission (FTC), summoned to evaluate the terms of the transaction, ordered Wal-Mart to sell four Amigo stores in different regions of the island to avoid local monopoly. A regional consumers' cooperative still filed a lawsuit in the Superior Court of San Juan seeking to block the deal, while Puerto Rico's Justice Department asked for an injunction in the same court until its Antitrust Affairs Office could review the case in full and negotiate certain conditions with Wal-Mart.

While the Puerto Rican government usually exercises considerable control over local economic policy, in this case the federal courts set sharp limits on the government's powers. The federal court judge immediately ordered the Justice Department to withdraw the injunction, ruling that because the transaction was already scrutinized by the FTC, Puerto Rico's secretary of justice "is potentially surpassing the boundaries of her authority into unconstitutional realms" (Berrios-Figueroa 2002, 68). The Justice Department appealed this ruling at the First Circuit Court of Boston, but in March 2003, before the case was heard, an agreement between Wal-Mart and the Justice Department was reached. The Justice Department dropped the antimonopoly allegations against Wal-Mart who in return agreed to maintain the current level of employment at its stores and of purchases from local farmers for ten years. Critics were unsatisfied, but the controversy and public discussion about global retail giants quickly subsided. By 2007 Wal-Mart operated six supercenters along with several Sam's Clubs and over thirty Amigo stores.

Beyond Wal-Mart, other global players and large U.S. corporations, such as White Rose, a major food distributor of the U.S. east coast, have recently considered entering Puerto Rico by buying into local chains (Rosa 2002a). Moreover, although Wal-Mart had already established a few smaller Neighborhood Markets tailored to the conditions of selected U.S. localities, the deal with Amigo is the company's first large-scale initiative of this kind in a relatively small geographical area (Saporito 2003). In this way an island with a per capita income about one-third that of the mainland United States is on the leading edge of a process that is now gaining steam in other Latin American and Asian countries (Coyle 2006).

Our analysis of the restructuring of food retailing since the 1960s underscores the importance of nutritional subsidies in enabling the emergence of supermarkets and reveals how the administration of those subsidies shapes the process of retail restructuring (Weisskoff 1985). Clearly it is the extent to which food consumption is subsidized in Puerto Rico and to which the island depends on the global food system that makes it an attractive target for global players. While most Puerto Rican families live on fairly low incomes, many can consume like higher-income households. In the United States the centralization of food retailing has been linked to the emergence of new kinds of outlets. Does the emergence of farmers' markets in Puerto Rico reflect the generally high food purchasing power of Puerto Rican consumers? Do consumers choose to spend more money at farmers' markets because of the PAN funds they receive? And can these markets make up for the lost marketing opportunities in supermarkets? Can they pave the way for a more independent Puerto Rican food system?

Emerging Farmers' Markets: Characteristics, Possibilities, and Limitations

New farmers' markets recently have been established in Puerto Rico with the hope of creating more direct connections between producers and consumers. By cutting out the intermediate sector, farmers' markets hold great promise for making fresh local produce available to consumers at affordable prices while offering farmers a new and more favorable market. However, it is important to understand the different kinds of farmers' markets in Puerto Rico and the different possibilities and limitations they face. In this section of the chapter we describe five markets in chronological order of inception and analyze their successes and challenges.

Mercado Agrícola de San Sebastián

Established in 1958 by the Puerto Rican Department of Agriculture to improve the distribution of local produce to middlemen and wholesalers, the Mercado Agrícola de San Sebastián (MASS) became a regional trading center that currently combines wholesale trade and livestock bartering with twice-weekly retailing to consumers. In 2000 it was the only outlet approximating a farmers' market in Puerto Rico. At present 80 of the 340 market stands (23.5 percent) are occupied by farmers who sell both local and imported produce to consumers (Armando Serrano, personal communication, May 2003). The rest of the market booths offer, among other things, prepared food, clothing, exotic birds, and used household goods, making the MASS more like a flea market than a farmers' market. However, the MASS is still the most popular "farm market" on the island. It has preserved a traditional outlet for fresh produce and has inspired more recent efforts. While the emphasis on local agriculture is lost, renewed support from the government may lead the market to convert one section so as to feature farmers selling exclusively local produce.

Mercado Orgánico Agrícola en la Placita Roosevelt

The Cooperativa Orgánica Madre Tierra, an organic consumer and producer cooperative, has established a market, the Mercado Orgánico Agrícola en la Placita Roosevelt (MOAPR), that similarly offers an array of products. However, the MOAPR's particular mission is to promote organic farming, serve as a distribution point for co-op members, and make healthy products and services more accessible to everyone. During the planning stage the cooperative had held two one-time markets at other events: a celebration in a small town square and an open house at a member's farm. The markets proved popular, and the cooperative's directors sought to create a more regular venue. They found a well-known, accessible location in San Juan, and the local resident's association allowed them to use it free of cost. The first MOAPR was held in April 2001, the second in June of that same year, and subsequently on the first Sunday of each month from 9 a.m. to 2 p.m. In addition to organic farm products, vendors also sell essential oils, natural soaps, massages, prepared vegetarian foods, and independently produced music and art. The cooperative's retail shop sells organic and natural staple products not produced in the island, items such as rice, beans, oils, and cereals. At present about fifty-five vendors participate, about one-third of whom are farmers.

The organizers consider the market very successful. The cooperative has rented a commercial property near the market location to serve as a warehouse and office and is currently raising capital for a more permanent shop. The cooperative's director feels that the principal limitation of the market initiative is that most cooperative members do not contribute to maintaining it. As the market and the cooperative grow, routine tasks multiply while the number of active volunteers has remained about the same.

Recent years, however, have also seen controversy associated with a preinspection program based on organic principles that the market instituted to screen the production practices of farmers. By July 2005 market personnel had inspected twenty-two farms. Four of these farms failed the inspection, and those vendors, some of whom had participated in the market from its inception, were no longer allowed to sell in the market (Adelita Rosa, personal interview, July 2005).

La Plaza del Agricultor Barranquiteño

La Plaza del Agricultor Barranquiteño (PAB), or Barranquitas Farmers' Market, is one of the recently emerging farmers' markets inspired by the commercial success of the MASS. While there was interest in establishing a farmers' market since at least 1994, the project did not move forward until 1998 when a group of farmers, the local Agricultural Extension Service, and the town's mayor collaborated to produce a three-day festival dedicated to celeriac (also called root celery, *apio* in Spanish), one of the region's most important crops. The festival, which combined farmers and food sellers, local craft persons and musicians was very successful and has since become an annual celebration in Barranquitas. That success convinced a group of farmers to pursue the farmers' market idea in 2000. The Agricultural Extension Service played an important role in the process by providing space for meetings, organizational assistance, and administrative and promotional support.

The first PAB market took place in April 2001 in the town's municipal parking lot with eight to ten farmers selling their products. It is still being held once a month on Fridays and Saturdays of the first weekend of the month. This market's principal goal is to promote local agriculture and to achieve better prices for farmers. Only Barranquitas farmers selling their own crops (usually plantains, roots and tubers, chayote, papayas, and ornamentals) are invited to participate.

At first the market was open on Saturday and Sunday of the selected weekends, which coincided with other special celebrations (religious and agricultural festivals). The initial markets enjoyed strong support by attendees from around the island, but later markets on nonfestival days had disappointing attendance. To boost sales, organizers held the market on Friday instead of Sunday in order to serve municipal government employees working in town. That change helped, but what saved the market, organizers assert, was the introduction of the WIC Farmers' Market Nutrition Program the following year, in which clients of the USDA's WIC program were provided with coupons that they may spend at approved farmers' markets on fresh fruits and vegetables. The participation of WIC clients boosted sales and renewed farmers' commitment to the market. PAB participants also anticipate another boost from the local implementation of the USDA FMNP for seniors.

Aibonito Farmers' Market

The Aibonito Farmers Market was initiated by a farmer who participated in the MOAPR and sought to replicate the direct marketing model in his own community (Raul Noriega, personal interview, March 2003). He saw that customers at the MOAPR were willing to pay a premium for a healthier product but felt that one market day per month would not be enough to benefit most farmers. Consequently, he convinced the mayor of Aibonito to allow a farmers' market in a small public square in the town's outskirts. Because there are few food crop growers in Aibonito (most farmers have converted to capital-intensive production of poultry and ornamentals), the organizer of the market invited farmers from neighboring Barranquitas and the MOAPR to participate. Five or six responded to the invitation, and the market opened in March 2002 and took place on the third Sunday of each month through November.

The market had some problems. While the market was certified to redeem WIC FMNP coupons, not all of the participating farmers had been individually certified, violating a requirement of the program. Meanwhile, the organizer found himself bearing the total burden of administrative and promotional tasks. Limited support from the community, he believes, was the major problem the market faced. Compared to the MOAPR, the Aibonito market lacked product diversity and failed to create a consistent customer base.

WIC FMNP Itinerant Markets

The U.S. Congress established the WIC FMNP in 1992 "to provide fresh, unprepared, locally grown fruits and vegetables to WIC recipients, and to expand the awareness, use and sales at farmers' markets" (USDA/FNS 2003, 1). The Puerto Rican government brought the program to the island in April 2002, although with some difficulties. First, only three of the existing markets could meet FMNP regulations regarding locally grown fruits and vegetables. Moreover, one market, the MOAPR, was not interested in being certified for the FMNP at the time. That left only the Barranquitas and Aibonito markets as prospective participants. At the time the Puerto Rico Department of Agriculture was organizing farmers into production nuclei to facilitate, among other outcomes, the marketing of their fresh produce directly to consumers, but the nuclei were not yet in place (Commonwealth of Puerto Rico 2002).[4] To solve this problem, the government created an itinerant (moving) market involving FMNP-certified farmers islandwide. Approximately thirty-five farmers were certified during FY 2002, and by March 2003 the number was close to seventy-five. On designated dates the market would travel to the selected municipalities, where the WIC clients then could purchase produce with the WIC FMNP coupons already provided by the local office. The Barranquitas and Aibonito markets could participate as well.

During the season, April to November, two or three markets were held weekly in different communities. Sales of imported and processed crops and the participation of nonfarmers were prohibited, but certified farmers were allowed to buy and resell part of the production of uncertified local farmers. Typically, farmers brought their own products—

usually roots and tubers grown in the mountainous interior—and also resold produce from other regions such as fruits from the coast (Carlos Alvarez, personal interview, March 2003). At every market personnel from WIC set up an educational booth with information on the nutritional value of local fresh fruits and vegetables.

The market has been a great success. With a program grant of $1 million in FY 2002, $20 coupons (the annual maximum) were distributed to 62,929 eligible recipients in thirty-eight municipalities (Commonwealth of Puerto Rico 2002, 4). The program grant rose to $1.634 million in FY 2006 (USDA/FNS 2006). Increased funding has enabled the incorporation of other urban areas into the program. The promising results with the WIC FMNP prompted the government to bring in the Senior Farmers' Market Nutrition Program for FY 2003. In 2006 the Senior FMNP alone contributed close to one million dollars in use at farmers' markets. The government hopes that guaranteed demand for several years will stimulate producers to increase and diversify their plantings, actions that would support more permanent year-round farmers' markets. In addition, the government has been successful in gaining certification for several new production nuclei as prospective farmers' markets. The principal challenges, at this point, are to get enough produce into the market to satisfy demand and establish farmers' markets on a more independent footing in Puerto Rico.

Analysis of Farmers' Market Information

While these markets differ in important ways, common themes emerge. In short, the new farmers' markets of Puerto Rico are promising but fragile, depending on the abiding commitment of a small number of organizers and for most, the WIC coupons that subsidize market purchases. The MASS, which enjoys long-standing government support, is the only market profiled here that does not experience staffing as a constraint. The organizers of the other markets are carrying a considerable burden that probably is not sustainable over the long-term. In addition, it is clear that at this point the FMNP plays the central role in enabling these markets to survive, a situation much less evident in other places under the U.S. flag. When the Aibonito farmer sought to replicate the model in a rural area, he found that the WIC FMNP was needed to make it possible. The markets that do not depend on WIC funds, the MOAPR and the MASS, are the ones where farm products are relatively deemphasized, accounting for only 20 to 30 percent of vendors.

Overall our analysis shows that nutritional assistance programs have played a major role in enabling the emergence of farmers' markets on the island, not because the funds increase the food purchasing power of Puerto Rican consumers but because the structure of those programs includes a minor, although exclusive, role for farmers' markets. Whereas farmers' markets in the United States succeed by serving relatively well-off consumers seeking an alternative to the globalized food system, most of the markets in Puerto Rico depend heavily on a small part of the system that undergirds the whole food system: nutritional assistance. Thus the emergence of farmers' markets in Puerto Rico is not an example of food system concentration creating small-scale economic opportunities on the margins. While the dual trends of concentration and innovation in the food sector in Puerto Rico seem to mirror similar trends in the United States, in reality they are driven by different forces.

In predicting the long-term impact of recent trends for farmers and consumers in Puerto Rico, it is important to recognize that nutritional assistance programs have determined the broad outlines of the retail landscape for decades. During the decades when the subsidies were administered as cash transfers with minimum spending restrictions, Puerto Rico became a heaven for all types of retailers. Industry sources report that by 2001 the island hosted the highest revenue stores for Kmart, J.C. Penney, and Sam's Club, and local Wal-Mart shops were among the chain's highest in revenue per square foot (*Puerto Rico Herald* 2001). Later, the 2001 introduction of the debit card mechanism and the 75/25 rule in the PAN program seem to have boosted the sales of food retailers, consequently drawing global retailers to Puerto Rico's food market. Now, with the introduction of the WIC FMNP and the Senior FMNP, nutritional assistance is contributing to the diversification of retailing.

It also is important to note that things will continue to change. Wal-Mart's agreement with the Puerto Rico Justice Department to continue purchasing local products is only for a term of ten years. In general, global retailers are much more likely to articulate with global suppliers, even for perishables (Hendrickson et al. 2001). Similarly, the WIC FMNP program in Puerto Rico is also based on a temporary (five-year) agreement. While there is no reason to think that the program will not be renewed, it may not last forever.

Does our analysis suggest that farmers and other organizers should abandon their efforts to create long-term viable markets independent of social subsidies? Certainly not. It may be that as the globalization of food retailing continues to constrain marketing opportunities, consumers drawn to markets sustained by the FMNP will become accustomed to and appreciative of high-quality Puerto Rican grown products. The resurgence of fresh produce markets as a cultural practice may help these markets gain a more independent footing over the long-term. In that time, as food retailing becomes more concentrated, the supermarket price of produce may rise to the point that farmers' markets are price competitive. The FMNPs may actually function as a kind of business incubator for farmers' markets that will be in place to attract more consumers if the global system no longer meets their needs.

Puerto Rican policy makers are concerned about the island's dependence on imports and subsidies. While food security became essentially a nonissue in Puerto Rico since nutritional subsidies gave virtually everyone access to an adequate food supply, the debate aroused by the Wal-Mart/Amigo deal brought about an incipient public awareness of the precariousness of such extreme dependence on a small number of major distributors purveying nearly all imported foods. Similarly, while the nutritional assistance programs are not immediately threatened, the block grant that funds that PAN program is a fixed value; meanwhile, local inflation continues to grow. The increasing cost of food and energy provide a solid rationale for formulating public policies that support local production and stabilize the conditions that make it possible.

The local Department of Agriculture recently has embarked on a series of such policies, designed to reinvigorate Puerto Rican agriculture. One key initiative is organizing commodity subsectors into a structure in charge of solving input problems and coordinating the sector's production in order to meet the market's demand. While the new initiatives seem headed in the right direction, they have not yet been fully implemented and may be too limited in scope to alter important declines in agriculture and make a significant difference for small farmers (Carro-Figueroa 2002). At present the expanded demand for fresh produce generated by the itinerant WIC market is being met by those producers with bigger transport facilities and more laborers to harvest and help in the market operation. Many of these midsized and large farmers are also the suppliers of local supermarkets and thus are in a better position to benefit from the current agreement with Wal-Mart.

If demand for local foods grows but local production is unable to expand to meet these new marketing opportunities, these new production and marketing arrangements may be weakened. Small farmers, who comprise a majority in the sector, could make the difference in the success of these projects, but they still face major obstacles in getting involved. The problems, including a lack of appropriate technology for production and adding value, poor access to capital, little support for organic practices, and a lack of research and education targeted to farmers' needs, require a more comprehensive approach. The creation of new marketing possibilities will not by itself ensure the viability of small-scale farming in Puerto Rico. The measures taken in the next five to ten years to address these issues will make a big difference in whether or not the new government initiatives will fulfill their potential to move Puerto Rico to a more vibrant and sustainable food system.

Notes

1. While most Puerto Rican towns have had since Spanish colonial times a central market, or *plaza del mercado*, these institutions were never fundamentally farmers' markets at which producers were involved in the retailing of their goods directly to consumers.

2. Programa de Asistencia Nutricional (Nutritional Assistance Program); the acronym PAN spells the word for "bread."

3. While regulations specify that 100 percent of the allotment should be spent on food, once the cash portion is retired, there is no way to trace how it is spent.

4. The nuclei are organizational structures in charge of solving input problems and coordinating and tailoring the sector's production to meet the market's demand.

References

Berrios-Figueroa, H. 2002. "Judge: Rodríguez engaged in 'vindictive prosecution.'" *The San Juan Star*, December 11.

Caribbean Business. 2002. "Amigo: Sale to Wal-Mart won't create monopoly." November 7. http://www.puertoricowow.com/html/cbarchivesdetail.esp (last accessed April 16, 2003).

Carro-Figueroa, V. 2002. "Agricultural decline and food import dependency in Puerto Rico: A historical perspective on the outcomes of postwar farm and food policies." *Caribbean Studies* 30 (2): 77–107.

Commonwealth of Puerto Rico. 2002. *WIC Farmers' Market Nutrition Program, Commonwealth of Puerto Rico, Plan of Operations*. Unpublished government document. Personal copy.

Coyle, W. 2006. "A Revolution in Food Retailing underway in the Asian-Pacific Region." *Amber Waves* 3:22–29.

Gigante, L. 1999. "Simply put, a local retail revolution occurred in the past decade." *Caribbean Business*, December 30. Archive detail: retail. http://www.puertoricowow.com/html/cbarchivesdetail.esp (last accessed April 16, 2003).

———. 2001a. "Food for thought." *Caribbean Business*, August 23. Archive detail: retail. http://www.puertoricowow.com/html/cbarchivesdetail.esp (last accessed April 16, 2003).

———. 2001b. "Independent supermarkets and 'colmados' bankruptcy filings up 58%." *Caribbean Business*, April 12. Archive detail: retail. http://www.puertoricowow.com/html/cbarchivesdetail.esp (Last accessed April 16, 2003).

———. 2001c. "PAN amendments take effect Sept. 1". *Caribbean Business*, July 19. Archive detail: retail. http://www.puertoricowow.com/html/cbarchivesdetail.esp (Last accessed April 16, 2003).

Guptill, A., and J. L. Wilkins. 2002. "Buying into the food system: Trends in food retailing in the U.S. and implications for local foods." *Agriculture and Human Values* 19:39–51.

Hendrickson, M., W. D. Heffernan, P. H. Howard, and J. B. Heffernan. 2001. *Consolidation in Food Retailing and Dairy: Implications for Farmers and Consumers in a Global Food System*. Report. Washington DC: National Farmers Union.

NACLA Report on the Americas. 1981. "Tax exemption: Basic terms." NACLA 15 (2): 33–34.

P.R. Departamento de Salud. 2004. *Caseload Management Report* and *Client Public Assistance Report*. Programa Especial de Nutrición Suplementaria para Manejo de Infantes y Niños 1-5 años WIC. Oficina de Divulgación y Promoción. Unpublished.

P.R. Junta de Planificación. 2003. *Censo de Población 2000*. http://www.censo.gobierno.pr/censo_de_poblacion_vivienda.htm (last accessed April 30, 2006).

Puerto Rico Herald. 2001. "Dense Puerto Rican market and American influence yield bounty of retail potential for WalMart." June 1. http://www.puertorico-herald.org/issues/2001/vol5n24/walmart-en.shtml (last accessed May 7, 2006).

Riley, H., C. Slater, K. Harrison, J. Wish, J. Griggs, V. Farace, J. Santiago, and I. Rodríguez. 1970. *Food Marketing in the Economic Development of Puerto Rico*. Research Report No. 4. East Lansing MI: Latin American Studies Center, Michigan State University.

Rosa, T. 2002a. "White Rose-Grande deal temporarily halted companies still doing business together." *Caribbean Business*, October 10. Archive detail: retail. http://www.puertoricowow.com/html/cbarchivesdetail.esp (last accessed April 16, 2003).

———. 2002b. "The debate about Wal-Mart's impact on the local economy rages." *Puerto Rico Herald*, Caribbean Business, October 10. http://www.puertorico-herald.org/issues/2002/vol6n41/cbwalmart-en.shtml (last accessed April 9, 2006).

Ruiz, A. L., and P. Choudhury. 1978. *The Impact of the Food Stamp Program on the Puerto Rican Economy: An Input-Output Approach*. Temas sobre Economía de Puerto Rico. Serie de Ensayos y Monografías (7). Universidad de Puerto Rico, Facultad de Ciencias Sociales.

Saporito, B. 2003. "Can Wal-Mart get any bigger?". *Time*, January 13.

USDA/FNS online WIC. 2003. *What is the WIC Farmers' Market Nutrition Program?* FNS online. WIC Farmers' Market Nutrition Program. http:www.fns.usda.gov/wic/fmnp/fmnpfaqs.htm (last accessed April 9, 2006).

———. 2006. *Grant Levels by State*. FNS online, WIC Farmers' Market Nutrition Program. http://www.fns.usda.gov/wic/fmnp/fmnpgrantlevels.htm (last accessed April 9, 2006).

Weisskoff, R. 1985. *Factories and Food Stamps*. Baltimore: Johns Hopkins University Press.

14. The Lamb That Roared: Origin-Labeled Products as Place-Making Strategy in Charlevoix, Quebec

Elizabeth Barham

Driving north along the St. Lawrence River from Quebec City to their rural village in the spring of 1994 (see map 3), four members of the Agro-tourism Roundtable of Charlevoix (Table Agro-Touristique de Charlevoix) excitedly discussed the interview that had just taken place with the French consul. He had listened to them quietly, seeming rather aloof, and they were uncertain as to his reaction. The group had asked him to bring experts from France in order to help them create in Quebec a label of origin system like the *appellation d'origine contrôlée* (AOC) system that protects traditional regional products in France (Barham 2003).[1] They needed a legal mechanism to stop the theft of the trade name of their product, the lamb of Charlevoix (*l'Agneau de Charlevoix*), which had become so well known that restaurants in Montreal and Toronto were listing it on their menus, even when in fact they had never purchased any. The last straw was a menu that a friend brought back from a restaurant in Paris that falsely claimed to serve l'Agneau de Charlevoix. Something had to be done. Could French models such as the AOC system designed to protect product names associated with places be adapted to the province of Quebec? Would the consulate bring someone to guide them in creating such a system?

Several years later during my first research visit to Charlevoix, the leader of the Agro-Tourism Roundtable, Lucie Cadieux, described that first meeting with the consul for me. She shook her head, laughing (an energetic, gregarious, optimistic woman, Lucie is always laughing): "He must have thought we were ridiculous! 'This little band of hayseeds driving down from the back country to propose a new set of laws

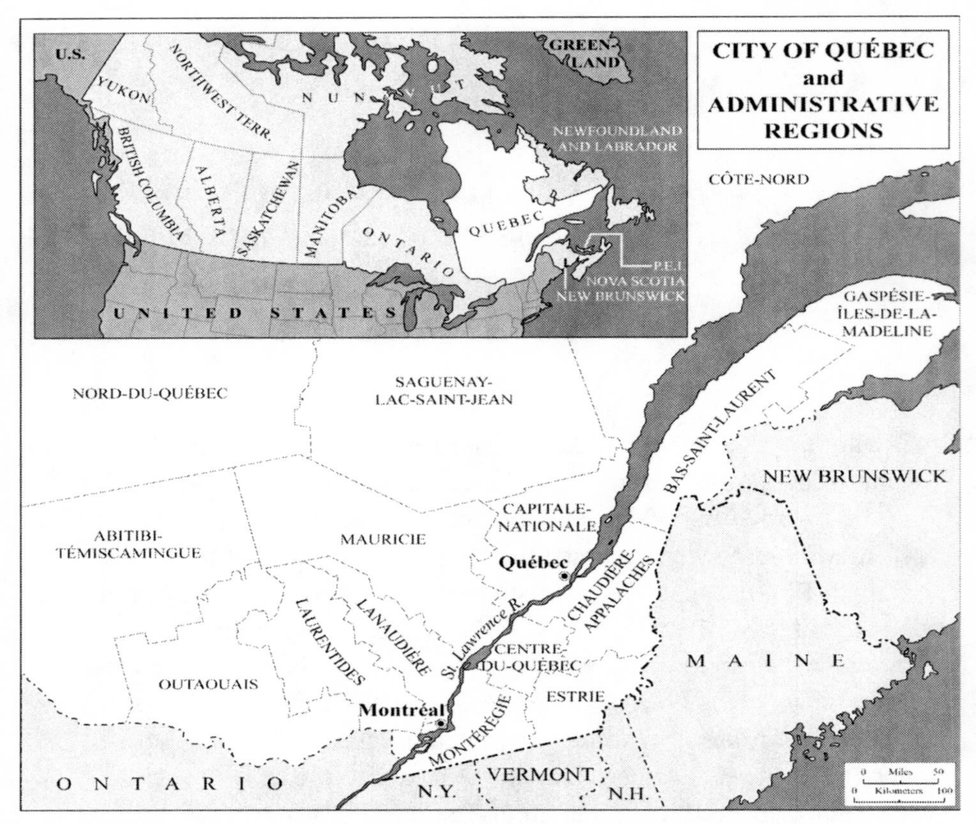

Map 3. The province of Quebec, Canada, showing the Capitale-Nationale (National Capital) region where Charlevoix is located. *Note: Map by Andy Dolan, Department of Geography, University of Missouri–Columbia*

for Quebec. Of all the nerve!' The more we thought of it, we couldn't stop laughing. We just laughed until we cried, thinking about how we must have looked to him. We were sure he would never call us!"

However, they were wrong. After their visit the consul had called them back. Furthermore, he had arranged for someone from the French Ministry of Agriculture to visit them in order to evaluate the situation. By the summer of 1994 a French agronomy student had been sent to Charlevoix to prepare a socioeconomic profile of the region. She had also begun the process of explaining to the members of the Roundtable how labels of origin were officially recognized in France and how the systems associated with the labels were administered. Labels of origin did not exist in Quebec at that time so the roles of producers and government agencies that stood behind the labels needed explanation.

In December 1995 top food-labeling experts from France traveled to Charlevoix to work with Roundtable members. Notably among them was Bertil Sylvander, head of a French research unit dedicated to quality products.[2] With their help and support from Quebec's regional offices of industrial development and the Ministry of Agriculture, the Roundtable laid plans for a conference in the spring of 1996 entitled "Protecting Product Origin and Guaranteeing Quality: The Future of Regional Food and Farming."[3] The purpose of the conference was to launch the first AOC system in North America.

Presentations at the conference detailed the politics and economics of quality food protection in Europe and France. European experts carefully explained the steps that would be required to build and sustain profitable markets for label of origin foods. Participants included representatives from Quebec's Ministry of Agriculture, Fisheries, and Food,[4] members of the powerful Agricultural Producers Union of Canada,[5] and academics from the University of Laval in Quebec City.[6] Members of the Agro-tourism Roundtable shared the history of their group, their vision for the agricultural future of their region, and the institutional barriers that they were encountering. In the end France agreed to lend its help to the effort to create a place-based labeling system in Quebec.

The conference set in motion an effort that would stretch across more than ten years and would be representative of a particular kind of local food endeavor described here as place-making, which is defined as the conscious use, construction, and reconstruction of social, historic, cultural, and ecological elements native to a particular location. Place-making is a reflexive effort to simultaneously preserve desirable aspects of a place and to enhance the economic viability of its inhabitants. It is therefore a kind of place marketing (Hinrichs 1996), but one that does not simply create a surface association with a place through a product in order to build sales. Instead, it reflects a concerted effort to literally create the social and economic basis for claims of uniqueness and place reputation for quality or high value-added products. From this perspective the choice of a local name to label a product indicates

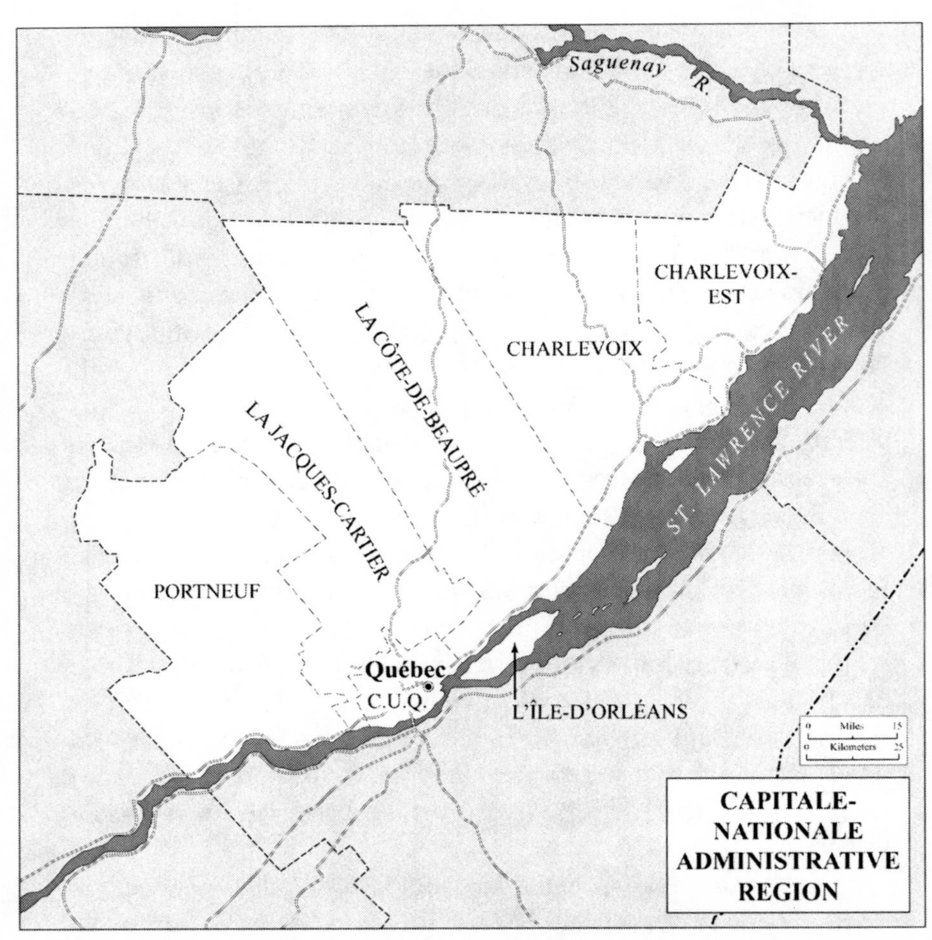

Map 4. The municipal districts of Charlevoix and Charlevoix-Est in the Capitale-Nationale region of Quebec. *Note: Map by Andy Dolan, Department of Geography, University of Missouri–Columbia.*

a commitment to the long-term development needs of the entire local territory because the quality of the product itself will be intimately associated with the place of origin.

There are other local food projects and labels in North America that resemble the approach described here and that generally fit Lyson's (2000) concept of civic agriculture, meaning "a locally-based agricultural and food production system that is tightly linked to a community's social and economic development" (p. 42). But these projects are less formal geographically, legally, and institutionally than a true appellation system. To begin with, awarding an appellation involves establishing the boundaries of production and value-added transformation to the degree that they can be drawn precisely on a map, thereby creating new places out of old. In the case of products with a long tradition of production, as in many European countries, appellation boundaries follow the history of production. As a result they typically delineate ecological and cultural rather than political boundaries, reflecting where a product could be made based on local inputs and attributes as well as on the traditional know-how of the inhabitants.

For example, Charlevoix as a political entity is made up of two adjoining municipal districts, Charlevoix and Charlevoix-Est in the Capitale-Nationale administrative region of Quebec (see map 4). But the label of origin for Charlevoix lamb actually will refer to a new place-within-a-place, bounded by the area of production of the lamb and reflecting the fact that the lamb must be raised within this area to carry the label.[7] This new area will be associated by name with the broader historic region that is Charlevoix in the mind of the public.

Another crucial difference between the place-making effort of Charlevoix and that of less institutionalized local food systems is that, when the work of the producers is completed, the label carrying the region's name will in fact be a new piece of intellectual property, both monitored and protected by the government. It will be available to be used by producers in the region willing to meet certification standards for their product. The producers as a group will have a large role in setting the standards, but control and use of the label will not belong to any single producer or, even to the group itself. It will in essence belong to the region, to be administered and protected by the state, in this case at the level of the province of Quebec because the law establishing the label of origin system covering Charlevoix was passed at the governmental level of Quebec Province. Similarly, laws recognizing labels of origin in Europe exist at the level of individual countries or states such as France or Italy. Certain of these laws are in turn recognized at the level of the European Union.

The result of the linkage of producer organizations to government protection of place-related trade names is a form of intellectual property protection that is of particular importance for small and midsized producers. While these producers may wish to export their products across state borders, they cannot be expected to bear the costs both in time away from their farms and in money that would be required to protect their product's name in international courts of law if it is abused or usurped. A label of origin system backed by the state through the Ministry of Agriculture removes this need when trading with other countries that respect such systems. If a name is misused, it is the state that intervenes. The story of Charlevoix's producers meeting the challenges demanded in the creation of this particular form of local food distinguishes their place-making efforts and becomes a remarkable example of how people in local places experience the global and in turn act upon it.

From Place Marketing to Place Names as Intellectual Property

The research that led to Charlevoix had begun as an investigation into the usefulness and prevalence of place-based labels for marketing local products in the United States. Research had been carried out on several cooperative place-based marketing efforts in Missouri, but it had revealed that producers did not seem concerned about defining the actual geographic area covered by their labels with any specificity. They were opting instead for the name of a nearby rural town ("Taste of the Kingdom" for Kingdom City), a general region of the state ("Northern Missouri Pecans"), or a local landscape depiction ("River Hills"). Similarly, the organizational frameworks or institutions being created for ongoing support of these marketing efforts were loose and informal, with no explicit connections to other entities that might prove supportive such as state government agencies like the Department of Agriculture. While this method certainly held some advantages, particularly in terms of the demands on producers' time required for start-up marketing and openness of the network created, it also might mean that joint marketing efforts could prove fragile over time.

The research in Missouri was complemented by an examination of labels of origin in France (Barham 2003) that provided a sharp contrast to the somewhat amorphous U.S. examples. French label of origin systems were quite institutionalized and dated back to the early 1900s.

Territories covered by the labels were precisely defined on the map, and producers organized themselves in relation to a shared label for a number of tasks, including establishing and enforcing quality standards for their production. Producer organizations were directly linked with the French state through the Ministry of Agriculture, which coordinated producer organizations at the national level, provided backstopping for some areas of label establishment and certification, and took charge of the legal registration of label names.

By the time Quebec was added as a case study location, the importance of the legal aspects of label of origin recognition and the role of the state had become more evident. I was searching specifically for a group of producers who were knowingly adapting the stronger, more European form of place labeling to a North American context. Charlevoix turned out to be the only location where an effort of this kind was underway. I also wanted to observe, if possible, how such place-labeled products would be treated under the North American Free Trade Agreement (NAFTA), and the proximity of this region to the U.S. border made international trade a marketing possibility in the future.

Initially, the lamb producers of Charlevoix, along with the Agrotourism Roundtable that supported them, had not really suspected the scope of the project that they had embarked upon. The type of label of origin scheme they envisioned—one legally protected by the government or state rather than by individually registered trademarks—was not only unknown in the province of Quebec, it was virtually unknown in all of North America except to certain specialists in international trade law. Furthermore, the United States, Canada's largest trade partner, objected strongly to this type of place-based labeling. Instead, the United States had taken a position in the World Trade Organization (WTO) in favor of an international trading system based on trademarks (Goldberg 2001). The Canadian government had joined the United States in this position.

At issue is a particular form of intellectual property protection that has become one of the most contentious unresolved trade issues between the United States and the European Union. For several decades these superpowers have been locked in dispute over protection of labels of origin, known in international trade as geographical indications, or GIs.[8] To most Europeans the dispute is a David and Goliath story of rich and powerful multinational corporations, many based in the United States, taking advantage of the historic reputations of some of Europe's most famous rural regions (Cabot 2003). While the issue is rather well known in Europe, it is barely discussed in North America where label of origin systems are unfamiliar. This ignorance is regrettable because the outcome of trade negotiations related to GIs could have important ramifications for rural regions in the United States, Canada, and Mexico, while the voices of people from these rural areas remain largely absent from the debate.

Given this situation, Charlevoix became emblematic for me of how small community-based efforts to promote local food can encounter and challenge the globalizing trends of the current industrial food system. Discussions and decisions taking place in Geneva could directly impact the chances of success for the lamb of Charlevoix labeling scheme, although the producers had no input into these decisions. This lack of input in turn raised a host of questions that fall under the rubric of local-global relations. Would the benefits of free trade promised by the proponents of NAFTA and the WTO apply to products strongly associated with particular places? Or should producers of such products focus primarily on creating and sustaining strongly localized markets and eschew trading at longer distances? The name of Charlevoix lamb had already crossed the ocean to France. Would the producers of the real Charlevoix lamb ever be able to benefit from the international reputation that they had built for themselves?

Origin-labeled products in Europe travel across national borders, but at least within the European Union producers are protected from the misappropriation of their product's name by their own state and by the European Union itself. It appeared this trade option would be closed to place-associated products in North America unless the producers were wealthy enough to hire lawyers with international trade expertise to file trademarks for them in any country with which they did business. Producers also would have to pay to file suit against trademark violators in other countries. In other words, international trade discussions underway at the WTO could lead to a system suitable for large corporations but not for small-scale production in rural communities.

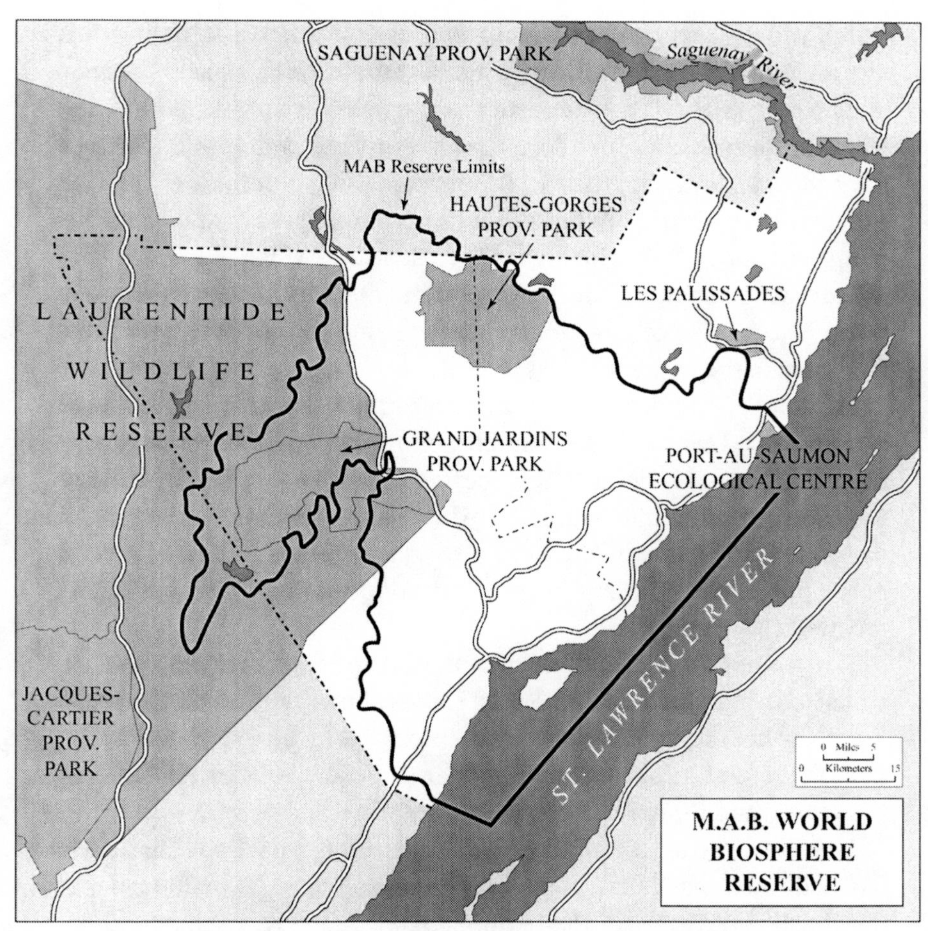

Map 5. Man and the Biosphere Reserve that encompasses much of Charlevoix. *Note: Map by Andy Dolan, Department of Geography, University of Missouri–Columbia.*

"Condemned to Quality"

When the lamb producers of Charlevoix approached the French consulate for help with their labeling efforts, they were not aware that they were pulling strings that would draw them into interaction with the international trading system. They wanted to survive, and they wanted to help their region face difficulties shared by many marginalized rural areas around the world: loss of local agriculture due to the pressure of agricultural concentration and international competition, a declining and aging population, difficulty in maintaining local services, and a struggling local economy (Perron and Gauthier 2000). Charlevoix once had been the site of a thriving dairy industry, but every year more farms went out of business. Two local packing houses and a milk distributor had shut down. In such a far northern climate public services are expensive to maintain. To become more competitive in a rapidly globalizing economy, Quebec had actually closed some villages in the 1960s, withdrawing support for utilities, road maintenance, and communications. The specter of this kind of rural decline haunts the region's inhabitants in the winter months.

Of course, Charlevoix has many resources, some of which have proven critical for the success of label of origin schemes in other countries. For example, because it is located largely within a Man and the Biosphere reserve and within proximity of several other provincial parks, the very name of Charlevoix evokes images of beautiful mountain landscapes and outdoor recreation opportunities in the mind of the public (see map 5).

The lower flanks of the mountains near the St. Lawrence River are still strongly marked by Charlevoix's agricultural traditions, with rolling fields interspersed amidst dense forests. And the region shares the rich history and colorful cultural influence of Quebec's French heritage (Bouchard and Courville 1993; Perron and Gauthier 2000).

However, in the face of progressive agricultural decline, Charlevoix has had to fall back on its environmental and cultural attributes, and they have made tourism the region's major industry, which accounts for more than 15 percent of overall economic activity (Jutras and Simard 2000). The Agro-tourism Roundtable was organized to take better advantage of this trend, and the viability of the Charlevoix label of origin may depend on it. But an economy overly dependent on tourism has its downside. The jobs it creates are seasonal and often do not pay well. Most visitors come during the summer, making it hard to accommodate the sudden influx. Additionally, winter unemployment in the region can range as high as 33 percent. In this context, building the profitability of local products appeared to be critical, and thus a label of origin might have an important impact.

I formed my first personal impression of Charlevoix in March 2000, driving north from Quebec City after dark to meet the local leaders of the Roundtable. The road was flanked by long stretches of fields and forests covered in three feet of snow, and more was falling. An occasional road sign indicated a moose crossing. The winter had been particularly cold and temperatures were still well below freezing during the day, dropping to 20° Fahrenheit at night. I passed several villages and finally arrived at my destination of Les Eboulements, whose center consisted of a few blocks of tidy homes and shops and a handsome Catholic Church, all built close to the road for access in the snowbound months of the year.

I knew from the map that the St. Lawrence River was off to my right, but I could not see it in the dark. Salt water from the Atlantic mixes with fresh water at this point and the river widens, making it possible to view whales nearby in the tourist season. Brochures I had seen carried pictures of the summer river view from the town—picturesque older homes and barns, green fields sloping down to cliffs at the river's edge, and shining blue water—the Charlevoix that attracts numerous artists. But it was hard to imagine a warm version of Les Eboulements on the night that I drove in, windshield wipers batting at the snow.

The degree of personal investment and self-reliance required to make a farm profitable throughout the year in such a difficult climate has earned the Quebecois their reputation as a hardy and tenacious people. Perhaps these characteristics lent energy and determination to the small group of local inhabitants who decided in 1993 to form the Agro-tourism Roundtable. They organized themselves as a nonprofit organization, bringing together local chefs and restaurant owners, producers of a variety of farm products such as lamb and vegetables, and makers of value-added specialty foods such as aged cheeses and smoked trout. All of their products were based on local inputs. A local chef, Eric Bertand, acted as first president for the group. He was of French origin and so brought a French perspective on the connection between quality food and local agriculture to the group. Along with grant-writing skills, Lucie Cadieux contributed research-based knowledge of the local economic situation gained through her job as a farm management counselor.[9]

The Roundtable members had shared their views and reached the same conclusions. Producers in Charlevoix were faced with a progressive decline in conventional agriculture. They already were already handicapped in terms of commodity production by their harsh climate and difficult soils, and NAFTA was pulling down trade barriers and intensifying competition from the United States for commodity markets. In addition, the government was reducing or eliminating aid to farmers. The farmers in the group saw that a continued attempt to compete on this basis would only exhaust the region and its people and lead to more depopulation and abandoned fields. Roundtable members from area restaurants didn't want further farm decline either. They preferred to buy quality items locally and knew their customers wanted foods from the region. Furthermore, tourism depended to some extent on the Charlevoix agricultural landscapes, which would deteriorate if fields were no longer cultivated and became overgrown. What the Roundtable thought was needed were new and viable family farms engaged in diversified production.

It seemed that the only solution was to transform the natural conditions that were their handicaps in the world of conventional agriculture into assets in a new vision based on exceptionally high-quality local products. As Lucie Cadieux put it, "We realized that we were condemned to quality—there was no other way out!" Like some traditional maple sugar producers in Quebec described by Hinrichs (1995), the farmers of Charlevoix were headed off the technological treadmill dictated by conventional agriculture to begin pursuing a more sustainable form of regional development.

Their new shared vision rested on three key goals, collected under the banner of quality regional products. First, they would reorient and consolidate a new agricultural sector based on specialties of the region. Such initiatives already existed within the group for lamb, veal, snails, hot house tomatoes, honey, wild boar, venison, baby vegetable production, smoked trout, butter, artisanal cheeses, and fruit liquors. Second, they would reinforce the regional tourism economy by building off of the existing gastronomic reputation of Charlevoix and diversifying the tourist offerings. By working closely with chefs to feature local products on restaurant menus, the Roundtable would give visitors to the region a chance to learn about local offerings. New opportunities to spread this awareness would be promoted through tastings, farm visits and farm stays, and direct purchasing at the farm and from other local businesses. Finally, this new agrifood economy of Charlevoix would be able to directly counterbalance the seasonal and low-paying work associated with tourism by creating opportunities for new young farmers to become established and for existing conventional farmers to diversify.

One important aspect of this new vision was the plan for an integrated agrifood center in Charlevoix that would combine test kitchen, food-related library resource, and food science laboratory along with a restaurant. It would serve several purposes: training for producers and makers of value-added products and for a variety of skills needed to support this effort (butchers, bakers, and so on), all in close relation to training for young chefs and wait staff. There would be ongoing relations with university departments to help them identify new breeds of animals and plant varieties needed for their products as well as new modes of value-added production such as aged cheeses. The center would incorporate a laboratory for testing soil and water, plants and animal feeds, and safety and quality of final products, all needed in support of certification procedures and quality control. It also would provide technical information and support to local producers and product developers. Several small treatment facilities for smoking and salting as well as a distillery would be included. An on-site store featuring local products would further enhance the marketing aspect, along with a strong commitment to certification procedures and controls that would guarantee the quality of the products and encourage continual improvement.

The Roundtable went into action by creating the Route des Saveurs, or the Charlevoix Flavor Trail. A regional driving itinerary linking together farms and restaurants, the Route des Saveurs opened a world of quality products and on-farms visits to tourists (see map 6). Participating establishments, open primarily in the warmer months, display a special logo (a chef's hat in a diamond) to indicate their participation in the network and the special quality of the local products to be found there.[10]

Through the sharing that takes place within the Roundtable and because of the larger regional market that it has helped to foster, local know-how is increasing, and new value-added products are constantly

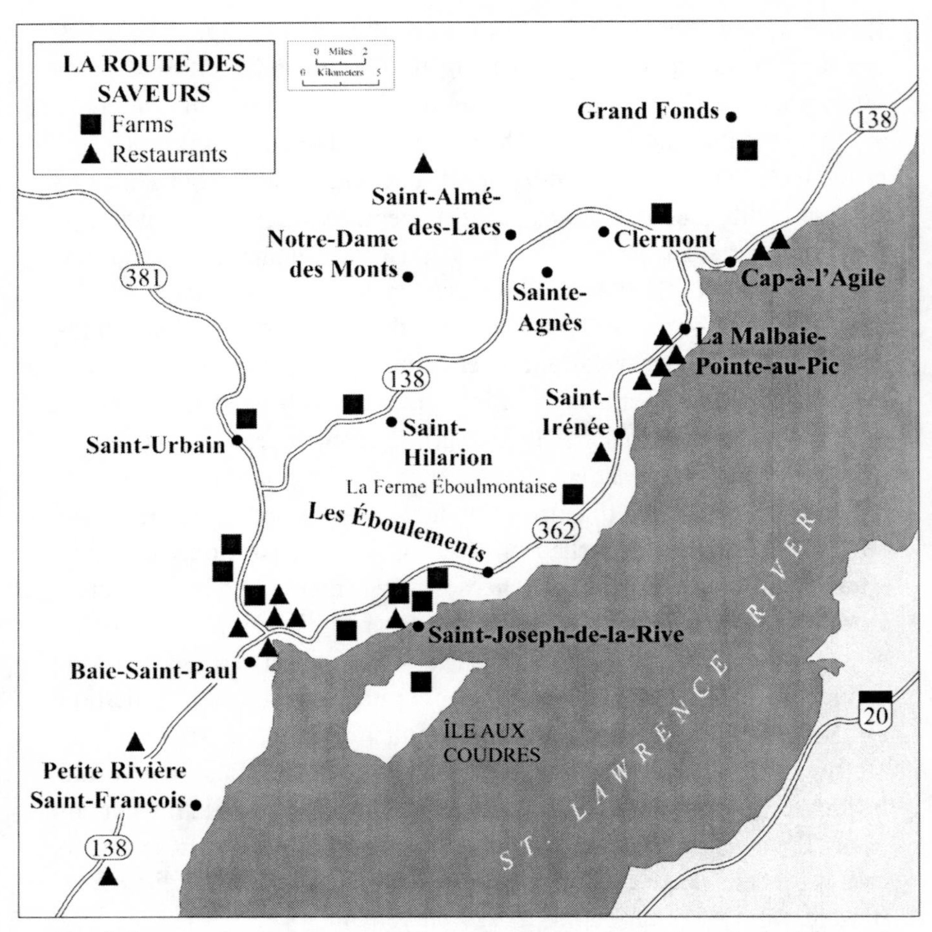

Map 6. Farms and restaurants on the Route des Saveurs circuit. *Note: Map by Dolan, Department of Geography, University of Missouri–Columbia, following a map prepared by Louise Paquin for the Charlevoix Regional Tourism Association.*

under development. In 2001 a restaurant featuring local food was built directly next to the Cadieux-Gagnon farmhouse. Named "Flavors of the Past" (Les Saveurs Oubliées), it is the first "country-style dining establishment" (*Table Champêtre*) of Charlevoix, taking part in an agrotourism system organized by the Federation of Agrotours of Quebec (Fédération des Agricotours du Québec).[11] Lucie obtained a grant to include a cold storage room for meats and other products below the restaurant, along with a butchering room and a test kitchen for developing new regional products. It was the start of the integrated restaurant-test kitchen-certification laboratory of which the Roundtable had dreamed.

As of this writing the lamb producers are in the process of forming a CUMA, a type of cooperative for sharing the use of farm machinery that is based on similar co-operatives in France.[12] As Gertler (2001) has noted, the strong local community and regional ties that are typical of cooperatives position them to both promote and to take advantage of sustainable practices. In Charlevoix this cooperation is further reinforced by very strong place-based identity related to the natural attributes and beauty of the area as well as the shared French heritage. Cheng, Kruger, and Daniels (2003) have recognized that this type of place identity can have positive effects on natural resource decision making, and it seems to be contributing to the ongoing spread of organic farming practices in the region.

In terms of economic sustainability, by 2003 five new family farms producing lamb had been established in the region. The Roundtable had explicitly targeted an increase in diversified small and midsized family farms as one of their goals, and they are proud of the recent growth. While there were no explicit rules prohibiting existing farms in the network from growing ever larger to meet increased market demand, such growth would go counter to the idea of keeping the region alive by attracting and retaining more young people.

The importance of this aspect of the Roundtable's work derives in part from the 1960s attempt by Quebec to slow its population loss by focusing resources on the province's cities and forcibly closing small villages considered to be in decline (Lavallée 1996). The people of Charlevoix had rebelled against this logic, assessing their local strengths and pulling together to reestablish a market niche for the region as a destination for tourism, outdoor enthusiasts, and artists. But as late as 1995 the idea of closing "non-profitable" regions had reappeared in the popular press (Lavallée, 1996, 34). Resisting this reasoning by making regional farm production economically viable as well as socially and ecologically sound was in many ways the crucial motivation of the Roundtable. While more concentrated production might satisfy the need for market profit, it would not meet the social and environmental goals shared by the group.

In line with the emphasis on creating regional synergy, the Roundtable focused much of its early efforts on developing local markets and loyal customers through direct sales. A large percentage of their production is sold to year-round residents. Each tourist season, more visitors return to buy directly as regulars. Clearly, the Roundtable has opened the possibility for some degree of longer distance marketing to urban areas, perhaps eventually even Internet sales, and they welcome this idea. Still the members anticipate that most of these purchases would be made by people who know or have visited the region because the Roundtable is focused on products linked specifically to the region and its resources, specialties that cannot be easily replicated elsewhere and that are generally available in limited quantities due to their family-farm production base. The Roundtable is also targeting consumers who know and appreciate the region. The result is a different kind of comparative advantage that seeks regional development based on resources that are "relatively unique, immobile, innovative and embedded in specific socio-cultural settings" (Jenkins and Parrott 2003, 53).

Fitting Institutions to Purpose

The determination of the Roundtable to make its region live resulted in a particularly sought-after product, the lamb of Charlevoix. But as Jenkins and Parrott (2003) and Parrott, Wilson, and Murdoch (2002) have noted, development models based on cultural, social, and territorial rootedness depend heavily for their success on the legal and regulatory regime that governs them. In this case the creation of a product with high market value initially was not complemented by governmental institutions to protect it. However, when members of the Roundtable traveled to Quebec City to ask the French consulate for assistance, they felt they knew what they needed. As Eric Bertrand put it at the conference held in 1996, "The [Round]Table will participate in developing new laws to protect the image and use of their name once in distribution on the market. This is not a protectionist measure nor a special favor, but rather a guarantee that one can make a living from one's work and reap the benefits of one's own labor faced with individuals or systems who only seek their personal profit. Models for this protection exist, as in other areas such as the arts and technological innovation, to inspire us."

While Bertrand was correct that models existed, the challenge of fitting them to the context of Quebec was in reality an enormous one for the Roundtable. The goal, of course, was to establish the first officially recognized label of origin in North America to be administered much like the AOC. Barham (2003) has traced the organization of AOC label of origin systems within France, and Parrott, Wilson, and Murdoch (2002) have examined the interplay of different state labeling systems in the context of the European Union.

Along the way to the Roundtable's goal there were laws to write and pass at the provincial level of Quebec government that would officially recognize labels of origin to be protected by the state. Government offices needed to restructure or create new departments in order to take on functions associated with protecting the labels as well as to oversee the accreditation of regionally specific products. Private organizations had to be identified and trained to take on the task of certifying products that might carry the labels. And naturally the producers themselves would have to define quality thresholds for production, standards to which they would be held if they intended to use a recognized regional name in their marketing. In other words, several public and private agencies had to be reworked or expanded, some new ones had to be created, and clear lines of authority had to be established in order to administer the new system. As these changes took place, questions had to be resolved about how each office would link with various preexisting authorities so that new functions could work properly with old, and at all levels from the municipal to the provincial and national and ultimately to the global.

In the end by pursuing a state-recognized label of origin, the Roundtable members had stumbled into the complex task of developing a new claim of identity in the form of intellectual property. It would make it possible for their collective work to be seen and recognized at the global level. Their strategy had brought them to the concept of local-global relations described by Dirlik (2001) as "place-based imagination," one that emphasizes how places attach to the global. Rather than seeing the local in juxtaposition to the global, as two universal opposites, he argues that place is place in relation to the global and has to attach to it. The question is, on what terms? In Dirlik's view both the local and the global are better understood as processes. The global implies connection to "change-bringing networks" (p. 16) that surpass the boundaries of nation states and can call national dominance over the future of a territory into question. However, the local can in turn coalesce into "a project that is devoted to the creation and construction of new contexts for thinking about politics and the production of knowledge" (Dirlik, 2001, 16). Such projects can become unexpected sites of resistance to homogenizing forces within the global, depending on the nature of their attachment to it.

Following Dirlik, the labeling efforts unfolding in Charlevoix provide good examples of how a local territory discovers a new way of attaching to the global. Success for the lamb producers depends on international trade negotiations, but the producers, with and through the Roundtable, can also influence those negotiations. As they formulate their labeling scheme, they speak to the need for recognition of regional specificity in the context of global trade. Recognizing the lamb of Charlevoix as a new form of intellectual property attaches the region to international trade systems in a way that goes beyond the exchange of material goods and touches the realm of shared meanings and values. The label asserts the meanings associated with the cultural understandings and practices of the producers and is shot through with their own experience of their place. To truly appreciate such a product and what its label means, the consumer would want to visit, know, and experience the place itself and become tied to it in some respects.

Thus place-making in the institutional sense discussed here is one logical response to the impacts of globalization on marginalized rural economies. It provides the possibility for nonmarket aspects of local places to be protected and valued as well as providing a modicum of economic and legal protection to that place's products when they are placed in a market economy. It helps retain capital in local places by reintegrating social, economic, and environmental goals and values and tying them to locality. Achieving this reintegration of market and nonmarket values is widely accepted as the basis for sustainable agriculture and for sustainable development more broadly.

Returning to the question posed earlier of whether free trade can benefit a region like Charlevoix, it would seem imperative that those concerned with the fate of rural places in North America consider the discussions and negotiations underway at the WTO relative to geographical indications. If the European approach to GIs is followed, place-based labeling could become a more viable regional development alternative, perhaps making an important contribution to Third World development. If the U.S. position wins the day, rural people everywhere, but particularly those in America, will have greater difficulty using this development tool. The ramifications go far beyond North America and Europe and touch on whether or not globalization will be able to accommodate the dreams and aspirations of local places in all their diversity. The creation of a global system of recognition for place-based labels would appear to be one way to move in that direction by helping regions organize around their specific strengths and use elements of their past achievements to open the door to a brighter future.

Acknowledgments

The author would like to thank the University of Missouri Research Council (grant URC-00-048) as well as the European Union Center at the University of Missouri for their support of this research. She also would like to thank Clare Hinrichs and the anonymous reviewers for their helpful comments on earlier versions. In particular, she is grateful to Janice Astbury, formerly of the Commission for Environmental Cooperation; to the Cadieux-Gagnon family; and to so many others in Charlevoix who in contributing their time and support made this research possible.

Note that terms in French, as well as quotations translated from the French, are represented in the text by italics. Translations are by the author.

Notes

1. *Appellation d'origine controlée* literally means that the name of the place of origin of the product is controlled, in this case by the state.

2. UREQUA (Unité de Recherches Economiques sur les Qualifications Agro-Alimentaires, or Economic Research Unit for Quality Food Products), located in Le Mans, France.

3. Protection d'Origine et Garantie de Qualité: L'avenir de l'agriculture et de l'agroalimentaire regional, held in Charlevoix, Quebec, April 3–4, 1996.

4. MAPAQ (Ministère de l'Agriculture, des Pecheries et de l'Alimentation du Quebec).

5. UPA (Union des Producteurs Agricoles).

6. The University of Laval shares aspects of the land-grant universities of the United States. The academics involved were primarily agricultural economists.

7. The kind of protection discussed here applies to food products that are made in an area using only inputs found in that area. It would not apply, for example, to craft items made with materials that came from outside the area.

8. The WTO and the World Intellectual Property Organization (WIPO), both located in Geneva, have established special working committees devoted to the topic of geographical indications and their role in the trade economy. Helpful documents on global trade negotiations related to GIs are available from their respective Web sites: http://www.wto.org and http://www.wipo.org (both sites last accessed April 29, 2006).

9. Lucie's position is funded by the Agricultural Council of Charlevoix, a trade union group. The group consists of thirty-four regional agricultural producers, who by virtue of belonging to a recognized Council are eligible to receive government assistance to hire an agrieconomist to advise them on farm management questions. More information on Quebec's Agriculture Management Unions in French and English is available at http://www.fgcaq.com/sitefgcaq (last accessed April 30, 2006).

10. An agrotourism link on the Charlevoix tourism Web site, http://www.tourisme-charlevoix.com (last accessed April 30, 2006), provides information on the individual businesses involved and includes a picture of the chef's hat logo.

11. Agricotours also organizes an accredited system of farm visits, farm stands, and bed-and-breakfast stays on farms. See http://www.agricotours.qc.ca (last accessed April 30, 2006).

12. CUMA stands for Coopérative d'Utilisation de Matériel Agricole (see Harris and Fulton 2000a, 2000b).

References

Barham, E. 2003. "Translating terroir: The global challenge of French AOC labeling." *The Journal of Rural Studies* 19 (1): 127–38.

Bouchard, G., and S. Courville. 1993. *La Construction d'une Culture: Le Québec et l'Amérique Française*. Sainte-Foy, Quebec: Les Presses de l'Université Laval.

Cabot, T. 2003. "Naming rights; Is America the home of the free but not of the brie?" *The Washington Post*, May 21, final edition, Food Section.

Cheng, A. S., L. E. Kruger, and S. E. Daniels. 2003. "'Place' as an integrating concept in natural resource politics: Propositions for a social science research agenda." *Society and Natural Resources* 16:87–104.

Dirlik, A. 2001. "Place-based Imagination: Globalism and the Politics of Place." In *Places and Politics in an Age of Globalization*, edited by R. Prazniak and A. Dirlik, 15–51. Lanham MD: Rowman & Littlefield.

Gertler, M. 2001. *Rural Co-operatives and Sustainable Development.*" Saskatoon, Saskatchewan: Centre for the Study of Co-operatives, University of Saskatchewan.

Goldberg, S. D. 2001. "Who will raise the white flag? The battle between the United States and the European Union over the protection of geographical indications." *University of Pennsylvania Journal of International Economic Law*, 22 (Spring): 107–51.

Harris, A., and M. Fulton. 2000a. *The CUMA Farm Machinery Co-operatives.* Saskatoon, Saskatchewan: Centre for the Study of Co-operatives, University of Saskatchewan.

———. 2000b. *Farm Machinery Co-operatives in Saskatchewan and Québec.* Saskatoon, Saskatchewan: Centre for the Study of Co-operatives, University of Saskatchewan.

Hinrichs, C. 1995. "Off the treadmill? Technology and tourism in the North American maple syrup industry." *Agriculture and Human Values* 12:39–47.

———. 1996. "Consuming Images: Making and Marketing Vermont as Distinctive Rural Place." In *Creating the Countryside: The Politics of Rural and Environmental Discourse*, edited by E. M. DuPuis and P. Andergeest, 259–78. Philadelphia: Temple University Press.

Jenkins, T., and N. Parrott. 2003. "The Commodification of Heritage and Rural Development in Peripheral Regions: Artisanal Cheesemaking in Rural Wales." In *Ecolabels and the Greening of the Food Market*, edited by W. Lokeretz, 51–61. Boston: Friedman School of Nutrition, Science, and Policy, Tufts University.

Jutras, G., and L. Simard. 2000. *Rapport Eté 2000, Ferme Eboulmontaise.* Stage Sainte-Foy, Quebec: en Entreprises Agricoles, Faculté de Sciences de l'Agriculture et de l'Alimentation, Laval University.

Lavallée, A. 1996. "Communautés d'adhésion et insertion dans les réseaux mondiaux." In *Le Québec des régions: vers quel développement?*, edited by S. Côté, J.-L. Klein, and M.-U. Proulx, 339-59. Rimouski, Quebec: Université de Québec à Rimouski (Groupe de recherche interdisciplinaire sur le développement régional de l'Est du Québec and Groupe de recherche et d'intervention régionales).

Lyson, T. A. 2000. "Moving toward civic agriculture." *Choices* 15 (3): 42–45.

Parrott, N., N. Wilson, and J. Murdoch. 2002. "Spatializing quality: Regional protection and the alternative geography of food." *European Urban and Regional Studies* 9 (3): 241–61.

Perron, N., and S. Gauthier. 2000. *Histoire de Charlevoix.* Sainte-Foy, Quebec: Les Presses de l'Université de Laval.

15. Be Careful What You Wish For

Democratic Challenges and Political Opportunities for the Michigan Organic Community

Laura B. DeLind and Jim Bingen

Early in January 2003 a meeting took place between members of Michigan's organic food and farming community and selected Michigan State University (MSU) faculty and administrators, the vast majority of whom came from the College of Agriculture and Natural Resources. This meeting was the second in what was expected to be an ongoing dialogue between proponents of organic agriculture and scientists, educators, and extension specialists. The three-hour meeting held at MSU, the state's land-grant institution, was an attempt by members of Michigan's organic community to review what the university had been doing to address their needs and once again to present their philosophies and concerns.

The meeting though polite was not without tension. University researchers and administrators outlined the existing (and imminent) programs and offices that would directly or tangentially impact organic agriculture around the state. Much was said about newly endowed chairs, research grants, dedicated funds, and professional expertise. University representatives also listened while several organic farmers acknowledged positive changes over the last twenty-five years but argued that these changes were neither sufficient nor timely enough, especially in light of the attention given to biotechnology. Organic advocates wanted to see more whole farm and on-farm research, more farmers incorporated into the university's program planning and research oversight committees, and more extension expertise in the area of organic production. They also asked how organic research and programming would fare in light of the state's budget deficit and the dramatic cuts in funding for the agricultural experiment station and MSU Extension. When no assurances were forthcoming, one well-respected, activist farmer argued that organic research should not take a back seat to biotechnology. Rather by strengthening the former, he contended, there would be far less need for the latter. If research on Roundup Ready crops, or Bt corn, or bST continued unabated, he challenged, "then I'm going to review *your* system. I'm going to call you out."

While no one else spoke as forcefully, there was an abiding sense on the part of organic representatives that the university and the state generally lacked any deep commitment to organic agriculture. Such disappointment and skepticism are neither new nor unfamiliar. In fact, it is their continued presence—the continued disconnect between the concerns of the organic community and those of the university—that has led us to ask a number of critical questions. These questions are not directed toward the university community per se but rather toward Michigan's organic community, consumers as well as producers. First, will the state's organic agenda be promoted simply by asking land-grant researchers to devote more attention and resources to organic agriculture? Second, what are the implications for Michigan's organic growers of having organic research become more established at MSU? And third, shouldn't the organic community be thinking more deeply about the type of relationship(s) it needs to cultivate if organic agriculture is to develop in ways that are consistent with its civic promise?

The purpose of this chapter is to use these questions to initiate a discussion about the political condition of organic agriculture within the state of Michigan. To this end we will discuss the way in which traditional organics contrasts with its newer, industrial counterpart, conceptually and pragmatically. Next, we will profile three emerging issues as Michigan organics is incorporated into land-grant research and becomes part of state-level agricultural policy. Finally, we will argue that in the process of reconfiguring Michigan organics to suit conventional scientific practice and the assumptions of economic rationalism, much of value is being lost. Confronting this loss as well as understanding what is happening and why is not only the responsibility of the organic movement, it is also essential for charting a more deliberate and democratically responsive course of action, locally and throughout Michigan.

The Changing Face of Organic

Industrial Organic

U.S. organic agriculture has experienced great changes in the last ten to twenty years. No longer seen as belonging to the counterculture, it has become a visible and viable alternative for mainstream farmers and consumers. In 2001 organic products, both processed and unprocessed, generated $7.7 billion in the United States. While still amounting to less than 2 percent of the national food industry, organic sales continue to grow at a rate of 20 percent a year. By 2010 it is anticipated that organics will represent 10 percent of the U.S. food economy (Organic Consumers Association 2001).

The discovery of organics has been energized by the proliferation of both consumer and farmer organizations, such as the Organic Consumers Association, the Organic Trade Association, the Ohio Ecological Food and Farm Association, and the California Certified Organic Farmers, all determined to improve the image, production, and accessibility of organic food and farming. Their efforts have resulted in small but noticeable increases in USDA Farm Bill appropriations for organic research and programming (Organic Farming Research Foundation 2002) as well as in farmland dedicated to organic production (Halweil 2001). Their efforts also shaped to a great degree the creation of national organic standards that now legally define the O-word, the allowable materials, and attendant processes. As part of these standards third-party certification and an organic label officially protect and attest to the consistency and purity of the organic promise.

Within this climate many food manufacturers and conglomerates have found themselves embracing rather than disparaging organics. As Michael Pollan observed, "agribusiness has decided that the best way to deal with that alternative [organics] is simply to own it" (2001, 32). Today, organic products are sold in discount outlets like Wal-Mart. Traditional food giants like General Mills, Heinz, Nestlé, and Danon have acquired organic farms and product lines. Even at a time when farmers' markets, community supported agriculture, and other direct marketing arrangements are growing increasingly popular, half of all organic sales take place in conventional supermarkets.

However, for all its economic success the new industrial organics has met with considerable criticism. There are those who feel that it has strayed from its ecological and spiritual roots (Anon. 1999; DeLind 2000; Norberg-Hodge, Merrifield, and Gorelick 2002; Pollan 2001, 2003; Reynolds 2000; Youngberg 1996). Not only do they argue that the O-word no longer embodies many of the values that originally gave it definition but that in its new persona it may be recreating the very conditions that it was initially designed to redress. For example, a few, large factorylike farming enterprises such as EarthBound and Cascadian Farms now dominate the industry (Anon. 1999; Hesser 2003; Severson 2002). Smaller organic producers, despite their wealth of diversity, from both a landscape and production standpoint, cannot compete. There is also the concern that by centralizing political authority over the O-word, it becomes easier for agribusiness to control organic standards and rework them to fit capital efficiencies rather than ecological and ethical principles (DeLind 2000; Kittredge 2003; Kneen 2001; Mendelson 2003; Pesticide Action Network 2003). Concerns, likewise, have been voiced that all organic foods are not created equal and that consumer demand for organically labeled products is not necessarily nutritionally, socially, or environmentally responsible—the organic Twinkie™ being a classic example.

The Organics of Place

If "[t]he term 'organic' describes a holistic approach to farming: fostering diversity, maintaining optimal plant and animal health and recycling nutrients through complementary biological interactions" (Halweil, 2001, 26), then the new industrial organics pays lip service to that notion but pulls hard in another direction. In fact, many scholars, farmers, and food analysts have noted that there are now really two organics at work—one small and civic, the other large and commercial (Klonsky 2000; Pollan 2001; Reynolds 2000; Wheeler and Esainko 1997). The difference between these two organics, however, is not merely a matter of adherence to ecologically based production practices. It is also a matter of something far less tangible, something we can call the "conscience of organics."

The recognition that organic agriculture is essentially something other than a commercial endeavor is central to understanding this conscience. As Henderson explained early on, "[w]e [organic farmers] are not an industry—we are a community with shared values that cannot be imposed by the regulatory process. We value stewardship of the land, cooperation, conservation of resources, sharing, and independence. We are a very diverse group—and that, too, is one of our values. . . . Although we want to make a decent living, we are not in this for the money. There are very few organic farmers who could not make a lot more money at some other kind of work" (1992, 21).

What Henderson is saying is that there is something else beyond the mechanical and the personally profitable that attracts many organic farmers to this manner of farming. This something is based in a set of core understandings or principles: diversity, place, democracy, and spirituality clearly among them. Together these principles describe and support a living system that breathes from the soil on up.

The soil, its fertility and health, is the first concern of organic farmers and organic agriculture. Without healthy soil, they allow, there can be no healthy plants and no healthy people. Soil is seen as a complex living system, built up over time through the interactions of a diverse and seemingly infinite cast of characters, humans among them. To know it (and them) requires firsthand experience and takes as much artful intuition as it does exacting science. It requires physical engagement, sensual interpretation, and a holistic way of knowing. Such work and such knowledge are not easily transferable but are grounded, tightly connected to an actual place on earth.

But soil for many proponents of organics is also a metaphor for culture, and as culture the same awareness applies (Berry 1990; Esteva and Prakash 1998; Kirschenmann 1997). As such, soil, literally and figuratively, embodies the work and wisdom as well as the physical remains of previous generations. Through soil we are fed by our ancestors as we will feed our descendants. It connects the past to the present to the future and holds a people in place. Soil then is the stuff to which we belong. It is us in trust.

Like all life forms, soil is greater than its component parts but diminished by the loss of any one of them. Given this bond, pests (microbes, insects, neighbors, and nations) cannot be simply and absolutely recognized as enemies—we against them—but as signs of shifting, and sometimes dangerously shifting, balances. It is not the outright elimination of the pest, or the removal of the inconvenience that will ensure safety and survival. Rather it is the restoration and maintenance of diversity and the dance of accommodation. Echoed in this commitment to "nature as measure" is the voice, or more accurately the voices, of grassroots democracy, that profoundly slow and messy process entered into by disparate, but connected, interests that when working well can produce tolerance, humility, forgiveness, and constructive change.

Still these are not the values or major principles that drive the new industrial organics, for which the scientization and rationalization of organic processes have begun to assert themselves. There is a growing need to simplify and codify. For industrial organics contradictions are seen to create opposition and inefficiencies are seen to create waste. Wants spin free of context and can only be enlarged. When approached from this perspective, it becomes quite logical for the notion of *quality* to be defined by cleanliness and spotlessness (Green 2001). It is quite reasonable for *fresh* to exist somewhere between 26 and 25 degrees Fahrenheit (Associated Press 1995) and for *healthy* to be measured by the absence of—or reduced presence of—trace pesticides. These attributes (and a thousand others like them) were once contextual, interpretive, and relational but now have become technical and absolute. Such simplification makes it possible to deliver an organic bagel, an organic chicken, or an organic tofu turkey anywhere, twenty-four hours a day, seven days a week, with the legal guarantee that it will not immediately make the consumer sick. But what has happened to such things as taste, history, timeliness, reciprocity, negotiation, diversity, democracy, spirituality, or place? The new organics is transforming a way of being into a purchasable life style. It is turning a philosophy into a formula.

This is the dilemma that faces organic farmers and the organic movement generally. Where do their values lie and what choices can and can't they make if they wish to follow a course of action consistent with their values? Here in large measure lies the source of the disappointment that Michigan organic farmers and consumers experienced when they met with land-grant scientists and extension staff. They had reason to be skeptical. At the same time they were conflicted themselves about what they wanted and so could not see or adequately address their own dissatisfaction.

The Science of Organics

This next section discusses what is happening in Michigan—how the organic course is being set and who is setting it—and documents the slow but steady transformation of organics from a deep commitment into an extensive commodity. The final section of this chapter suggests possible ways in which those who value a deeper organic may reclaim the process.

Three different and only slightly overlapping research programs attest to MSU's commitment to certified organic production. At the MSU W. K. Kellogg Biological Station (KBS) near Kalamazoo, eight certified organic acres are devoted to cover crops and especially the rotation of soybeans, winter wheat with frost-seeded red clover, and corn interseeded with red clover. In response to requests from the station's organic farmer advisory council, several weed control experiments also are underway.

At a second research site, the Clarksville Horticulture Research Station, a five-acre orchard with over 2,500 apple trees is dedicated to helping large-scale conventional apple growers transition to organic production. This research specifically examines soil fertility and biology, tree vigor and ground floor management, pest and disease strategies (plum curculio, coddling moth, apple scab, and fire blight), and the costs of transitioning to, as well as beginning, an organic orchard.

At the Northwest Horticultural Research Station outside of Traverse City, a farmer-industry-researcher group has designed and continues to support research on alternatives to monoculture tart cherry cropping and conventional pest management. Since tart cherries are the backbone of the Michigan cherry industry, cherry research is designed to help large-scale conventional growers overcome constraints on the transition to organic. By contrast, no research is underway on sweet cherries, a less economically significant crop intended primarily for roadside and local markets and one that organic growers frequently incorporate into their diversified production and marketing strategies.

These research programs confirm the observation by the Organic Farming Research Foundation (OFRF) that the "official taboo against scientific study of organic systems" has been cracked at land-grant institutions like Michigan State (Sooby 2001, vi). At issue, however, is whether the OFRF vision that "an expanding organic knowledge base will only be realized through significant investments by the USDA and others in new research at our Land Grants" (Sooby 2001, viii) represents an effective way to attend to the concerns of small-scale growers. Is this adequate for and consistent with the organic conscience? While we agree that it is necessary to demand expanded and diversified financial support for organic research, we also caution that MSU's organic research programs, embody characteristics that should prompt organic growers to look beyond the land grant and embrace more pluralistic and collaborative approaches to their research needs.[1]

Research Protocols

The KBS, Clarksville, and Northwest programs all apply standard agronomic research protocols to address key and specific component problems of concern to organic growers and especially to conventional growers transitioning to organic. This research is consistent with an "ecological approach to farming that affects the entire production and processing system" (Greene and Kremen 2003, 1). Nevertheless, many small-scale organic growers remain critical of the ways in which conventional research protocols overlook the production and pest management complexities that are inherent in their highly diversified operations as well as their need to understand the implications of biointensive management.

Instead of thinking about discrete research topics and activities, researchers need to enter the worlds of organic growers as they confront issues arising from multifaceted relationships (that is, soil health and the use of predator food for pest management) (Landis et al. 2002). They need to design protocols in active partnership with farmers and around the rhythms and practices of actual working farms. In other words, there is a need to contextualize research and to recognize the particularities of production and of place. To this end we have identified three structural features inherent in most university-based organic research that prevent the consideration of systemic, integrative, and whole farm issues as they relate to cropping, cultivation, fertility, and pest and disease management.

First, most research protocols assume that organic growers will benefit from the activities of university researchers. At the same time they also acknowledge that researchers must respect and integrate indigenous knowledge into their work. While this perspective reflects the common delivery mentality, it also embodies a paradox that most researchers seek to resolve by involving growers in their research through visits of the latter to research sites and discussions of research findings. Nevertheless, grower input, like the research design itself, is carefully controlled, and knowledge and authority continue to reside principally with the researcher.

A second, closely related assumption raises more serious concerns. Most research is based on the belief that there is a fundamental and important difference between the scientific foundations of production that university researchers provide and the traditional organic knowledge of farmers. Organic knowledge and the solutions it engenders are permeable and changeable. By contrast, the knowledge and solutions of conventional research, even when applied to organic concerns, tend to be regarded as definitive, stoically literal, and testable. This way of thinking, characteristic of Berry's rational mind, reinforces the delivery mentality, precludes any consideration of farmer knowledge as equal to that of the researcher, and reflects another way in which researcher control is maintained in the relationship (Berry 2002).

Third, standard research protocols for organic research often pay lip service to their holistic or integrated features, but the research design by definition precludes emergent properties or different ways of knowing. Conventional approaches are not built around, nor do they allow for, the amazing process of discovering (Nabhan 1997) or for considering different ways of understanding validity and reliability. This inability to accommodate emergence is one reason why conventional researchers, even those committed to organic, find it difficult to accept alternative approaches based on biodynamic principles.

Organic as Commodity and Market Niche

Land-grant agricultural research has a long history of courting and nurturing commodity constituencies. Its recognized power stems from serving what have become mutually beneficial economic and professional interests. Consequently, it should come as no surprise that land-grant research administrators approach and try to deal with organic agriculture as just another commodity. This mentality compartmentalizes widely diverse situations and interests and makes them disappear within a neatly defined category or package. In fact, this thinking easily shifts from product to persons. With few exceptions university administrators as well as extension personnel tend to think of organic growers as a uniform group that represents and fills a niche market instead of regarding them as a group of individuals made whole by their commitments to place, a way of life, and a way of connecting with the soil. In addition, by assuming a measure of uniformity over a diverse group of growers, administrators can set the terms of discussion, that is, language, timing, content, and location.

Without a new frame of reference for considering how local, small-scale, and organic can become integral to our food and farming system, research administrators will conveniently assume that any organic grower can speak for the state's organic community. It is hard to discern whether university administrators deliberately depoliticize meetings by regarding the organic community as simply one big happy family or whether they really do not recognize important differences and voices within this community. Many land-grant research administrators have grown up on modern, large grain, cattle or dairy farms or spent their professional careers working on problems presented by this sector of the society. Consequently, it is difficult, if not impossible, for them to think seriously about becoming smaller and getting better, an important principle for many in the organic community.

Regulatory Co-optation

The Michigan Organic Products Act (Act 316 of 2000), which was written and passed with lightening speed, embodied a strategic set of policy and drafting decisions. First, those representing the broader organic community agreed not to tie state certification to a separate set of organic standards. It was felt that establishing state standards would have significant and unacceptable budgetary implications and would take years to draft, as had been the case elsewhere. Moreover, since the

National Organic Program (NOP) was close to being approved, separate Michigan standards appeared redundant.

Second, and perhaps more important, those involved in developing the Michigan Organic Products Act agreed that two separate initiatives were needed: one focused on *protecting* the organic standard and another on *promoting* organics throughout the state. A clear and unbreachable organic standard, they reasoned, would protect consumers against fraudulent products and marketing claims. It also was necessary to uphold the integrity of organic growers and to encourage others to transition to organics. Thus from the beginning there has been a conceptual as well as a functional divide between organic regulation and organic development. In fact, not only has regulation received far more administrative attention, but promotion frequently is regarded as a natural outgrowth of regulation. This orientation has had decided consequences for the organic community in Michigan.

From a legal standpoint the state act is more restrictive than the NOP. Certification is now required of all persons or concerns representing their agricultural products as organic. No exceptions are made for growers who have less than $5,000 in gross sales. This certification, according to the Michigan Department of Agriculture, "give(s) meaning and understanding to the term and use of the word. . . . It will also discourage small farmers who are growing conventionally from mislabeling product as organic."[2] Once again, organic has been legally enclosed, reduced to a discrete and ownable thing. In exchange for uniformity and predictability, context and diversity have been sacrificed. Farmers, and especially small organic farmers, are granted no space within which to interpret or to construct relationships suited to particular soils or particular lived environments.

From an administrative standpoint similar difficulties exist. While the act insists on a unique set of organic provisions, the state does not have the power to suspend or to revoke the accreditation of any NOP-approved certifying agent. This lack of authority means that all such issues must be handled through a federal agency or in the federal court system, yet another level removed from place-based negotiation. Likewise, assigning the administration of the act to the Michigan Department of Agriculture's Pesticide and Plant Pest Management Division makes it difficult, if not impossible, for broader development and food system issues to arise. Almost by definition, organic oversight becomes a matter of bureaucratic and conventional scientific authority.

Broadly speaking, the Michigan Organic Products Act and the decision to establish a State Organic Program have not created, and may well have hindered, opportunities for realizing the civic nature of organic agriculture. An act based on the assumption that "protection enhances promotion" squeezes and narrows discussions related to the role of organic agriculture in local and community food systems. An act wedded to regulation can only uphold and simplify the roles of producer and consumer—roles consistent with faceless and increasingly narrow market transactions. Ultimately, it will deny the vitality as well as the messiness of dialogues and relationships sensitive to place(s) and will provide little room to consider how organic agriculture might contribute to a public consideration of food citizenship.

Over the last ten years both Michigan State University and the Michigan Department of Agriculture have recognized the term "organic" and have incorporated it into production research and marketing programs. This incorporation has been handled in ways consistent with both the tenets of conventional research practice and economic development. In this way research sites in monocultural orchards yield data on effective management strategies for overcoming specific pests and for increasing production. Such an orientation allows researchers to flip from organic to molecular—from ecology to biotechnology—without bothersome internal contradiction. State policy manifests a similar commitment to organics. By focusing on the promotion of organic labels and product sales and by fully commodifying organic foods and processes, the state is paving the way for the largest and most consistent suppliers and for a prescriptively applied consumer culture.

The issue is not that such an orientation exists, because we all know that it does. Rather, the issue is that knowing this orientation exists and knowing how radically it departs from the organic conscience, we must also recognize that to embrace it is to lose much, if not most, of what is meant by organic. If the organic community is to follow its own heart and deeply held principles, it cannot hanker after more institutional science and more commercial legislation and then be disappointed when the movement as well as its purpose and practices are politically diminished and economically co-opted. It could hardly be otherwise. Or stated a bit differently, "we cannot get there from here."

Certainly MSU's approach to education and research must change if it is ever to actively honor its land-grant mission and its motto for the twenty-first century, "people matter." A little of this change is already happening as restless insiders begin to push established boundaries and mindsets (such as with the Student Organic Farm and the Office of Campus Sustainability).

However, addressing the land-grant university's needs and anticipating how it must change to better serve the interests of the organic community is to miss the point we are making. It is the organic community that needs to refocus its efforts. Simply put, it is not the job of the organic community to change the university, its science, and its thinking. It is not in the best interest of organics to expend what are still modest energies and resources to deliver place-based wisdom to well-paid and distant experts and simultaneously to ask for their legitimization.

Organics, despite its present-day insecurities and inconsistencies (or possibly because of them), is already quite legitimate. It already belongs to many people and to many places across Michigan. It does not need experiment station approval or a "Buy Organic" campaign to make it so. What is needed is for advocates—farmers and eaters alike—to recognize the power that underlies the concept and the full extent to which organic provides a living, as well as a growing, alternative. If organic forces us to think differently, it also requires us to act differently. Here at the nexus of thinking and acting is where Michigan's organic movement might best put its energies.

What is wanting, we feel, are not additional organic products or new market niches but expanded discussions at home for the purpose of connecting residents to their food and to the ecology and the culture of their places. Organic food and farming is particular; it is contextually specific. It is not a one-size-fits-all proposition. Rather it embodies and is embodied in the wisdom that emerges from dwelling in a place, close to natural systems, over extended periods of time. Eating organic food, supporting organic activity, and creating a set of shared understandings as well as a cuisine—in Mintz's (1996) relational and nontransferable sense of the term—is a source of belonging and identity. It is what Kirschenmann (2002) refers to as "in dwelling." It is also, as people like Nabhan (1997) recognize, a source of biodiversity, cultural diversity, and real homeland security.

To tap this strength, Michigan's organic community might consider working with and through many less traditional agencies, organizations, and nongovernmental organizations, such as sports clubs, chambers of commerce, labor unions, and historical societies, as well as more traditional environmental and consumer groups. Conversations might emerge through the sorting out of such contentious issues as hunting and farming, organic prices and fixed income, and farm labor and union representation. Conversations might be catalyzed through the creation of great good places (that is, cafes, exercise clubs, bookstores, and farmers' markets) and through the dedication of common sites for daily as well as celebratory public work (public parks, community farms and kitchens, parades, theaters, and memorials) (Loren 2003; Oldenburg 1989).

The opening up of organic conversations moves them out of the exclusive domain of agricultural scientists and practicing farmers. It makes them the property of ordinary people, people who are also Boy Scouts, librarians, merchants, mothers, drain commissioners, school teachers, ministers, doctors, musicians, and storytellers. These expanded voices in turn are the source of new words and metaphors in a language that is not wholly informed by controlled experiments, replicated plots, or the statistical calculation of acceptable risk. Likewise, the shape or color of a native plant, the location of a rocky outcropping, the taste of a local meal, a story told and retold across generations all might contribute to the symbol and substance of a place and of an organic conscience. These, no less than drip irrigation and BRIX assays, need to become the tools of Michigan's organic movement.

Understandings and wisdom of this kind are not provable. Rather they are sensual, embodied, and shared; they emerge from lived experience, from being there. Being there is also a source of discovering, which Nabhan explains, "[is] a process far different from the heroic act of *discovery*. Through the process of *discovering*, we seldom achieve any hard-and-fast truth about the world, its cornucopia of creatures, or its cultural interactions with them. Instead, we are inevitably assured of how little we know about that on which each of our lives depends" (1997, 98). We suggest that the Michigan organic community needs to remain comfortable with a healthy measure of not knowing, of sharing uncertainty, and of remaining humble.

But humility is not at all the same thing as fatalism or the willingness to self-exploit. There is no beauty in despair and no honor in hunger. Humility, the sense that we are part of something greater than ourselves, needs time and space within which to grow. Here, then, is another recommendation for Michigan's organic community: find (and tutor) champions in township and county government and in the Michigan legislature. These champions do not necessarily need to be agriculturalists. They do need to be defenders of diversity, decentralization, and democratic engagement. They need to be persons who understand and can delight in the partialness of human wisdom and the paradox of place.

It is by protecting multiple forms of expression, multiple resources and landscapes, and multiple ways of being that we will find the time and space within which to grow and thereby protect the complexities, interconnections, and natural processes that sustain and enrich our lives on earth. It is time Michigan's organic movement began grooming "organic" leaders (not scientists, not data managers, not career politicians) by physically and financially underwriting their education over the course of several years via workshops, internships, conferences, and place-based dialogue.

Becoming organic and a champion of organic will mean knowing more than how to grow or access organic food; it will mean more than passing legislation to legalize specific techniques, labels, and market savvy. It will mean knowing how to speak (and translate) the languages, tell the stories, and recount the embodied experiences that nurture an organic conscience. These abilities are the skills that will bring local, place-based voices into state and national discussions. This place-based and vernacular context will allow organic farmers to grow secure in their respective communities, and as one Michigan farmer wrote, "to look to each season not to make a million dollars, but to simply [be able to] do it all over again."

Notes

1. The six-acre Student Organic Farm/CSA, located on the Horticulture Department research farm just south of the main MSU campus, may be an exception to the land-grant pattern. Less than a year old, it evolved from a five-year project on greenhouse organic crop production and is supported by a three-year grant from the W. K. Kellogg Foundation with additional funding from North Central USDA-Sustainable Agriculture Research and Education. The farm has the potential to become a diverse learning community committed to local, year-round diversified food production.

2. From the "Michigan Department of Agriculture Organic Program Application" (p. 5), submitted by Ken Rauscher, director, Pesticide and Plant Pest Management Division, to Rick Mathews, acting program director, USDA National Organic Program, October 19, 2001. Program application received as a result of a Freedom of Information Act request by Organic Growers of Michigan.

References

Anon. 1999. "The true meaning of 'organic.'" *Acres USA* 30 (9): 24–27.

Associated Press. 1995. "Frozen chicken to lose fresh labels." *Lansing State Journal*, August 25.

Berry, W. 1990. *What Are People For?* San Francisco: North Point.

———. 2002. "Two minds." *The Progressive*, November.

DeLind, L. B. 2000. "Transforming organic agriculture into industrial organic products: Reconsidering national organic standards." *Human Organization* 59 (2): 198–208.

Esteva, G., and M. S. Prakash. 1998. *Grassroots Post-Modernism: Remaking the Soil of Cultures*. London: Zed.

Green, E. 2001. "Gone for good?" *Los Angeles Times*, January 3.

Greene, C., and A. Kremen. 2003. *U.S. Organic Farming in 2000-2001: Adoption of Certified Systems*. Washington DC: Economic Research Service, Resource Economics Division, USDA.

Halweil, B. 2001. "Organic gold rush." *WorldWatch*, May/June.

Henderson, E. 1992. "Some comments and concerns about NOSB." *Natural Farmer*, Fall.

Hesser, A. 2003. "Salad in sealed bags isn't so simple, it seems." *New York Times*. January 14.

Kirschenmann, F. 1997. "On Becoming Lovers of the Soil." In *For All Generations: Making World Agriculture More Sustainable*, edited by J. P. Madden and S. G. Chaplowe, 101–14. Glendale CA: World Sustainable Agriculture Association.

———. 2002. *What Constitutes Sound Science?* Ames IA: Leopold Center for Sustainable Agriculture.

Kittredge, J. 2003. "Advocates say organic program yielding to industry pressure." *Organic Farming Research Foundation Information Bulletin* 12:4–5.

Klonsky, K. 2000. "Forces impacting the production of organic foods." *Agriculture and Human Values* 17 (3): 233–43.

Kneen, B. 2001. "A matter of context." *The Ram's Horn*. 187 (January): 1–3.

Landis, J. N., J. E. Sanchez, G. W. Bird, C. E. Edson, R. Isaacs, R. H. Lehnert, A. M. C. Schilder, and S. M. Swinton. 2002. *Fruit Crop Ecology and Management*. Michigan State University Extension Bulletin E-2759.

Loren, B. K. 2003. "Got tape?" *Orion*, May/June.

Mendelson, J. 2003. "Who's watching the USDA's organic 'henhouse?'" *Organic Farming Research Foundation Information Bulletin* 12:1,8.

Mintz, S. 1996. *Tasting Food, Tasting Freedom*. Boston: Beacon.

Nabhan, G. P. 1997. "Beyond the Zipper: Discovering the Diversity around Us." In *Cultures of Habitat: On Nature, Culture, and Story*, edited by G. P. Nabhan, 97–106. Washington DC: Counterpoint.

Norberg-Hodge, H., T. Merrifield, and S. Gorelick. 2002. *Bringing the Food Economy Home: The Social, Ecological, and Economic Effects of Local Food*. Bloomfield CT: Kumarian.

Oldenburg, R. 1989. *The Great Good Place: Cafés, Coffee Shops, Community Centers, Beauty Parlors, General Stores, Bars, Hangouts, and How They Get You through the Day*. New York: Paragon.

Organic Consumers Association. 2001. "The politics of food: Moving beyond USDA organic." *Organic View* 1:3,6.

Organic Farming Research Foundation. 2002. *Hidden Organic Gems Included in the 2002 Farm Bill*. Santa Cruz CA: Organic Farming Research Foundation.

Pesticide Action Network Updates Service. 2003. *Action Alert: U.S. Organic Standards Threatened*. San Francisco: Pesticide Action Network North America.

Pollan, M. 2001. "Naturally: How organic became a marketing niche and a multibillion-dollar industry." *New York Times*, May 13.

———. 2003. "Getting over organic." *Orion*, July/August.

Reynolds, P. C. 2000. "Organics at the crossroads: Future for runaway industry is community-level systems." *Acres USA* 30 (9): 1,8–11.

Severson, K. 2002. "Agribusiness goes organic." *San Francisco Chronicle*, October 13.

Sooby, J. 2001. *State of the States: Organic Farming Systems Research at Land-Grant Institutions, 2000-2001*. Santa Cruz CA, Organic Farming Research Foundation.

Wheeler, V., and P. Esainko. 1997. "Purity and Danger: The Social Evolution of National Standards in Organic Farming." In *Life and Death Matters: Human Rights and the Environment at the End of the Millennium*, edited by B. R. Johnston, 151–72. Walnut Creek CA: AltaMira.

Youngberg, G. 1996. "The future of rural communities in the United States: A shared vision of organic agriculture or 'bowling alone?'" *Organic Farming Research Foundation Bulletin* 2 (Winter): 1, 9–10.

16. The Social Foundation of Sustainable Agriculture in Southeastern Vermont

Matthew Hoffman

An increasing awareness of industrial agriculture's negative impact on human health, the environment, and rural communities has prompted many to call for an alternative system of agriculture, one that is sustainable. However, the development of such an agriculture entails more than just perfecting an alternative set of growing techniques. If industrial techniques of production have evolved in response to a certain type of economy, the success of sustainable agriculture will depend on the development of a different type of economy. Thomas Lyson (2000, 2002) has claimed that whereas industrial agriculture is driven by a competitive market system, sustainable agriculture is based on cooperative community relationships. Using data collected through in-depth interviews with farmers in southeastern Vermont, this chapter reports on a study testing Lyson's theory and describes a variety of cooperative practices that function in support of sustainable agriculture.

The Industrial Model of Agriculture

The dominant trend in U.S. agriculture is toward increasing industrialization and consolidation. Following 4,100 mergers and buyouts in the U.S. food industry between 1982 and 1990 (Korten 1995, 224), it was reported that "corporate agribusiness manufactures and markets over 95 percent of the food in the United States" (Lehman and Krebs 1996, 123). After the Second World War the number of American farms declined by two-thirds. By 1994 two companies were responsible for 50 percent of U.S. grain exports, and three packers controlled the slaughter of more than 80 percent of beef (Korten 1995, 224; Lehman and Krebs 1996, 125–27). In a report to the National Farmers Union

William Heffernan explained that "[the] continuing concentration of ownership and control of the food system . . . [is like] an hour glass in which farm commodities produced by thousands of farmers must pass through the narrow part of the glass that is analogous to the few firms that control the processing of commodities (1999, 2)."

Not only has the number of producers and processors declined (representing an increase in scale and concentration of ownership), the food and agriculture system has become more vertically integrated as individual firms acquire control over each stage of production for particular commodities (Heffernan and Hendrickson 2002). As a result it has become common for farmers, in this case called growers, to work under contract for a company that controls the entire process from seed to shelf. One of the earliest examples of this vertical integration is the poultry industry, in which contract farmers raise hundreds of thousands of birds in warehouse-like facilities for companies that dictate precise management practices and deal in everything from chicks and medicated feed to processing and marketing. According to a page on the United States Department of Agriculture Web site, "[t]he broiler industry was one of the success stories in American agriculture during the last century and is an example of how the use of technology, improvements in production practices, and product marketing can change the basic structure of agriculture. . . . The broiler industry has evolved from millions of small backyard flocks of dual-purpose (eggs and meat) chickens in the early 1900s to less than 50 highly specialized, vertically integrated agribusiness firms" (USDA-NASS 2005).

Dennis Avery credits industrial agriculture with high levels of productivity that is achieved through the intensive use of inputs: "[a]dvanced farming methods utilize monocultures, potent new seed varieties, irrigation, fertilizer and pesticides to minimize land needs; medicines keep livestock and poultry healthy and productive; and the best genetics help herds and flocks convert feed more efficiently" (1997, 10). He and other proponents of industrial agriculture claim that these techniques are necessary in order to feed the hungry and to preserve the landscape in an increasingly crowded world.

Criticism of the Industrial Model

This industrial food and agriculture system has come under increasingly severe criticism. The use of pesticides has been shown to be a serious health hazard for humans and a persistent problem in the environment (Moore 2002). Soil degradation and erosion resulting from the use of synthetic chemical fertilizers and other industrial techniques have cost the United States half its topsoil since 1960 (Kimbrell 2002; Ponting 1991). The emergence of antibiotic-resistant pathogens has been linked to the use of antibiotics in animal feed (White et al. 2001). Increasing incidences of food poisoning and an epidemic of bovine spongiform encephalopathy, transmittable to humans as a variant of Creutzfeldt-Jakob disease, have contributed to public mistrust of industrial farming (Nottingham 1999). Many scientists and consumers are alarmed by the rapid and largely unheralded introduction of genetically modified organisms into the food system, a move that they feel is being made without adequate understanding of the health and environmental consequences (Commoner 2002; Regal 2000; Rifkin 1998).

As an increasingly industrialized and highly consolidated food system erodes the economic foundation of rural communities (Geisler and Lyson 1991; Heffernan 1999), we see the emergence of those conditions Goldschmidt warned about: "with industrialization will come an increasing concentration of economic power in the hands of fewer and fewer men at the head of great organizations, and an end to the broad diffusion of social and economic benefits that has long been characteristic of American rural communities" (1978, 280). With agrarian landscapes continuing to change or vanish beneath commercial real estate development, the disappearance of small-scale farming, once central to U.S. society, is seen by many as a deepening cultural tragedy.

The Increasing Popularity of Organic Techniques

The last several decades have seen the emergence of a significant countertrend to industrial agriculture. Many farmers are eschewing the use of herbicide, pesticide, and chemical fertilizer in their fields; they are raising poultry and livestock without hormones or antibiotics in feed. Organic farmers focus on maintaining healthy soil and animals through cultural practices based on natural biological processes. The demand for organic products is very high: "Organic farming is one of the fastest growing segments of U.S. agriculture during the 1990s. . . . The number of organic farmers is increasing by about 12 percent per year and now stands at about 12,200 nationwide, most of them small-scale producers. According to a recent USDA study, certified organic cropland more than doubled from 1992 to 1997. Two organic livestock sectors, eggs and dairy, grew even faster" (USDA 2000).

These figures might understate the number of organic farmers since many of these farms are not officially certified and market their products locally through informal channels, as suggested in the chapter in this volume by DeLind and Bingen. Recent mergers and acquisitions in organic food processing and distribution as well as the emergence of large producers indicate that this segment of the food industry is also consolidating (Jacobs 2000; Starke 2001). Several popular brands of organic food are now owned by conventional agribusiness corporations (Starke 2001).

A Sustainable Model of Agriculture

Criticism of the industrial model has prompted many to call for an alternative system of agriculture. The term sustainable agriculture has been used widely to denote a system of farming that functions without depleting the natural resources on which it depends. It has also come to mean a system that is beneficial for its participants and society as a whole. The organic food industry, which is supposed to be based on sustainable agriculture, has come under criticism for defining itself exclusively in terms of growing techniques. Because the ill effects of industrial agriculture are not just environmental but also social, organic farming techniques cannot by themselves be an adequate remedy. Moreover, if industrial farming techniques have evolved in response to a certain type of economy, it seems unlikely that sustainable agriculture can enjoy more than limited success within that same framework. In the words of Patricia Allen and Carolyn Sachs, "[A tendency] . . . to rely on technology as the solution . . . does not examine the overarching structural forces that have contributed to the adoption of resource-intensive farming practices. Technologies and social relations are inseparably linked, both in terms of their inspiration and their consequences" (1991, 5). Marty Strange writes, "[t]o sustain itself, commercial agriculture will have to reorganize its *social and economic structure* as well as its technological base and production methods" (1984, 116, emphasis added).

Lyson has argued that the gulf separating sustainable agriculture from industrial agriculture is more than a difference in technology. He claims that these two systems represent two paradigms that are "fundamentally different" and "essentially incompatible." Whereas industrial agriculture has its foundation in neoclassical economics, "[s]ustainability is framed by an emerging community-centered, problem-solving perspective. . . . The underlying social science paradigms are portrayed by Beus and Dunlap [1990] as competition versus community" (Lyson, 2002, 193–95).

The association of industrial farming with a market economy and of sustainable agriculture with community is a common theme. Murray Bookchin writes, "[t]he contrast between early and modern agricultural practices is dramatic. Indeed, it would be very difficult to understand the one through the vision of the other, to recognize that they are united by any kind of cultural continuity. Nor can we ascribe this contrast merely to differences in technology. Our agricultural epoch—a distinctly capitalist one—envisions food cultivation as a business enterprise to be operated strictly for the purpose of generating profit in a market economy" (1976, 3).

Agricultural economist Steven Blank exemplifies the neoclassical viewpoint: "[w]e need to strip away the romance and nostalgia surrounding agriculture and see it for what it is: a business" (1999, 25). Because Americans can buy cheaper imported food, he advises farmers to sell their land to developers: "People trying to hang on to the outdated version of farming are simply holding onto a bad investment. . . . The prudent thing is to manage their portfolio of assets with their head, not their heart" (Blank, 1999, 25–27).

From the community problem-solving perspective, however, agriculture is "integrally related to the social and cultural fabric of the community" (Lyson, this volume). Lyson uses the term civic agriculture to refer to "the embedding of local agriculture and food production in the community. . . . Civic agriculture embodies a commitment to developing and strengthening an economically, environmentally, and socially sustainable system of agriculture and food production that relies on local resources and serves local markets and consumers. The economic imperative to earn a profit is filtered through a set of cooperative and mutually supporting social relations" (Lyson 2001, 41–42).

"We therefore find ourselves," Andrew Kimbrell observes, "in the midst of a historic battle over two very different visions of the future of food in the 21st century" (2002, xiii). The success of sustainable agriculture in its broadest sense depends not only on the right technology but more fundamentally on how the food system is structured. If Lyson and others are correct, proponents of sustainable agriculture must seek alternatives to the market-driven neoclassical model; they must seek to reembed farming in community, that is, to reestablish agri*culture* where it has been displaced by agri*business*. According to Wendell Berry, "[f]armers must understand that this requires an economics of cooperation rather than competition" (1995, 5).

Investigating the Community-Based Model

Is community rather than the market a realistic basis for a sustainable model of agriculture? In the summer of 2002 I conducted a study of sustainable agriculture in southeastern Vermont with two principal aims. The first was to test Lyson's theory that "community problem-solving, rather than economic competition, is the social foundation of sustainable agriculture" (2002, 195). The second was to describe how certain cooperative practices function in support of this type of agriculture.

Community problem solving is being contrasted here with the neoclassical model of economic behavior. In the neoclassical model profit-seeking individuals respond to price signals determined by supply and demand. In doing so, it is claimed, they are led to behave exactly as if they were acting out of a well-informed sense of civic duty. To the extent that market mechanisms fail to produce satisfactory outcomes, the neoclassical model relies on government intervention. In the community problem-solving model people are motivated by their relationship to their work, which can also mean by their relationship to other people. I call this attitude being task oriented, or problem solving in contrast to profit seeking. In this model, dense networks of social interaction generate enough "social trust" (see Putnam 1993) that economic behavior may be integrated through voluntary cooperation instead of by market mechanisms or government control.

The data for this study were gathered by means of loosely structured in-depth interviews on sustainable farms, all located in the vicinity of the same large town in southeastern Vermont. In order to meet my definition of sustainability, the farms had to be small, they had to produce for local consumption (although not exclusively), and they had to employ ecologically attuned growing practices. I did not require that they meet the standards for organic certification, only that they be committed to farming in a way that protects the environment and approximates the organic standards. I tried to err on the side of inclusiveness in order to have the largest possible study population. Fifteen farms met my definition; and thirteen of them are included in the study.

Observations in the Field

Table 28 presents a summary of the on-farm interviews. Under the top category, relationship to work, the first variable, labeled "task oriented," is based on farmers' responses to questions about their choice of techniques, why they produce what they do, and (most importantly) why they farm. The next two variables, "has chosen to limit scale" and "does not rely on hired labor," are not in themselves indicative of task orientation, but the feelings expressed by some farmers regarding scale and hired labor helped to illustrate their relationship to their work. As shown in the matrix, every farmer but one appears to be task oriented. Each of these farmers made statements indicating that their farming activities are motivated primarily by their relationship to the work rather than by calculations of how to maximize profit. While the choice of which crops to grow is sometimes determined by price, the decision to use sustainable practices—indeed the decision to farm at all—is mostly determined by farmers' feelings about the work, other people, and the land.

The fourth characteristic in the relationship to work category shows that eight of the thirteen farm operations are on land that is protected by one of three private nonprofit organizations. The largest of these or-

Table 28. Summary of on-farm interviews

		Farms												
		1	2	3	4	5	6	7	8	9	10	11	12	13
Relationship to work	Is task oriented	X	X	X	X	X	X	X	X	X	X	X	X	
	Has chosen to limit scale			X			X	X		X		X		
	Does not rely on hired labor	X		X	X			X	X	X	X	X		
	Land is protected				X	X	X		X	X	X	X	X	
Production practices	Assisted at times by unhired labor	X		X	X		X		X	X	X		X	
	Uses others' land rent free	X	X				X	X	X				X	
	Belongs to support group		X	X	X					X			X	
	Engages in balanced reciprocity	X	X	X	X		X			X	X	X	X	X
	Engages in generalized reciprocity	X		X		X	X		X	X	X		X	X
Distribution practices	Sells at farmers' market	X	X					X	X	X	X	X	X	X
	Does community supported agriculture				X	x			x			X	x	X
	Sells to one or more cooperative food stores		X			X		X	X	X	X	X	X	X
	Belongs to VT Fresh Network		X			X			X				X	
Relationship to customers	Knows most customers	X	x	X	X		X				x	x		
	Would not need official certification	X		X	X				X	X	X	X	X	X

Note: Characteristics shared by at least one-third of studied farms.

ganizations is the Vermont Land Trust, which holds conservation easements on two of the farms in this study. Three of the farms in this study are part of a community land trust. The farms each have a eighty-nine-year renewable and inheritable lease from the community land trust, which holds title to the land in order to ensure proper stewardship. The third organization is not a land trust but rather a nonprofit whose mission is to protect and restore important historic properties. Different parts of a 571-acre farm acquired by this organization in 1995 are being used by three of the farmers in this study.

In the next category, production practices," five trends emerge that indicate cooperation in production. Looking at the first variable in "Production Practices," one sees that eight out of thirteen farms are occasionally assisted by unhired labor, usually at times of exceptional need. This help does not include the labor of family members. Sometimes this assistance is with labor-intensive seasonal activities such as sugaring or lambing, while at other times it is in response to such unexpected events as personal injury, a sick animal, or a broken machine. Barn raisings are infrequent but important occasions when friends and neighbors provide valuable assistance. One farmer has put up three buildings with barn raisings, and provides similar help to other people "all the time." Several farmers also have received much-needed help following the arrival of a new baby in the family. Members of the community land trust have workdays on which they gather to provide assistance at one of the member's farms as a matter of rotation.

One important aspect of production for six of the farms is the ability to use other people's land rent free, especially for grazing, haymaking, and tapping maple trees. The farmer's use of this land is helpful to landowners because in the case of grazing and haymaking the land gets fertilized and mowed and because putting land to agricultural use allows landowners to enroll in Vermont's Current Use Assessment Program, which assesses the value of their land differently for tax purposes. One farmer taps maple trees on ten different properties and has his sugarhouse located on a neighbor's land. Some of these landowners are given syrup, but nobody receives any money.

There are three networks that I have labeled support groups. One of these is made up of a dozen families who have participated in a workshop on sustainable pasture management. They meet to take pasture walks on each other's farms and occasionally to help each other with certain tasks. One farm in this study is a member of this group. There is a cheese guild, made up of five to six shepherds, one of whom is included in this study. All of these shepherds work with a particular sheep's milk dairy and cheese cave. However, they are independent producers, not a cooperative, and meet four times a year in order to exchange information, set standards, and occasionally to make bulk purchases. There is also a women's group that meets periodically to discuss agricultural and personal issues. The community land trust and especially the farmers' market also serve as support groups, but membership in these two organizations is not included in this variable.

The distinction I am making between balanced and generalized reciprocity in the next two variables is as follows. Balanced reciprocity involves a quid pro quo, usually in terms of cash value (that is, $50 worth of firewood for $50 worth of shingles). In contrast, generalized reciprocity involves uncalculated giving in a context in which the giver knows that the receiver will do, has done, or *would* do something in return, although such a bargain is not specifically agreed upon. When interviewing farmers, it was sometimes hard to distinguish balanced from generalized reciprocity because most Vermonters in my experience take pains to make all exchanges appear balanced. A typical and humorous example of this attitude is someone selling a car to a relative for $10 or a six-pack, and as often as not, upon receipt of payment declaring, "There, now we're even."

The casualness with which farmers give and receive help often causes them to understate the importance of such activity. Some farmers said that they never cooperate with neighbors, and then when numerous examples of them doing so came up during the interview, they said, "Oh, well, I suppose, if you count *that*." In the words of one farmer, "Not a lot of money changes hands, but people get what they need." The norms of balanced and generalized reciprocity that exist among these farmers form an important part of many of their operations. Their mutual willingness to lend out machinery or to spend time and effort on each other's projects significantly reduces the amount of money that they need to spend and offers some assurance against the uncertainties of farming. Often the exchange of cash for professional services is embedded in long-term relationships, the price of a given service taking into account previous and future transactions. Asked why a local expert was willing to help some people at all hours, one farmer replied simply, "It's called being good neighbors."

There is another form of cooperation in production that was not common enough to be called a trend but which appears to be an effective strategy for small farmers. This strategy is collaboration between a farmer and a craft processor. In one case five to six shepherds (only one of whom is included in this study) age their cheeses together in the same cave. The dairy where the cave is located buys cheeses from the other farmers, hires staff to supervise the aging process, and markets the cheeses under the same label. Another farm rents space to a winery, for which it is raising bittersweet apples and elderberries. In the near future the farm will provide all the apples to the winery for estate-bottled wines and cider. A small dairy located on one of the farms in this study is jointly owned by the farm and by a cheese maker. Milk from the farm is used in the making of specialty cheeses. All three of these collaborations are known for the exceptional quality of the resulting products.

Four trends emerge in the category covering distribution of farm products. Noteworthy about each of these trends is the engagement and cooperation of local *consumers*. Nine of the farms in this study are members of the local farmers' market, a nonprofit organization whose primary goal, to quote the bylaws. is "to further the local production of agricultural products, prepared foods, and crafts." The market takes place every Saturday throughout the summer and fall in a glade west of town, where farmers, craftspeople, and vendors of prepared food set up rustic stalls arranged in a wide circle. Usually there is live acoustic music under a large tree in the middle. The market is also held every Wednesday on a smaller scale and for a somewhat shorter season along a sidewalk that intersects Main Street. Farmers and craftspeople may only sell goods that they themselves have produced. When I asked 100 customers why they buy produce at the farmers' market, 42 said that they came to support local organic agriculture. This answer was the most common. More than a third came to socialize and enjoy the general atmosphere. Only two said that they came because of price.

Another distributive practice that supports sustainable farming through the cooperation of farmers and consumers is community supported agriculture (CSA). Three of the farms in this study are CSA operations. Three more (each indicated by a small italicized "X") sell produce to a CSA. Each of the CSA farms in this study differs—and one of them is quite unique—but they each distribute produce based on the same central concept: members pay up front for a share at the beginning of the year and receive a quantity of produce each week throughout the growing season. In most cases CSA members share in the risk of a bad crop, but joining a CSA also can be more convenient and much less expensive than produce shopping in a supermarket. Selling shares provides the farmer with money for expenses in the spring and creates a personal connection with his or her customers. Consumers enjoy this connection to the farm, the farmer, and other members. Some CSA farms have occasional potluck dinners and parties. When I asked seventeen CSA members why they joined, twelve said that they did it in order to support local agriculture. Others said that they were friends with the farmer.

Nine of the farmers in this study sell their produce to a natural food store in town as well as to other stores like it in nearby towns and around the state. The store is a consumer cooperative, and some of the farmers are members. It is not unusual for the produce manager at the cooperative to buy something from several different farmers at different prices in an effort to spread his purchasing around. "I try to help everyone out," he told me. I asked him why he doesn't just buy from whoever is selling something the cheapest. He pointed out that, if such were his purpose, he would buy everything from California. When asked why he doesn't do that, he explained that supporting local agriculture is part of the co-op's mission, and such support is very important to its members. When selling to the co-op, farmers sometimes cooperate with each other on an informal basis. The produce manager explained, "Sometimes [one farmer] will call up and say, 'Yeah, I've got lettuce; but [another farmer] has really got a lot. You should buy it from him.'"

The Vermont Fresh Network (VFN) supports local agriculture by making connections between chefs and local farmers. In addition to a catalog listing all the members and their products, the VFN puts out a weekly "Fresh Sheet" telling chefs what products are currently available. This information is submitted by the farmers. Restaurants advertise their participation in the program, and the VFN publishes a dining guide and does other publicity. Four of the farms in this study are part of this network.

Farmers distribute their produce in a variety of ways beyond the trends mentioned above, but the most important remain the farmers' market, community supported agriculture, and sales to consumer co-ops. It is through one or a combination of these methods that the farmers in this study sell most of their produce. Only two farms do not use any of these ways: one of them has a farmstand and sells to two nearby inns; the other sells lamb and wool products, mostly by word of mouth and at wool and fiber festivals.

While these three main distribution practices each involve a certain amount of cooperation between farmers, they rest essentially on the willingness of consumers to support practitioners of sustainable agriculture. The security of sustainable farmers therefore depends on their ability to form committed relationships with consumers. I suspect that this is best done through structures in which those commitments are made formal. Community supported agriculture is an example, and it will be interesting to see how this model evolves in the future, particularly in light of issues raised in Marcia Ostrom's chapter in this volume about community supported agriculture. The cooperation of multiple farms, the establishment of food-processing facilities, consumer investment through the formation of community corporations, and further decommodification of land could each contribute to the advancement of this model.

The last category deals with relationship to customers. Seven farmers claimed that they knew most of their customers. The ones marked with a small italicized "X" sell to the co-op. Their claim must be interpreted as counting the produce manager at the co-op as one of their customers but not all of the shoppers at the co-op. In the case of some farmers that do not know most of their customers, most of those customers still know them. This customer recognition turned out to be very important.

One idea of theoretical interest to this study is the relationship between the ability to cooperate and the need for authoritarian intervention. At the time that I was interviewing farmers, a new federal law regulating use of the term organic was about to go into effect. Whereas in the past certification was not necessary to sell produce as organic, now certification under the new national standards is required of anyone marketing produce under this label. This change is intended to protect the value of the term and to protect consumers from its fraudulent use. Proponents of this legislation believe that government intervention is necessary in order for the public to be able to trust farmers' claims about how they grow food. I have hypothesized, however, that in the context of community such intervention is unnecessary.

The conditions among farmers in this study afforded the opportunity for an experiment. I have seen that their production and distribution practices are embedded in social relationships that generate trust and a willingness to cooperate. This trust and cooperation should mean that for them the federal legislation is unnecessary. Indeed, nine of the farms in this study said either that they do not need to be certified or that they would not need to be certified were it not for the law. The other four farms did not comment on the issue. Those nine farms that did comment all said the same thing, almost word for word: "We don't need certification; our customers know us." When I spoke with the produce manager at the co-op, I was told that the store is required to comply with the new federal law about labeling. But would he require certification if it weren't for the new law? He said no. They never had up until now. "We took them for their word; and we know who most of them are anyway."

Because their farming practice is embedded in community, the farmers in this study do not need the new legislation. And perhaps neither do consumers, for there is another question to be asked: could the new legislation be harmful? Kropotkin (1902) believed that the intervention of authority leads to a decline in cooperative institutions. Without a certifying authority people would *have* to build trust by personally engaging with the people and processes on which they depend. The new legislation encourages people to shift their trust away from relationships and experiences that are accessible to them and to place trust in an authority that is not so accessible—and perhaps not so trustworthy. When trust is no longer something negotiated between neighbors but something embodied in a label produced by legislation and when the role of farmers is only to compete in the production of commodities and in the ability to influence or evade legislation, then the foundation is being laid anew for the type of agriculture from which organic farming is supposed to be breaking away.

The observations made during the course of my study in southeastern Vermont tend to support Lyson's claim that "[c]ommunity problem-solving, rather than economic competition, is the social foundation of sustainable agriculture" (2002, 195). The farmers and customers that I interviewed appear much more greatly motivated by their relationship to the practice of farming, to the landscape, and to each other than they are by a desire to maximize profit or to get the most for each of their dollars. Production and distribution activities are characterized more by cooperation than by competition, and for this reason are not in need of government regulation. Cooperation is facilitated by a setting in which people interact frequently enough to develop strong social ties and by institutions in which people have sufficient emotional and practical investment that informal social controls can operate effectively. It is this kind of social context that I am referring to when I use the word "community." Because sustainable agriculture requires an economics of cooperation, it needs to be embedded in community.

The practices described in this study are good strategies for small farmers, and many of them have been discussed by other authors in this volume and elsewhere. In describing them here, I have endeavored to show that they are indicative of a certain model of economic integration that I call community problem solving. I cite these practices as evidence that sustainable agriculture is based on a community problem-solving model rather than a neoclassical economics model. The conclusion I draw is that sustainability requires us to look beyond the market paradigm. Recognizing that different techniques of agricultural production are rooted in different forms of economic integration, proponents of sustainable agriculture need to focus their efforts on supporting community problem-solving institutions and the conditions under which these institutions flourish.

References

Allen, P., and C. Sachs. 1991. "What Do We Want to Sustain?" *Sustainability in the Balance*. Issue Paper No. 2. Santa Cruz: University of California.

Avery, D. T. 1997. "Environmentally sustaining agriculture." *Choices*, 12 (1): 10–14.

Berry, W. 1995. *Another Turn of the Crank*. Washington DC: Counterpoint.

Beus, C. E., and R. E. Dunlap. 1990. "Conventional versus alternative agriculture: The paradigmatic roots of the debate." *Rural Sociology* 55 (4): 590–616.

Blank, S. C. 1999. "The end of the American farm." *The Futurist* 33 (4): 22–27.

Bookchin, M. 1976. "Radical Agriculture." In *Radical Agriculture*, edited by R. Merrill, 3-13. New York: Harper & Row.

Commoner, B. 2002. "The spurious foundation of genetic engineering." *Harper's*, February.

Geisler, C. C., and T. A. Lyson. 1991. "The cumulative impact of dairy industry restructuring." *BioScience* 41 (8): 560–67.

Goldschmidt, W. 1978. *As You Sow*. Montclair NJ: Allanheld, Osmun.

Heffernan, W. 1999. "Consolidation in the Food and Agriculture System." Report to the National Farmers Union. http://www.nfu.org/wp-content/uploads/2006/03/1999.pdf (last accessed April 22, 2006)

Heffernan, W., and M. Hendrickson. 2002. "Multi-National Concentrated Food Processing and Marketing Systems and the Farm Crisis." Paper presented at the annual meeting of the American Association for the Advancement of Science, Symposium: Science and Sustainability, Boston. http://www.foodcircles.missouri.edu/paper.pdf (last accessed April 22, 2006).

Jacobs, L. 2000. "Organics: Yesterday, Today, and Tomorrow." Paper presented at the Agricultural Outlook Forum. http://agecon.lib.umn.edu/cgi-bin/pdf_view.pl?paperid=3034&ftype=.pdf (last accessed April 22, 2006)

Kimbrell, A., ed. 2002. *Fatal Harvest: The Tragedy of Industrial Agriculture*. Washington DC: Island.

Korten, D. C. 1995. *When Corporations Rule the World*. San Francisco: Berrett-Koehler; Bloomfield CT: Kumarian.

Kropotkin, P. 1902. *Mutual Aid*. Reprinted. Montreal: Black Rose, 1989.

Lehman, K., and A. Krebs. 1996. "Control of the World's Food Supply." In *The Case Against the Global Economy*, edited by J. Mander and E. Goldsmith, 122–30. San Francisco: Sierra Club.

Lyson, T. A. 2000. "Moving toward civic agriculture." *Choices*, 15 (3): 42–45.

———. 2001. "The promise of a more civic agriculture." *Catholic Rural Life* 43 (2): 40–43.

———. 2002. "Advanced agricultural biotechnologies and sustainable agriculture." *TRENDS in Biotechnology*, 20 (5): 193–96.

Moore, M. 2002. "Hidden Dimensions of Damage: Pesticides and Health." In *The Fatal Harvest Reader*, edited by A. Kimbrell, 130–47. Washington DC: Island.

Nottingham, S. 1999. *Eat Your Genes*. Cape Town: University of Cape Town Press.

Ponting, C. 1991. *A Green History of the World*. New York: Penguin.

Putnam, R. D. 1993. *Making Democracy Work*. Princeton NJ: Princeton University Press.

Regal, P. 2000. "A brief history of biotechnology risk debates and policies in the United States." An occasional paper of the Edmunds Institute. Washington DC: Edmunds.

Rifkin, J. 1998. *The Biotech Century*. New York: Tarcher/Putnam.

Starke, A. M. 2001. "New rules to govern organic label." Newhouse News Service. http://www.newhouse.com/archive/story1c072601.html (last accessed April 22, 2006).

Strange, M. 1984. "The Economic Structure of a Sustainable Agriculture." In *Meeting the Expectations of the Land*, edited by W. Jackson, W. Berry, and B. Colman, 115–25. San Francisco: North Point.

USDA. 2000. "Glickman announces national standards for organic food." Release No. 0425.00. http://www.usda.gov/news/releases/2000/12/0425.htm (Last accessed April 22, 2006).

USDA-NASS. 2005. "Broiler Industry." Trends in U.S. Agriculture: A Walk through the Past and a Step into the New Millennium. http://www.usda.gov/nass/pubs/trends/broiler.htm (last accessed April 22, 2006).

White, D. G., S. Zhao, R. Sudler, S. Ayers, S. Friedman, S. Chen, P. F. McDermott, S. McDermott, D. D. Wagner, and J. Meng. 2001. "The isolation of antibiotic-resistant salmonella from retail ground meats." *New England Journal of Medicine*, 345 (16): 1147–54.

17. Community Food Projects and Food System Sustainability

Audrey N. Maretzki and Elizabeth Tuckermanty

Former U.S. House of Representatives Speaker Thomas P. "Tip" O'Neil is credited with saying, "All politics is local," by which he meant that public decisions are influenced by situations that affect citizens in the communities in which they reside, raise their families, and earn their livelihoods. When the United States was an agrarian society, the politics of food was largely driven by the interests of local landowners who controlled the key food system assets of land, labor, and capital. Today, the historical relationship between food and local communities no longer exists in the United States. The politics of the food system has shifted, on the one hand, to national and international levels at which agricultural subsidies and global trade issues dominate the political agenda and, on the other, to considerations of household food security for which federal food and nutrition programs are intended to provide a safety net for those in need.

Food system issues other than land use seldom play a role in the election of local officials. In fact, food systems rarely even appear on the agenda of city planners (Abel 2000; Pothukuchi and Kaufman 2000). We could assume therefore that food is largely absent from the local public agenda in the United States. Yet food systems have come to the attention of a broad swath of citizens who have initiated many distinctive efforts to retain or restore a portion of a locality's food system for the economic, sociocultural, aesthetic, and health benefits of all who live in that geographic area (Biehler et al. 1999; Pothukuchi et al. 2002).

Community Food Systems

Since food has long been a catalyst to bring people together, using the term "community food systems" reflects the central role that food plays in the lives of people who want to secure food locally for humanitarian, ecological, and economic reasons as well as for social reasons. The community food systems movement is challenged by the need to bring together many distinctly different local food system efforts into a definable community food mosaic. This mosaic illustrates the vitality of what is emerging as a social movement as well as a fledgling political agenda. The concept of local community food systems incorporates ecological principles of energy conservation, land stewardship, social capital building, and economic development, all of which are considered to be equally important goals. In community food systems food is the organizing tool for improving nutrition and health by increasing access to fruits and vegetables as well as supporting local economic development and promoting collective local environmental action (Ashman et al., 1993).

Local food system initiatives have grown out of both farmers' needs and consumers' demands. Some local food system initiatives are economically viable because increasing concentration and integration in the global food system allows space for niche markets to thrive. This success is most evident when the economy is expanding. But consumer demand for locally produced food is also promoted by a waning trust in the safety and nutritional quality of the global food supply as well as by an increasing awareness of the energy cost of transporting food from distant producers to local eaters. The policy environment, however, presently is not organized to support local food systems or to enable these systems to respond to the needs of the currently expanding sector of the population that cannot afford the premium prices generally charged for locally produced agricultural products.

The Community Food Security Movement

The founding in 1994 of the Community Food Security Coalition (CFSC) marked the beginning of a concerted effort to bring political visibility to the reality of community food systems. The following definition of community food security, written by Michael Hamm and Anne Bellows and employed by the coalition, reflects the way CFSC views a food-secure community: "[c]ommunity food security is a condition in which all community residents obtain a safe, culturally acceptable, nutritionally adequate diet through a sustainable food system that maximizes community self-reliance and social justice."[1] The CFSC definition thus goes beyond traditional antihunger approaches to encompass the need to protect and promote local family-based agriculture as essential to a food-secure community (Joseph 2000).

There was, and still is, a bifurcation of the U.S. food system that enables those with financial means to acquire high-quality locally grown organic produce while those without adequate financial resources oftentimes are unable to avail themselves of sufficient food, much less high-quality locally grown food. The small corner markets and convenience stores that serve disadvantaged urban communities usually offer only a very limited choice of fresh produce, and the fruits and vegetables they sell are expensive and often of inferior quality. Ironically, these urban food deserts may lie within a relatively short distance of rural areas where small farmers, unaware of possible demand for quality produce, proclaim an inability to reach customers. At the lowest rung of the economic ladder are individuals and families who frequently resort to emergency foods available through local food pantries. These pantries, typically run by the faith community, have very limited storage facilities for perishable foods and consequently provide primarily shelf-stable processed items. Many of these processed products are donated by food manufacturers in exchange for federal tax benefits. The observation that many of these processed foods are high in fat, sugar, and salt and may therefore be contributing to chronic health problems such as obesity, cardiovascular disease, and diabetes—diseases that disproportionately affect the poor in the United States—has led food banks and pantries to search for ways to improve the variety and nutritional quality of foods included in their offerings. This quest has become especially important in light of increased reliance on emergency sources of food by families whose dietary needs are not being met by the public safety net (Poppendieck 1998).

The CFSC initially had a distinctly urban orientation. However, in collaboration with a number of agricultural groups and rural organizations, it quickly became concerned with bringing visibility to the plight of small family farmers, many of whom were unable to remain economically viable in a food system in which concentration and integration increasingly resulted in their marginalization (Joseph 2000).

Community Food Projects

A 1996 amendment to the Food Stamp Act of 1977 (SEC. 25) was enacted as a direct result of very active lobbying by a coalition of groups led by CFSC that were concerned with the economic plight of small farmers, access to healthful foods by low-income citizens, and the importance of food in building cohesive, caring local communities. This legislation established the Community Food Projects (CFP) Competitive Grants Program and authorized USDA's Cooperative State Research, Education, and Extension Service (CSREES) to award $16 million in grants over a seven-year period. The program provided funds to private non-profit organizations that had experience in the area of community food work so that they could conduct projects that would be self-sustaining after a one-time infusion of federal assistance. A maximum grant of $250,000 for a three-year period was awarded to organizations that could demonstrate that a similar amount in matched funds would be provided locally. These CFP projects were intended to highlight innovative approaches by which communities could "meet the food needs of low-income people . . . , increase the self-reliance of communities in providing for their own food needs, and promote comprehensive responses to local food, farm and nutrition issues."[2]

An initial group of thirteen organizations received grants in 1996. The total number of projects funded grew to over one hundred by 2002, with several organizations receiving second grants to address new objectives. As a result of legislative action in 2002, the CFP Competitive Grants Program was reauthorized, its approved funding level being doubled and the maximum grant being increased to $300,000. By 2007 over two hundred and fifty grants had been funded.

Because of their combined focus on food and farm and nutrition issues, the Community Food Projects collectively constitute a national incubator in which comprehensive, but relatively small-scale, food system innovation is taking place community by community. Even if limited in their scope, the ability of these highly visible projects to raise communities' understanding of the value of "thinking and acting locally" to become more food-secure should not be underestimated. These local experiments, taken individually, cannot be expected to transform the prevailing food system, but viewed collectively, they draw attention to the interacting social, political, and economic forces that support the concepts of a community food movement.

The innovative strategies implemented with CFP support have had a demonstrable impact at the local level, and their combined success may contribute to the long-term viability of local food systems. A particular approach that is successfully employed in one community may or may not be employable elsewhere, but a national incubator affords the opportunity to try out various methods in diverse settings and to draw attention to the importance of maintaining viable local food systems. Even small changes can make a big difference in a community if a few working farms remain economically viable, with the stewardship of the land being maintained and poorly fed households gaining better access to high-quality food (Tauber and Fisher 2002).

The new directions that emerge from successful community food project models can engage the minds, the hearts, and the imaginations of leaders who are committed to bringing about change in public and corporate policies. Such policies have the potential to more comprehensively address local food, farm, and nutrition issues and to mitigate the effect of current policies that promote cheap food for those who can afford to pay while failing to recognize that marginalized farmers, the poor, and taxpayers in general are required to pay in other ways for the luxury and the efficiency of a global food system.

Sustainability and Impact of Community Food Projects

A study, funded in part by CSREES and conducted by the senior author of this chapter, was begun in 2000 to examine selected aspects of several USDA community food projects that were identified by the federal program staff as successful in addressing the congressional intent expressed in the CFP legislation. In-depth discussions were conducted on-site in late 2000 and again by telephone in early 2003 with individuals associated with these projects (Maretzki 2001). All of the sponsoring organizations were continuing to address the community food security objectives for which they received their initial CFP grant. None of the successful strategies were abandoned when federal funding ended, and innovative approaches frequently were adopted to maintain those activities that increased food self-reliance in the community while meeting the food needs of low-income people. Replicating these site-specific approaches elsewhere, however, may prove difficult, largely because successes are associated with dedicated, charismatic leaders whose vision drives the project and inspires trust on the part of other local collaborators.

The private nonprofit organizations experienced in community food work when the CFP grants were initiated included those established either to help food-insecure individuals and families with urgent needs for quality food or to aid small family farmers with operations that were not economically viable in an increasingly cost-competitive agricultural economy. Small farmers were searching for markets but did not see low-income consumers as a marketing target. Meanwhile, low-income consumers and the food banks and pantries that were attempting to respond to the needs of a very vulnerable population were unlikely to see local farmers as a part of the answer to their food problems. Consequently, the organizationally preferred approach to a community's food-security tended to emphasize either the situation of resource-stressed farmers or that of households with inadequate food resources, but not both.

The challenge faced by organizations submitting CFP proposals was to develop both innovative and comprehensive approaches to a community's food, farm, and nutrition issues. In this environment the scene was set to identify creative ways to allow local farmers to capture a portion of the federal funds that were being spent to improve the nutrition of the nation's poor. Projects conducted by organizations in urban areas were likely to focus on a number of goals. These goals included creative strategies to get locally grown fruits and vegetables into the offerings of food pantries, strategies to identify ways to defray the cost of community supported agriculture (CSA) subscriptions for low-income consumers, strategies to encourage the formation of farmers' markets in low-income areas where WIC and senior farmers' market coupon programs were in place, strategies to enable low-income clients to purchase locally grown produce, and strategies to promote farm-to-school initiatives that encouraged school districts with a high proportion of low-income children to buy locally.

Community food programs conducted by organizations with a rural focus were likely to include ways to identify upscale local markets in which resource-stressed farmers could sell their fresh produce individually or collectively. These markets included restaurants, resort hotels, and specialty food outlets. Rural projects also created mechanisms such as cooperatives or limited liability corporations through which a group of small farmers could market high-value agricultural products locally or create new value-added products for wider distribution. In these projects the link to low-income consumers was through skill building and job creation at the community level.

Some community food projects began with the formation of policy councils or boards that were seen as ways to generate public support for high-quality locally produced foods or to create a mandate that might redirect a portion of the public funds spent on institutional food purchases to contracts with local farmers. In California a concerted policy strategy was launched under a CFP led by the Center for Ecoliteracy in which the activities of the Berkeley Food Systems Council resulted in the adoption of a comprehensive food policy by the Berkeley Unified School District.[3] The policy had as its goal to provide children in every school in Berkeley with breakfast, lunch, and afternoon snacks, with much of the food being purchased locally or produced in school and community gardens linked to a hands-on food education curriculum in the classrooms. Implementation of Berkeley's food policy is monitored by a Child Nutrition Advisory Committee, an action that is seen as vital to the successful implementation of the plan. The extensive press coverage that resulted when the Berkeley policy was adopted gave impetus to similar initiatives being promoted in many other locations across the country.

Whether sponsoring organizations are more urban or more rural in their focus, they all find it both necessary and desirable to form partnerships with other organizations and agencies in order to address local food, farm, and nutrition issues comprehensively. These collaborations were identified by every project director as a key element in successfully bridging the gap between resource-stressed farmers and food-insecure eaters. Sponsors such as Practical Farmers of Iowa initiated collaborative activities under their CFP that were continued by other organizations when the value of the activity to the community and its relevance to the mission of the collaborating organization had been convincingly demonstrated (Tauber and Fisher 2002). When such transitions occurred in this and other CFPs, the organizations continued to work together even though the specific activity might have evolved into something quite different under new management.

People working on many community food projects are satisfied to know that in their community, small farmers can sell their products locally and receive a reasonable return for their labor while food-insecure residents have greater access to nutritious fresh fruits and vegetables. The most successful projects, however, have more ambitious goals. The Missouri Rural Crisis Center (MRCC) represents a sponsor with roots firmly planted in the agricultural sector, yet with a clear mission to help struggling Missourians feed themselves with dignity. The mission of MRCC is to preserve family farms, promote stewardship of the land and environmental integrity, and strive for economic and social justice by building unity and mutual understanding among diverse groups, both rural and urban.

MRCC challenges corporate domination of the food supply, advocating for fair food and farm policies at the state and federal levels. In 1992 MRCC created Patchwork Family Farms, an economic development project that buys hogs from its fifteen member farmers and markets the meat under the Patchwork Farms label. MRCC also supports a network of food purchasing cooperatives. The collaboration between Patchwork Family Farms' small sustainable hog farmers and the members of the Crisis Center's network of food purchasing cooperatives resulted in a doubling in one year of the amount of Patchwork Family Farms pork products purchased by the local cooperatives (Tauber and Fisher 2002). In that same year Patchwork's family farmers received fourteen cents more per pound for their animals than the going rate for hogs on the open market.

A successful project on the Tohono O'odham Reservation in Sells, Arizona, is taking its traditional agriculture very seriously as a strategy to bring an epidemic of Type II diabetes under control).[4] Tribal elders and youth are working to regenerate their traditional dryland agriculture, a holistic system that produced foods that protected these ancient agriculturalists from chronic diseases. For the Tohono O'odham, as for the Hopi, farming is not seen just as an economic necessity but as a religious duty. So in 1999 resurrection of the Nawait I:I (saguaro wine ceremony) to "sing down the rain" became an important step in bringing people back into the fields to produce healthful traditional crops. The community is now also working to increase local production of such crops as tepary beans, squash, melons, chiles, and sorghum in an effort to meet an increasing tribal demand. The Tohono O'odham Community College, meanwhile, is training students to conduct participatory research in an effort to determine how best to encourage the transition from government-donated commodities to a system that promotes local production and gathering of foods that can help control the rampant diabetes that now affects more than half the Tohono O'odham population.[5]

Rethinking Success

One might assume that successful community food projects succeed at everything that they attempt, but this outcome is certainly not the case. The following observations, gleaned from the interviews conducted, suggest that, when a project is conscientiously addressing the mission of the federal program, its leaders appear to be able to see when a particular strategy is not working and to find alternative ways to achieve the project's stated goals. These leaders are also quick to observe what works in their specific situation and to rapidly shift resources to take advantage of their project's successes. Additionally, successful project directors are constantly scanning the environment for further sources of funding that will support food system objectives that build upon the experience acquired through both their USDA project and other activities of their organization. They also establish a level of understanding of the complementary missions of other community organizations with which they are partnering. This understanding enables the community as a whole to move very rapidly in response to emerging funding opportunities.

Successful projects are almost always led by individuals who are not only visionary and opportunistic but also realistic about what can be accomplished in their communities and by whom. They are calculated risk takers who are not afraid to learn by trial and error when necessary.

The successful community food projects and their leaders have employed a number of strategies to achieve that success. For instance, when community gardens did not prove successful on the Tohono O'odham reservation, the local project acquired a cultivator that could be transported in the project truck in order to assist families who wanted a garden plot close to their house and were willing to grow their own produce once the soil was initially turned. In another example, local retail markets were not a totally successful outlet for Patchwork Family Farms pork products. However, sales to restaurants and from the Crisis Center's office in a low-income neighborhood far exceeded expectations; accordingly, these efforts, along with sales to members of food-buying cooperatives, became the backbone of an expanded marketing system. Elsewhere, the people of Kauai wanted to be assured that food bank clients would receive the same high-quality produce that was being purchased from Hui Mea'ai's growers by a resort hotel on the island, so a rigid produce grading system was initiated. Hui Mea'ai members, the island's retail outlets, and the food bank now pride themselves on providing top-quality locally grown produce on the Garden Island that is not just for tourists. In rural Hampshire County, West Virginia, a community food incubator and contract processing facility struggled to build volume in order to utilize its processing capacity, but a complementary marketing effort initiated through the same limited liability corporation was able to focus, at least temporarily, on selling local agricultural products bearing the Highland Harvest label. In summation, flexibility is clearly a characteristic of successful community food projects.

Policies That Support Community Food Security

Community food projects provide exciting examples of federal dollars being used creatively at the local level to increase food security in U.S. communities, but local efforts need to be supported by a federal policy environment that simultaneously supports both small family farmers and marginalized consumers. Local farmers typically face unit production costs that cannot be recovered directly from low-income purchasers unless these purchases are publicly or privately subsidized. WIC and Senior Farmers' Market nutrition coupon programs are current examples of the successful transfer of public sector food dollars to local farmers. Meanwhile, efforts to enable food stamp recipients to use their electronic benefits transfer (EBT) cards in farmers' markets or to use them for the purchase of CSA shares are being investigated.

Additional policy changes in the administration of programs that constitute the federal nutrition safety net for low-income consumers could directly improve dietary and food quality in these programs while expanding markets for local farmers. National efforts to encourage local sourcing of foods used in the National School Lunch, Breakfast, Summer Feeding, and After School programs as well as in the Child Care and Adult Feeding Program are slowly making headway, but local action remains the key when decisions are made independently in the more than 13,000 school districts in the United States (Azuma and Fisher 2001; Harmon 2003).

Less well understood than farmer incentives that could be provided through food and nutrition programs are those disincentives to local production created by such factors as the disappearance of independent food processors who formerly added value locally to grains, horticultural crops, and livestock. Much of the public interest in reinvigorating local food systems has been focused on high-value produce, specialty crops, and value-added processed foods such as jams, jellies, and salsas that have a relatively limited dietary impact. Consequently, local food systems may exist, and even flourish, outside the agricultural policy and agribusiness context that drives the players in the dominant food system. In the existing system policies operate effectively to keep the minimum wage as well as the price of food at the retail level quite low, enabling food stamp benefits pegged to the Thrifty Food Plan to remain low as well.[6] This situation generally pleases more affluent taxpayers, many of whom are also supporters of local farmers' markets, subscribers to CSAs, eaters at restaurants that highlight local foods, and online buyers of exotic food products shipped directly from farms and boutique processors. All of these activities can effectively support local farmers, but they do little to bridge the gap between local growers and those who earn too little to feed themselves and their families without federal or charitable assistance.

Through the USDA community food projects, an incredible number of site-specific insights have been gained into the challenge of linking the economic interests of small farmers and resource-stressed citizens, those groups in any community that are the least well served by the existing food system. These insights have served to raise the bar for CFP proposals that will be funded in the future. Enough lessons have been learned to allow policy makers, planners, citizen eaters, and human-service providers throughout the country to seriously consider what it means for their community to be food secure and whether food-farm-family linkages that encompass these marginalized groups can contribute to the social and biological as well as the economic security of our nation. Community food project success stories clearly need to be told but so also do stories that explicitly illustrate what was tried and failed and how that failure was used by determined leaders as a stepping stone to move toward the goal of meeting the food needs of low-income people and increasing the self-reliance of communities in providing for their own food needs.

Acknowledgments

This project was supported by the Community Food Projects program of the USDA Cooperative State Research, Education, and Extension Service under Cooperative Agreement no. 90-CSA-PA1–144. The views expressed herein are those of the contributors and do not necessarily reflect or represent the opinions or policies of the U.S. Department of Agriculture or its Cooperative State, Research, Education, and Extension Service.

Notes

1. Community Food Security Coalition. http://www.foodsecurity.org/views_cfs_faq.html (last accessed April 30, 2006).

2. CFP Competitive Grants Program. http://www.csrees.usda.gov/nea/food/in_focus/hunger_if_competitive.html (last accessed April 30, 2006).

3. Center for Ecoliteracy. http://www.ecoliteracy.org (last accessed April 38, 2006).

4. TOCA Community Food System Program. http://www.tocaonline.org/homepage.html (last accessed April 30, 2006).

5. For further information, see Tohono O'odham Community Action (2002).

6. Center for Nutrition Policy and Promotion. http://www.usda.gov/cnpp/foodplans.html (last accessed April 30, 2006).

References

Abel, J. L. 2000. "Assessing the involvement of Pennsylvania professional planners in food system activities." Master's thesis, Pennsylvania State University.

Ashman, L., J. de la Vega, M. Dohan, A. Fisher, R. Hippler, and B. Romain. 1993. *Seeds of Change: Strategies for Food Security for the Inner City*. Venice CA: Community Food Security Coalition.

Azuma, A., and A. Fisher. 2001. *Healthy Farms, Healthy Kids: Evaluating the Barriers and Opportunities for Farm-to-School Programs*. Venice CA: Community Food Security Coalition.

Biehler, D., A. Fisher, K. Siedenburg, M. Winne, and J. Zachary. 1999. *Putting Food on the Table: An Action Guide to Local Food Policy*. Venice CA: Community Food Security Coalition and California Sustainable Agriculture Working Group.

Harmon, A. 2003. *Farm to School: An Introduction for Food Service Professionals, Food Educators, Parents, and Community Leaders*. Venice CA: National Farm to School Program, Community Food Security Coalition.

Joseph, H., ed. 2000. *CFS: A Guide to Concept, Design, and Implementation*. Venice CA: Community Food Security Coalition.

Maretzki, A. N. 2001. "Our Food—Our Future: Sustainability and Impact of Selected Community Food Projects." Unpublished report to USDA/CSREES. University Park PA: Department of Food Science, Pennsylvania State University.

Poppendieck, J. 1998. *Sweet Charity*. New York: Penguin Putnam.

Pothukuchi, K., H. Joseph, A. Fisher, and H. Burton. 2002. *What's Cooking in Your Food System? A Guide to Community Food Assessment*. Venice CA: Community Food Security Coalition.

Pothukuchi, K., and J. Kaufman. 2000. "The food system: A stranger to the planning field." *Journal of the American Planning Association* 66 (2): 113–24.

Tauber, M., and A. Fisher. 2002. *A Guide to Community Food Projects*. Venice CA: Community Food Security Coalition.

Tohono O'odham Community Action and Tohono O'odham Community College. 2002. *Community Attitudes Toward Traditional Tohono O'odham Foods*. http://www.tocaonline.org/programs/food%20system/images/usdareport.pdf (last accessed April 30, 2006).

Conclusion: A Full Plate

Challenges and Opportunities in Remaking the Food System

C. Clare Hinrichs and Elizabeth Barham

Efforts to remake the North American food system have become more widespread and diversified through the 1990s and early 2000s. As observers, analysts, and participants, we recognize the growing hum of initiatives by farmers, educators, activists, researchers, and citizens who are now inspired and concerned enough to work on redirecting particular components of the food system toward greater sustainability. Harmonies certainly can be detected across these efforts but so can distinctive melodies, born of circumstances, resources, interests, and needs in particular places and regions.

As the chapters in this book have shown, new institutions and practices linking food production and consumption provide the most obvious social organizational evidence of change in the food system. Fueled by a potent mix of disenchantment with what is and optimism about what might be, new institutions and practices emerge from hard work and sometimes from hard-won lessons. This book has presented the flourishing, but also the fragility, of farmers' markets; the striking innovation, yet sometimes sobering reality, of community supported agriculture; the democratic experiment but also the operational constraints of local and state food policy councils; and the potential revalorization of place in origin-labeled products, something both prompted and challenged by new global relations of economy and trade. Emerging from a multidisciplinary base of empirical research and practical engagement, the picture provided by this book is ultimately one of informed, cautious hopefulness.

Our hopefulness is grounded in a clearer understanding of how food system initiatives draw their strength from common ways of working. We have organized this understanding into a series of conceptual frameworks that we believe provide key insights into the shared dynamics of a varied set of efforts to create change. The most encompassing frameworks developed relate to the values and goals that tie food system actors together. Thomas Lyson's notion of civic agriculture, for example, pictures locally based agriculture as relinking food production to a community's social and economic development and food system actors as engaged in a form of democratic practice emanating from the grass roots (this volume; 2004). Laura DeLind, on the other hand, has called for more "discussion of the nature and potential of civic agriculture not only as an alternative strategy for food production, distribution and consumption, but also as a tool and for grounding people in common purpose—for nurturing a sense of belonging to a place and an organic sense of citizenship" (2002, 217). These differing interpretations provide evidence of the evolving conversation on food system change. Both resonate in some degree with the notion shared by this book's contributors, including Hoffman, Barham, Gillespie and colleagues, and Thompson and colleagues, that agricultural and food issues must be reembedded into broader community values and guided by honest and equitable accounting of community needs.

Taking their cues more from actions than values and motivations of food system actors, Stevenson and colleagues develop the concept of coalitions weaving initiatives across food sectors and platforms as a powerful way to foster change, as with food policy councils. In another notable manifestation of weaving, researchers themselves become actors in these chapters through participatory research strategies that link academics and practitioners together, as in Ostrom's CSA studies or Lev and colleagues' work to enhance farmers' markets. The brief biographies included in the section "About the Contributors" testify to the engaged research approaches of the authors included here and should sensitize readers to the contribution of academic and practitioner partnerships in research and practice aimed at understanding and guiding food system change.

Most of the practices and institutions we have documented turn on deepening and dovetailing traditional producer and consumer roles.

In many respects this deepening and dovetailing calls for new flexibility and creativity in identities, skill sets, time management, and social relations. The small-and medium-scale family farmers featured in our research cannot content themselves with knowing their farm machinery, their livestock, or the needs of their fields. They also must grapple with marketing strategy, consumer relations, and more, as Marcia Ostrom makes clear in her chapter on community supported agriculture. Similarly, consumers must do a good deal more than mindlessly shop. As customers they are now also associates of particular farmers with whom they interact and whom they support; they are expected to act as food citizens, vigilant about monitoring and participating in both public and private arenas that have bearing on their desired food and agricultural system. In the scenarios presented here, participating in the food and agricultural system as either a producer or a consumer may well be more demanding but also potentially more rewarding. Continued and careful documentation of how those demands and rewards are allocated remains important.

The new practices and the institutions that they create also have the potential to promote systems orientations, if not full-blown systems thinking. For example, being a consumer at a farmers' market initially may be motivated by the desire for fresh food. But interacting with farmers on a personal basis also may stimulate recognition of the need for farmland preservation if those farmers are to continue their enterprises. It may increase appreciation of seasonal constraints on food availability and of crops and varieties particular to the region. Developing a deeper set of roles and concerns thus opens up the possibility of more connections and linkages and a wider view on food and farming, countering the common tendency to compartmentalize interests and problems.

As this book has been assembled, certain trends and patterns in food system work have become more pronounced. Emerging issues and new developments affecting the social, political, economic, and environmental context of food systems also have come into view. We now consider some of the elements that we believe are most likely to pose important challenges and opportunities for remaking the food system through the first decade of the twenty-first century.

Challenges

One vexing challenge for efforts to remake the food system, particularly in relation to garnering public and private support, is the crucial issue of definition. The best initiatives and actions to remake the food system are comprehensive and integrated, typically involving multiple sectors, various disciplines, diverse skills, and different types of participants. Is a community supported agriculture (CSA) farm primarily a farm business or a community resource, a commitment to landscape or a source of recreation, a setting for public education or a promoter of public health? It is, of course, in various ways all these things. But multiple roles, functions, and identities can complicate the ability to draw on what is generally more compartmentalized support.

Similarly, as initiatives such as those described in this book mature, many recognize that they must integrate their multiple strands of action. However, limitations of capital and human resources affect their ability to achieve their desired comprehensiveness and reach. Not matching standard or homogeneous program definitions, they do not fit into the usual boxes for public funding or private financing. Thus some initiatives to remake the food system may, not so much by choice as by circumstance, retain a piecemeal quality due to the fragmentation of resources available to launch and sustain their work. One solution would seem to be more flexible sources of funding, which will require concerted and continuing education and lobbying of those holding the purse strings and of those seeking benefits in just one area about the advantages of a more comprehensive, integrated approach.

A further and related challenge is the ongoing problem of action at the local level. Local efforts to remake the food system often remain just that—local. Whether doing builder or warrior work, as Stevenson and colleagues put it, people and groups working for a different, more sustainable food system face issues both in scaling out and scaling up (Johnston and Baker 2005). Some replication (or near replication) of models does occur in new locations and contributes to the sense of gathering momentum shared by many food system activists. In the U.S. context this replication tends to happen more through cultural diffusion and information exchange than through the influence of any universalizing governmental program or mandate. Indeed, networking and learning across small similar projects clearly has led to some powerful outcomes. Less ideal, however, are missed opportunities to train a more deliberate analytical eye on the experience of similar efforts in other places. Our effort to integrate the theme of place in this book is one response to the need for a more holistic and comparative approach in food system studies.

The issue of scaling up may be the more challenging. Related to the issue of definition discussed above and to the short-term, project-based nature of much food systems work, organizers and activists are often understandably drawn to, and arguably confined to, feasible local actions (Allen et al. 2003). These actions more typically center on entrepreneurial work (that is, direct and niche marketing) than on national or international level policy changes. Despite its economic and even civic contribution, such local action focused on entrepreneurship may also limit the potential for addressing the structural basis of current problems in the mainstream food system (Allen 2004).

The weaver work described by Stevenson and associates addresses this problem in principle and offers some hope for reinforcing local achievements and learning from local missteps and failures. Certainly the growing proliferation and vitality of federations and conferences providing forums for individuals and groups to work together on similar problems offer encouragement. Several states now have established state-level farmers' market federations. Further, the North American Farmers' Direct Marketing Association, including its Farmers' Market Coalition, provides education, training, and networking. These organizations are joined by several state and regional associations of CSAs such as the one recently organized in Michigan and efforts to bring together various food policy councils for conferences and workshops. Such groups act in effect as new trade and professional associations for a future food and agricultural system. By organizing in this way, separate local initiatives are better able to follow and potentially influence policy at higher levels, in much the way farmers' market associations have recently engaged with U.S. national policy governing the Farmers' Market Nutrition Program.

Increasingly, there is recognition of the need to lobby and act at highly significant policy junctures, such as the periodic engagement over the U.S. Farm Bill. Many grassroots organizations have been hesitant to engage in policy work, fearing that it might become a black hole for their scarce time and resources. But it is only by engaging with policy to some extent that these organizations will gain the insight necessary in order to devise solutions to persistent impasses in the larger food and agricultural policy arena.

Many such challenges in the United States and to a lesser degree Canada can be understood within the framework of global neoliberalization (Buttel 2002). Such efforts to remake the food system as those described in this book are inescapably shaped by the ascendancy of free-trade rhetoric and resolutely market-oriented federal agricultural policy. This shaping translates into the primacy of funding for commodity export production at the international level and a national food policy geared to conventional production and marketing channels. The role of government funding for the types of alternatives described here remains modest. In the United States the federal government provides some funding through comparatively small pots of USDA money, such as that distributed through the Sustainable Agriculture Research and Education grants program, the Community Food Projects program (discussed by Maretzki and Tuckermanty in this volume), and the Risk Management Agency. Some individual states have modest funding streams to support small food system projects like those through the Leopold Center for Sustainable Agriculture based at Iowa State University and the Sustainable Agriculture Research and Education program based at the University of California, Davis. Innovative federal programs such as the USDA's Fund for Rural America or its Initiative for Future Agriculture and Food Systems have offered and delivered greater resources, coupled with a valuable emphasis on integrated projects, but have lacked consistent funding over the years. Overall an underlying presumption persists that markets can best address the sustainability of the food system, and commitments from government can and perhaps should be more limited.

Given limited and sometimes fragmented government funding, the initiatives we have explored display a remarkable strength and vitality. One could argue in fact that lack of substantial governmental support has stimulated self-reliance at the grassroots level. Still the backdrop of neoliberalism also reduces space for many creative possibilities and reinforces a troubling need to compartmentalize efforts as new initiatives are invariably called upon to meet conventional market definitions of the public good in the first instance.

A remaining thorny challenge for efforts to remake the food system is the meaningful integration of social justice criteria. The neoliberal context described above in underscoring a certain primacy of the market also subordinates other concerns and interests that include meaningful participation by diverse populations and benefits sharing with disadvantaged groups. Particular types of food system initiatives, such as CSA and farmers' markets in some places and regions, have been innovative and sometimes dogged in their efforts to incorporate and address social justice concerns. However, public attention to food system change still tends to focus on elite enthusiasms for, say, pricey heritage tomatoes or the lifestyle statement made by purchasing artisanal bread from a particular farmers' market vendor. Far less attention is directed to spreading the presumed—and more varied—benefits of local food to all populations.

Opportunities

In the United States societal acquiescence in the matter of agricultural production subsidies may now be eroding. The conditions of U.S. agricultural production and trade can fluctuate significantly from year to year. There are also deep-seated cultural understandings and norms that are slow to change. The 2004 decision within the World Trade Organization (WTO) against the acceptability of U.S. subsidies to cotton offered a warning to the whole of American commodity production that is similarly structured (such as for corn, wheat, soy, and sugar) that major production subsidies are no longer sacrosanct. Information recently made public on the Web site of the Environmental Working Group about the amount and destination of U.S. agricultural subsidy dollars also has fueled public awareness and questioning about equity issues and environmental impacts of mainstream agriculture.

While cuts to major commodities such as corn and soybeans may be slow to come, conversations are already being engaged as to what U.S.

agriculture and rural landscapes might look like if production subsidies for a few large program crops either are reduced or eliminated. On the one hand, the disproportionate number of urban and suburban voters versus their rural counterparts and the potential pressures from the WTO suggest a possible reconfiguration of longstanding subsidy patterns to U.S. agriculture. On the other hand, recent intensified government and industry interest in addressing energy concerns by producing more fuels from subsidized agricultural crops such as corn and soy could serve to reinforce the present system. While contradictory forces appear in play, it is clear that for the first time since World War II, new lines of public debate are opening over the Farm Bill that governs so much of U.S. food and agricultural policy. Potentially dramatic changes are newly imaginable.

Sudden reductions or radical changes in agricultural subsidies would have a negative influence on the well-being of many U.S. rural communities. The American government has not shifted assistance to WTO-compliant categories such as rural development and direct payments to farmers to the same extent as European Union countries. On the one hand, this comparative lack of assistance makes U.S. rural communities more vulnerable to major dislocations in the form of farms in foreclosure, lost jobs in rural towns, and an accelerated exodus of the rural population, in particular young persons. On the other hand, the situation may present an opportunity for rural communities if leadership emerges in Congress to call for a redirection of agricultural funding toward rural development. This redirection would mean that rather than leaving the agricultural sector altogether the dollars that have gone to crop-related subsidies might be reallocated to rural development, including more sustainable models of farm production and marketing integrated with community and landscape (Cochrane 2003).

It is not clear whether we will face a situation of having to react, for example, to WTO rulings or of initiating proactive changes. At the time of this writing, environmental and agricultural coalitions have come together in anticipation of the need to press for more attention to what might be called a multifunctionality agenda in the 2007 Farm Bill. The influence of such developments on the kinds of practices documented in this book surely depends on deeper American attitudes and patterns of politics that have influenced every Farm Bill. But the 2007 Farm Bill, whatever its content, will also have emerged through important and intensifying topics of debate. In the past this debate has been fueled simultaneously by pressure to reform agricultural subsidies, shifting perceptions of national interests, the deepening challenges for much of rural America, and new food-related concerns of urban and suburban citizens. Each of these factors will probably continue to play a role in this ongoing debate.

Another opportunity supporting efforts to remake the food system in directions we have described is growing public and media attention to the potentially negative public health implications of the conventional food and agricultural system. As increasing rates of obesity and diabetes become a matter of public record, links are being suggested to both the high quantity and the poor quality of the mainstream American diet as promoted by much of the food industry (Nestle 2002). As subsidized corn and soy now figure prominently in processed foods, including soft drinks, baked goods, snacks, and desserts, some have gone so far as to claim that the present system of agricultural subsidies in part explains why Americans are obese (Pollan 2006). Recognizing the growing salience of health as an issue, regional sustainable agriculture organizations recently have organized their annual conferences around the theme of health, precisely because this topic speaks so compellingly to people across the food system.

The sudden popularity of another type of food system initiative, farm-to-school projects, has been driven as much by a desire to improve the diets and health of children as by the need to develop new market outlets for sustainable family farmers (Vallianatos, Gottlieb, and Haase 2004). Similarly, consumer interest in grass-fed beef has been spurred by some concern to avoid meat from animals given antibiotics, just as interest in organic produce stems significantly from consumer concern about the health implications of pesticide use on conventional fruits and vegetables. Given consumer perception that directly sourced local foods are fresher, freer of agrichemicals, and hence healthier, there are opportunities to mesh at least some consumers' interest in wellness and health with food choices centered on local food. But experience tells us that caution is in order, as without strong scientific evidence, an emphasis on health can devolve into a matter of claim-counter claim and continuing food wars.

Efforts to remake the food system also will find new opportunities emerging from the growing evidence of potential vulnerabilities in a food supply sourced at great distances, often from around the globe. While international trade in food—be it wine, cheese, grapes, or soybeans—is unlikely to disappear (nor should it necessarily in every case), technological and natural disasters as well as the specter of large-scale acts of terrorism have raised fresh public concerns about the risks of undue reliance on distant food sources. For the United States hurricanes Katrina and Rita in 2005 brought home the vulnerabilities of trade dependence and the massive disruption of everyday services and essentials likely to occur with such catastrophic events. These anxieties are only exacerbated by rising energy prices in the early twenty-first century, the growing salience of discussions about peak oil, and inevitable shifts in energy priorities, which are likely to reconfigure the spatial organization of the food economy.

In short, the vulnerabilities of a long-distance, industrial food and agricultural system have become more noticeable and worrisome with recent events. This worry fuels public perception of and receptivity to potential advantages in eating closer to home, knowing at least some of the producers of one's food, and supporting their efforts to sustain their enterprises. Beyond these perceived benefits, it suggests more grounds for renewed public interest in the preservation of farms and farmlands. In brief, reminders in the daily news of potential perils in relying on food from an unspecified anywhere will prompt more people to consider the provenance of their food and to seek to know how at least some of that food might be obtained from nearby sources.

Despite the many accomplishments of efforts to remake the food system, it seems clear that these efforts do not as yet present us with a comprehensive, coordinated plan of action for the future. Some have made the case that a national green plan for sustainability is needed, and food systems would presumably figure importantly into such planning (Johnson 1995). Many European countries offer examples of attempts at national planning of this sort, but nothing quite this comprehensive can be identified in the North American context. Given the political moment in the United States, it seems unlikely that anything along these lines could be proposed, let alone negotiated and implemented.

But should the varied initiatives that we have detailed be held to task for the lack of an overarching national or regional plan or vision for the entire food system? Perhaps Jules Pretty's advice on "connecting up the promising cases" (2002, 187) is the better way to approach this challenge, with a careful eye to more systematic accounting of the impacts over the longer term of promising cases and examples. This volume makes a contribution to what we hope will continue to be an ongoing research effort at documenting, monitoring, and evaluating new directions in the North American food system. While such research efforts can appear a luxury in the context of often underfunded and overstretched grassroots organizations, it will be crucial for assuring that their accomplishments and discoveries are retained and gathered together to inform more systematic planning efforts that may yet emerge in the future.

In the meantime, individuals and groups concerned with the food system will continue to work to make changes intended to improve environmental and human health, economic viability, and social equity. They may deliberate and sometimes argue about the details of those changes. They may try out strategies, adapting some and abandoning others. We will know whether and to what extent they are succeeding as we make progress together toward these goals, looking back with better understanding at what has been accomplished and looking forward with greater insight to the next wave of creative initiatives toward meaningful, sustained change in the food system.

References

Allen, P. 2004. *Together at the Table: Sustainability and Sustenance in the American Agri-food System*. University Park PA: Pennsylvania State University Press.

Allen, P., M. FitzSimmons, M. Goodman, and K. Warner. 2003. "Shifting plates in the agrifood landscape: The tectonics of alternative agrifood initiatives in California." *Journal of Rural Studies* 19:61–75.

Buttel, F. H. 2002. "Continuities and Disjunctures in the Transformation of the U.S. Agro-food System." In *Challenges for Rural America in the 21st Century*, edited by D. L. Brown and L. E. Swanson, 177–89. University Park PA: Pennsylvania State University Press.

Cochrane, W. 2003. *The Curse of American Agricultural Abundance: A Sustainable Solution*. Lincoln: University of Nebraska Press.

DeLind, L. B. 2002. "Place, work, and civic agriculture: Common fields for cultivation." *Agriculture and Human Values* 19:217–24.

Johnson, H. D. 1995. *Green Plans: Greenprint for Sustainability*. Lincoln: University of Nebraska Press.

Johnston, J., and L. Baker. 2005. "Eating outside the box: FoodShare's good food box and the challenge of scale." *Agriculture and Human Values* 22:313–25.

Lyson, T. A. 2004. *Civic Agriculture: Reconnecting Farm, Food, and Community*. Medford MA: Tufts University Press.

Nestle, M. 2002. *Food Politics: How the Food Industry Influences Nutrition and Health*. Berkeley: University of California Press.

Pollan, M. 2006. *The Omnivore's Dilemma: A Natural History of Four Meals*. New York: Penguin.

Pretty, J. 2002. *Agri-Culture: Reconnecting People, Land, and Nature*. London: Earthscan.

Vallianatos, M., R. Gottlieb, and M. A. Haase. 2004. "Farm-to-school: Strategies for urban health, combating sprawl, and establishing a community food systems approach." *Journal of Planning Education and Research* 23:414–23.

Contributors

Elizabeth Barham is nationally and internationally known for her research on labels of origin as catalysts for rural development. She holds both MS and PhD degrees in development sociology from Cornell University and currently teaches courses on the sociology of food, agriculture, and globalization at the University of Missouri–Columbia.

Jim Bingen is a professor of community, food and agriculture in the Department of Community, Agriculture, Recreation, and Resource Studies at Michigan State University. He currently works with organic, small, and family farmers and farmers' markets in order to promote local and organic food and farming that contributes to rural development in Michigan.

Troy C. Blanchard is an associate professor of sociology at Louisiana State University. His research interests include stratification, social demography, community, and rural sociology. His current research focuses on the effects of community and organizational ecology on labor market outcomes.

Holly Born has been the economics and marketing specialist for the National Center for Appropriate Technology and the Appropriate Technology Transfer for Rural Areas National Sustainable Agriculture Information Service since 1998. She holds an MA in agricultural economics from Washington State University and an MBA from the University of Arkansas.

Linda Brewer works with Oregon State University's Extension Service as a technical writer and project manager, principally with Extension Small Farms and Extension Horticulture. Her marketing research with Extension Small Farms promotes sustainability of and enhances risk management for Oregon's small farms.

Viviana Carro-Figueroa is research rural sociologist and assistant dean for research at the University of Puerto Rico Agricultural Experiment Station. Her research interests include the impact of global and local forces on Caribbean agrifood systems and emerging community agricultural development alternatives in Puerto Rico.

Kate Clancy is a consultant on a variety of sustainable agriculture and food system issues. She was previously a senior scientist at the Union of Concerned Scientists, director of the Wallace Center for Agricultural and Environmental Policy, and a faculty member of Cornell and Syracuse universities. In 1984 she was instrumental in launching the second food policy council in the United States.

Laura B. DeLind is a senior academic specialist in the Department of Anthropology at Michigan State University. As an advocate of more placed-based and democratized food systems, she studies the impact of alternative food and farming on communities, culture, and identity. She is the former editor-in-chief of *Agriculture and Human Values*.

Gail Feenstra is a food systems analyst with the Sustainable Agriculture Research and Education Program at the University of California, Davis. She has a doctorate in nutrition education with an emphasis in public health. Her research has been dedicated to integrating human, environmental, and community health through sustainable food systems.

Gilbert Gillespie is a senior research associate in the Department of Development Sociology at Cornell University. Formerly he was with the Community, Food, and Agriculture Program at Cornell. His primary research and teaching interests are food systems. Current work includes studies of dairy farmer knowledge about and views on anaerobic digestion of dairy manure and investigations of how small and midsized dairy farms can be made more viable.

Amy Guptill is an assistant professor of sociology at SUNY College at Brockport. Her research focuses on sustainable agriculture and rural development. Her recent projects have investigated the alternative agricultural movement in Puerto Rico, organic grain marketing, and organic dairies in the northeast United States.

Michael W. Hamm is the C. S. Mott Professor of Sustainable Agriculture at Michigan State University. He is affiliated with the Departments of Community, Agriculture, Recreation, and Resource Studies; Crop and Soils Sciences; and Food Science and Human Nutrition. He directs the C. S. Mott Group for Sustainable Food Systems, which conducts research, teaching, and outreach to aid the development of community-based food systems throughout Michigan.

Janet Hammer has been teaching, researching, and working on sustainable food system issues for more than fifteen years. She has helped establish collaborative regional food system initiatives, farm link programs, and a student-run sustainability oriented cafe, as well as served on the task force that created the Portland/Multnomah Food Policy Council. She is presently a doctoral candidate at Portland State University.

Alison H. Harmon is an assistant professor of food and nutrition at Montana State University, at which she also directs the dietetics program. Her research is related to incorporating sustainability concepts into teaching about food and nutrition. She teaches undergraduate and graduate courses on the food system and develops resources for educators on food system topics.

Duncan L. Hilchey is a senior extension associate with the Community and Rural Development Institute in the Department of Development Sociology at Cornell University. He was formerly with the Cornell Community, Food, and Agriculture Program. He has degrees in agricultural education and city and regional planning. His applied research and extension focuses on agricultural economic development, food system policy and planning, and microenterprise development.

C. Clare Hinrichs is an associate professor of rural sociology at Pennsylvania State University. She holds a PhD in development sociology from Cornell University and served previously as a faculty member at Iowa State University. Her research, teaching, and practice focus on transitions to sustainability in the food and agricultural system.

Matthew Hoffman is a doctoral candidate in development sociology at Cornell University. He currently lives in southeastern Vermont, where he has been restoring an eighteenth-century farm to working condition. His upcoming research examines the role community ownership plays in protecting the noncommodity benefits of farmland in Norway, Scotland, and Vermont.

Raymond A. Jussaume, Jr., is professor and chair of the Department of Community and Rural Sociology at Washington State University. His research has examined global and local dimensions of agrifood system structure, with a particular emphasis on the role of alternative agrifood systems in sustainable development.

Larry Lev is a professor and extension specialist in the Department of Agricultural and Resource Economics at Oregon State University. His work focuses on developing innovative research and marketing programs that address the needs of Oregon's small-scale farmers and communities seeking to foster the growth of local food systems. He teaches undergraduate courses in agricultural policy, agricultural marketing, and comparative world agriculture.

Sharon Lezberg is a sustainability studies scientist at the Environmental Resource Center, University of Wisconsin–Madison. She has a PhD in land resources from the University of Wisconsin–Madison and an MA in international development and social change from Clark University.

Debra Lippoldt, MS, directs Growing Gardens, a Portland, Oregon nonprofit organization that addresses root causes of hunger through supporting home food gardens and youth in low income neighborhoods. She cofounded the Coalition for a Livable Future Food Policy Working Group and was formerly a food policy analyst with the Hartford Food System. She is currently a member of the Portland/Multnomah Food Policy Council and chair of its food access committee.

Thomas A. Lyson (1948–2006) was the Liberty Hyde Bailey Professor of Development Sociology at Cornell University. He received his PhD in sociology from Michigan State University and was previously on the faculty at Clemson University. His earlier work had examined the career plans of rural and farm youths, rural labor markets, farm family work roles, public apathy about the farm crisis, dairy restructuring, and the impact of biotechnology. His later research had come to focus on civic agriculture, population health, and community problem solving.

Audrey N. Maretzki, PhD, is an emeritus professor of food science and nutrition in the College of Agricultural Sciences at Pennsylvania State University. She has worked both domestically and internationally to promote local food systems that are responsive to the nutritional and economic circumstances of low-income households.

Todd L. Matthews is an assistant professor in sociology at LaGrange College. His research interests include stratification, environmental sociology, and social movements. His recent research utilizes spatial analytical methods to assess environmental equity in the United States.

Marcia Ruth Ostrom is an associate professor in the Department of Community and Rural Sociology and the director of the Small Farms Program at Washington State University. Her research interests include social movements in the agrifood system, land-grant university accountability, and new immigrant farmers.

Kathryn Ruhf is the coordinator of the Northeast Sustainable Agriculture Working Group and has led public education, advocacy, food system development and community organizing activities since 1992. She specializes in farm entry, tenure, and transfer and in farm and food policy. She holds Master's degrees in administration and natural resources management.

Garry Stephenson is an associate professor and extension small farms specialist in the Crop and Soil Science Department at Oregon State University. His research and outreach interests include farm direct marketing (especially farmers' markets), local food systems, small-scale agricultural production systems, and livestock and water quality.

G. W. (Steve) Stevenson is a senior scientist emeritus and a rural sociologist at the Center for Integrated Agricultural Systems at the University of Wisconsin–Madison. His research focuses on alternative and values-based food supply chains, farm structure, and intergenerational farm transitions.

Joan S. Thomson is a professor of agricultural communications at Pennsylvania State University. She has been recognized for her communications research by the Association for Communication Excellence. In addition to strengthening local food systems through community engagement, she has in collaboration with others explored print media coverage of agricultural biotechnology.

Elizabeth Tuckermanty, PhD, is a national program leader for the United States Department of Agriculture's Cooperative State Research, Education, and Extension Service. She has managed the Community Food Projects Program since 1996 and has worked to support alternative, sustainable food systems that improve environmental, economic, social, and nutritional health in communities.

Jennifer Wilkins is a senior extension associate in the Division of Nutritional Sciences at Cornell University for whom she directs both the Farmer's Market Nutrition Program and the Cornell Farm-to-School Program. Her research and extension focus on food system implications for health and consumer food choices. She writes a monthly column, "The Food Citizen," for the *Albany Times Union*.

Index

Page numbers in italics refer to illustrations.

Abel, J., 137, 188
agribusiness corporations, 20–21, 33; challenging or replacing, 52–53; fragmentation and, 67–68; industrial model of agriculture and, 315–17; organic foods and, 300–301
agrifood systems: agenda for changing, 50–55; agribusiness corporations and, 20–21; builder work in, 43, 45–46, 52–53, 56; challenging or replacing corporate-dominated market structures in, 52–53; civic, 20, 23–30, 33, 186; coalitions and linkages, 46–47, 54–55, 56–57; community supported agriculture as a model for changing, 116–18; contract farming in, 21–23; conventional/commodity model of, 20, 24; dualistic nature of, 144–45; elite globalization and, 47–50; government subsidies and, 351–52; history of change activities within, 38–40; inclusion in, 40; mobilizing master frames and, 53–55; public policy change and, 53; remaking, 7–10, 40–41; research on, 4–7, 33–35; shared visions in, 51–52; social movement theory and, 35–38; space and, 10–13; warrior work in, 42, 44–45, 51–52, 55–56; weaver work in, 43, 46–47, 56, 349. *See also* farmers
Agro-tourism Roundtable of Charlevoix, 277–79, 286, 287–90
Aibonito Farmers' Market, 269–70
Albertson's, 203
Allen, Patricia, 38, 239, 318
Amigo, 264–65
antibiotic-resistant pathogens, 317
associative economies, 52
attendance at farmers' markets, 85–86
Avery, Dennis, 316

Baker, L., 47
Barber, B. R., 25
Barham, Elizabeth, 12–13, 179, 293, 346
Bay Friendly Chicken, 156–60, 162n19
Bellows, Anne, 333
Benbrook, C., 51, 53
Berry, Wendell, 320
Bertrand, Eric, 287, 292
Better Together: Restoring the American Community (Putnam and Feldstein), 94
Bialas farm, 72–73
Bingen, Jim, 13, 318
Blanchard, Troy, 12
Blank, Steven, 319
Born, Holly, 10, 179
Briggs, Suzanne, 89
Britain's National Food Guide, 176–77
Britten, P., 163
builder work, 43, 45–46, 52–53, 56
Buttel, F., 53

Cadieux, Lucie, 277–78, 287, 288, 290
California Certified Organic Farmers, 300
Campbell's, 67
Carro-Figueroa, Viviana, 12, 81
Cascadian Farms, 301
Charlevoix, Quebec: Agro-tourism Roundtable of, 277–79, 286, 287–90; consultation with France, 277–79, 285–86; demographics, 286; geography of, 286–87; lamb producers of, 277, 283, 285; place marketing by, 292–95; political organization of, 281; quality of local products in, 285–92
Cheng, A. S., 291
Christy, R. D., 27
citizenship, food, 33
City of Hartford Advisory Commission on Food Policy, 129–36
civic agriculture, 20, 33, 186, 280, 319–20, 346; agenda for, 29–30; civic community, socioeconomic welfare and, 25–27; new, 27–29; theory of, 23–25
civil rights, 38

civil society, 57n3
Clancy, Kate, 9, 167
Clean Water Act, 38
Cleverly, Larry, 75–76
coalitions in agrifood systems, 46–47, 54–55, 56–57
Cobb, J., Jr., 238
Coca-Cola, 67
community: -based food systems, 216–19; -based model of sustainable agriculture, 320–21; civic agriculture and, 25–27, 29–30; diversity, 185–86; environmental and social issues affecting, 183–84; food projects, 335–43; food security, 36–37, 47, 54, 80–81, 134, 333–34, 341–43; food systems, 194–95, 216–19, 332–33; forums, 189–92; initiatives for food systems, 194–95; media coverage of, 184–85; strategies for building common ground, 186–94
Community Food Security Coalition, 333–34
community supported agriculture (CSA), 9, 28, 57n6, 71, 229; brokers and networking in, 115–16; characteristics of successful, 114–15; defined, 99; demographics of farmers in, 104–5; demographics of members in, 109–10; farmer goals and objectives in, 105–6; food security and, 105–6; growth of, 99–100, 217–18; handbooks and manuals on, 101; levels of participation by members, 111–14; literature on, 100–102; management of, 106–8, 114–15; members/shareholders, 99–100, 108–14; as a model for changing agrifood systems, 116–18; participatory research methods and, 102–4; reasons members leave, 110–11; research on, 101–4; sustainable model of agriculture and, 326–27
Cone, C., 102
confined animal feeding operations (CAFOs), 44
Connecticut Food Policy Council, 129–36
consumers: acceptance of regional dietary guidance, 174–76; counting, 85–86; direct food purchasing by, 250–54; dot surveys of, 87–92; food expenditures among, 219–23, 227, 252–53; -producer linkages, 227–31; relocalization of relationships between producers and, 238–40; support for family farms, 253–54, 347; support for Label Rouge, 147–48, 151. *See also* dialogue, public
Continuing Survey of Food Intake of Individuals, 219
contract farming, 21–23
conventional/commodity model of agriculture, 20, 24
Cooley, J. P., 229
Cornell Cooperative Extension, 79
Corvallis-Albany Farmers' Markets, 94
craft processors, 325
credibility, empirical, 35–36
Creutzfeldt-Jakob disease, 317
CSA. *See* community supported agriculture (CSA)
customers. *See* consumers

Dahlberg, K., 126, 127, 139
Daly, H., 238
Daniels, S. E., 291
Danon, 300
Davis, C. A., 163
Davis Farmers' Market, 76, 78
DeLind, Laura, 13, 23, 102, 318, 346
democracy and civic agriculture, 25, 40
deserts, food: classification of, 206–7; emergence of, 202–4; location of, *208*, 208–13, *209*; measuring, 205–7
de Tocqueville, Alexis, 23
Dewey, John, 7
dialogue, public: benefits of, 188; community forums for, 189–92; focus groups for, 193–94; strategies for, 186–89; study circles for, 192–93. *See also* consumers; research
dietary guidance (U.S.): expanding the purpose of, 167; farm-to-school programs and, 28, 353–54; publications, 163–67; regional food guides, 167–77; seasonally varied, locally based diets and, 170–77; usability, 176–77
Dietary Guidelines for Americans, 163, 164–65, 166, 169–70, 177, 180n1
Dillman, D. A., 241
direct food purchasing by consumers, 250–54
direct marketing by farmers, 242, 242–50, *243*, 255–56, 349
Dirlik, A., 293–94
distribution, food, 204, 327
diversification, farmers' markets and, 72–74, 80
Dole, 67
dot surveys, 87–92

Downtown Des Moines Farmers' Market, 71, 73, 75
Drabenstott, Mark, 22
dualistic nature of agrifood systems, 144–45
Duxbury, J. M., 174

EarthBound, 301
economics: interaction and farmers' markets, 77–81; justice, 36, 52–53
Edible Connections (Nunnery, Thomson, and Maretzki), 189–90, 191, 192
elite globalization, 47–50, 57n1
emergency preparedness and food policy councils, 132
empirical credibility, 35–36
environmentalism and civic agriculture, 30
environmental sustainability, 36, 39
experiential resonance, 36

Fair Trade Fair, 50
Fals-Borda, O., 103
family farms, 105, 144–45, 253–54, 339, 347
Farm Bills, 48, 300, 350, 353
Farm Bureau, 246
farmers: African American, 38; collaboration with craft processors, 325; contract, 21–23; direct marketing by, 242–50, 255–56, 349; diversification by, 72–74; economic justice and, 36; family, 105, 144–45, 253–54, 339, 347; goals and objectives of, 105–6; government subsidies to, 351–52; income, 106–8, 158; Label Rouge, 149–55; local and regional, 68–69, 70–71, 79–80; Michigan, 223–27; modern, 2, 19; organic, 301–5, 312, 317–18; organizations, 246–47; pesticides used by, 303, 308–9, 316; selling direct to the public, 28–29; small and midsize, 144–45, 347; support groups, 323–25; sustainable model of agriculture and, 321–29; third-world, 49–50; Washington state, 240–42; women, 104–5. *See also* agrifood systems
Farmers' Market Coalition, 349
Farmers Market Nutrition Program (FMNP), 81
farmers' markets, 9, 41; business community support for, 90–92; cash flow and profitability at, 77–78; as community social and economic institutions, 66; counting customers at, 85–86; diversification and, 72–74, 80; dot surveys of, 87–91; facilitating social and economic interaction, 77–81; food system transitions and, 67–69; in low-to-moderate income areas, 78–79, 80–81; making local food more visible, 70–71, 79–80; new civic agriculture and, 27–28; in Puerto Rico, 261–62, 266–74; rapid market assessments of, 85, 92–95; reasons for popularity of, 68–69, 84, 347; resurgence of, 65–66, 68; study of, 70; supporting local and regional farms, 68–70; supporting small businesses, 75–76; WIC vouchers and, 54
Farms of Tomorrow (Groh and McFadden), 101
farm-to-school programs, 28, 353–54
Ferguson, A., 102
focus groups, 193–94
food: citizenship, 33; community projects involving, 335–43; co-ops, 78; deserts, 202–14; direct marketing of, 242–50, 255–56; and direct purchasing, 250–54; distribution industry, 204, 327; emergency, 132; expenditures by consumers, 219–23, 227, 252–53; genetically modified, 254; globalization and, 10–11, 13; labeling, 277–79; making practice, 8; pesticides and, 303, 308–9, 316; poisoning, 317; public interest in, 1–2; safety and health, 37, 44, 147–49; security (community), 36–37, 47, 54, 80–81, 105–6, 134, 333–34, 341–43; supplies (regional), 171–74. *See also* agrifood systems; organic agriculture; regional food; retailers, food
Food Commodity Intake Database, 172, 180n4
Food Guide Pyramid: Michigan food systems and, 218, 222, 223, 224, 225; nutritional guidelines, 163, 166, 168, 168–69, 173, 176–77, 180n1
food policy councils: advice for new, 136; challenges to, 134–35; documentation of food systems by, 132–33; emergency preparedness and, 132; engaging government, community, and visibility of, 133; factors in success of, 134; functions of, 126, 132; future of, 135–36; government sanction of, 127, 129–30, 138–39; history of, 122–25; longevity of, 129; membership and leadership, 130–32; reasons for emergence of, 125; research on, 127–40; and role in shaping policy agendas, 121–22, 139–41; staffing, 134–35, 136; state, 123–24

foodsheds, 47
Food Stamp Program, 263, 335, 342
food systems: agenda for changing, 50–55; civic agriculture and, 23–30, 346; community-based, 194–95, 216–19, 332–33; community food security movement and, 333–34; consumption patterns, 219–23; diversity within, 54; documentation of, 132–33; dualistic nature of, 144–45; gaps, 178–80; lack of consensus and clarity on, 51–52; local, 333, 348–49, 351; mapping, 194–95; public interest in, 1–2; remaking, 4–7, 345–55; research, 2–3; transitions and farmers' markets, 67–69
forums, community, 189–92
fragmentation of food systems, 67–68
framing processes in social movement theory, 35–37, 53–55
France: Charlevoix, Quebec and, 277–79. *See also* Label Rouge poultry system
Franz, L., 203
Future Search, 191–92

Gatenby, S., 177
General Agreement on Tariffs and Trade, 48
General Mills, 300
genetically modified crops, 254
Geographic Information Systems (GIS), 194–95, 206
Gertler, M., 291
Gillespie, Gilbert, 9, 46, 346
globalization, 10–11, 13, 217–18, 354; agribusiness corporations and, 20–21; contract farming and, 22; elite, 47–50, 57n1; food retailers and, 203; fragmentation and, 67–68; trade agreements and, 48; warrior, builder, and weaver work in, 47–50; Washington state agriculture and, 235–37, 254–55
Goldschmidt, Walter, 6, 26, 27
Goldsmith, E., 238
The Good City and the Good Life (Kemmis), 77
Gould, K. A., 39
government support: Farm Bills and, 48, 300, 350, 353; of food policy councils, 127, 129–30, 138–39; legislative reforms and, 38, 48, 349–51; of red label poultry, 146–49; subsidies and, 351–52
the Grange, 246–47
Green, J., 27, 29
grocers. *See* retailers, food
Groh, Trauger, 101, 114
Guptill, Amy, 12, 81, 260
Gussow, J. D., 167, 217

Hamm, Michael, 12, 333
handout/mail-back surveys, 87
Hart, P., 22
Hassanein, N., 240
Heffernan, William, 316
Heinz, 300
Henderson, E., 302
Hendrickson, M., 260
Hilchey, Duncan L., 27, 29
Hillman, James, 39
Hillsdale Market RMA, 94
Hinrichs, C. Clare, 39, 102, 239, 288
Hoffman, Matthew, 346
Hollywood Farmers' Market (Portland), 89
homogenization and national brands, 67–68
Hungry for Profit (Magdoff, Foster, and Buttel), 240
Hunt, P., 177
Hurricane Hugo, 123

income of farmers, 106–8, 158
incubation, small business, 75–76
industrial model of agriculture, 315–17
Institute for Agriculture and Trade Policy, 49
intellectual property, place names as, 281–84
Iowa Food Policy Council, 129–36
Irwin, M., 27

J. C. Penney, 272
Jenkins, T., 292
Johnston, J., 47
Just Food, 79

Kantor, L. S., 173
Kaufman, J. L., 137
Kelvin, R. E., 185
Kemmis, Dan, 77
Keon, T., 203
Kimbrell, Andrew, 320
Kirschenmann, F., 311
Kmart, 264, 272
Kneen, B., 238
Knoxville Food Policy Council, 122, 125, 126, 129–36
Korten, D. C., 238

Kropotkin, P., 328
Kruger, L. E., 291

Label Rouge poultry system: consumer, cultural, and government support for, 146–49; costs of products produced under, 148, 151–52, 161n5; farmer-centered economic organizations and, 149–55; farming standards guidelines, 148–49; food safety standards, 147–49; member farmers, 148, 149–52; as a model for the U.S. poultry system, 155, 158–60; reasons for establishing, 145–46, 149; supermarkets and, 151–55. *See also* poultry industry, U.S.
labels of origin. *See* Charlevoix, Quebec; place marketing
Lass, D. A., 229
Laytonville Certified Farmers Market, 74
legislative reforms, 38, 48, 52, 349–51
Lev, Larry, 75
local food. *See* regional food
Los Angeles Food Security and Hunger Partnership, 129–36
Lyson, Thomas A., 6, 236, 315, 329; on centralization and consolidation in the food system, 178–79, 203; on civic agriculture, 23, 27, 186, 280, 319–20, 346; on farmers' markets, 68, 84; on sustainable agriculture, 319

MacNear, Randii, 78
MacRae, Rod, 139
Maguire, P., 103
management of community supported agriculture, 106–8, 114–15
Mander, J., 238
mapping, food system, 194–95, 207–14
Maretzki, Audrey, 13, 188, 192, 217
market experiments, 38, 46–47
marketing: direct, 242, 242, 243, 243–50, 255–56, 349; organic agriculture, 306–7, 310–11; place, 281–84
La Marqueta Communitaria, 79
Marsden, T., 39
Marshalltown Farmers' Market, 74
Mast farm, 73
Matthews, Todd, 12
McAdam, D., 35
McCarthy, J., 35
McDonald's, 68
McFadden, Stephen, 101, 114
Mead, Margaret, 183
media coverage: of communities, 184–85; of community supported agriculture, 100–101
Melucci, A., 117
membership: community supported agriculture, 99–100, 108–14; food policy council, 130–32
Mercado Agrícola de San Sebastián, 267
Mercado Orgánico Agrícola en la Placita Roosevelt, 267–68
Michigan: community supported agriculture in, 229–30; consumer-producer linkages, 227–31; demographics, 218–19; food consumption patterns, 219–23; food production in, 223–27; marketing of organic products in, 306–7, 311–12; organic food and farming community, 298–99, 304–12
Michigan Organic Products Act, 307–9
Michigan State University, 298, 309–10
Mills, C. Wright, 6, 26, 27
Mintz, S., 311
Missouri Rural Crisis Center, 339
mobilizing structures in social movement theory, 37–38, 53–55
Murdoch, J., 292, 293
Murray, James E., 26
Myers, B., 163
Myhre, A., 102
MyPyramid, 163, 166, 180n1

Nabhan, G. P., 311–12
National Agricultural Statistics Service, 240
National Farmers Union, 260, 315–16
National Organic Program, 308
Native Americans, 339–40, 341
Nestlé, 300
New Mexico Food Policy Council, 127
North American Farmers' Direct Marketing Association, 349
Northeast Regional Food Guide, 166–69, 177, 179, 180n1; local and seasonal food knowledge and, 178; nutritional adequacy, 171
nutritional adequacy of locally based diets, 171

obesity, 219
Ohio Ecological Food and Farm Association, 300
O'Neil, Thomas P., 332
on-farm processors, 29

Onondaga Food System Council, 122, 125, 126, 129–36
organic agriculture: as commodity and market niche, 306–7; farmers, 301–5, 312; food retailers and, 300; growing popularity of, 317–18; marketing of, 306–7, 310–11; in Michigan, 298–99; principles of, 301–4; regulation, 307–9, 328–29; research on, 305–12; science of, 304–5. *See also* food
Organic Consumers Association, 300
Organic Farming Research Foundation, 304–5
Organic Trade Association, 300
Ostrom, Marcia, 46, 102, 327, 346
Our Food—Our Future, 190–91, 192

Parrott, N., 292, 293
participatory action research model, 93–95, 102–4
Payne, T., 229
pesticides, 303, 308–9, 316
Petzen, Joan, 76
Philadelphia Community Farm, 102–3
Philadelphia Food and Agriculture Task Force, 123
Pike Place Market Basket, 116
place marketing: in Charlevoix, Quebec, 292–95; fitting institutions to purpose in, 292–95; in Missouri, 282–84; and place names as intellectual property, 281–84; quality products and, 285–92
La Plaza del Agricultor Barranquiteño, 268–69
poisoning, food, 317
political opportunities in social movement theory, 38, 39, 44–45
Pollan, Michael, 300
pollution, 38
Portland/Multnomah Food Policy Council, 127
Pothukuchi, K., 137
poultry industry, U.S.: Bay Friendly Chicken and, 156–58; dualism, 145; farmer-centered strategies for, 155–58, 158–60, 161n17; Label Rouge as a model for, 155, 158–60; specialty producers, 160n1; Wholesome Harvest and, 155–56. *See also* Label Rouge poultry system
Pretty, Jules, 355
protest, public, 38, 39
public dialogue. *See* dialogue, public
public policy change, 53, 349–51; community food projects and, 336–40. *See also* food policy councils
public protest, 38, 39
Puerto Rico: economic factors and the agrifood system in, 261; farmers' markets in, 261–62, 266–74; food retailing in, 262–66
Putnam, Panna, 76
Putnam, Robert, 23

Rapid Market Assessment (RMA), 85, 92–95
Rayner, M., 177
Raynolds, L., 39
regional food: community supported agriculture and, 229–30; consumer acceptance of, 174–76; gaps, 178–80; guides, 167–77; labeling, 277–79; local and seasonal food knowledge and, 178; nutritional adequacy of, 171; relocalization and, 238–40; supplies, 171–74; systems, 333, 348–49, 351. *See also* food
relevance, subjective, 36
relocalization of producer-consumer relationships, 238–40
research: community supported agriculture (CSA), 101–4; dot surveys, 87–92; food desert, 205–14; food policy councils, 127–40; organic farming, 305–12; participatory action research model, 93–95, 102–4; rapid market assessments, 85, 92–95; sustainable model of agriculture, 321–29; on Washington (state), 240–42, 254–57. *See also* dialogue, public
resonance, experiential, 36
retailers, food: distribution and, 204; food deserts and, 202–14; globalization and, 203, 260–61; Label Rouge, 151–55; locations of, 208–13; organic foods sold by, 300–301; in Puerto Rico, 262–66; quantity vs. quality of, 201–2; rural, 201. *See also* food
Rhaman, M. A., 103
Ritchie, Mark, 49
Robb, E., 203
Robinson, K. L., 27
rural areas: community food projects in, 339–40, 341; food deserts in, 202–4; food retailers in, 201–2; industrial model of agriculture and, 317
Rural Enterprise Association of Proprietors (REAP), 76

Sachs, Carolyn, 38, 318
safety, food, 37, 44, 147–49

Sam's Club, 203, 264, 272
Saturday Stockton Certified Farmers' Market (SSCFM), 76
scaling up and out, 47
Schnaiberg, A., 39
security, food, 36–37, 47, 54; community, 333–34, 341–43; community supported agriculture and, 105–6; farmers' markets and, 80–81; food policy councils and, 134
shareholders, community supported agriculture (CSA), 99–100, 108–14
Shreck, A., 39
Sklair, L., 236
small business incubation, 75–76
Smith, J., 192
social movement theory: civil society and, 57n3; community supported agriculture and, 102; direct marketing of food and, 255–56; framing processes in, 35–37, 53–55; mobilizing structures in, 37–38, 53–55; political opportunities in, 38, 39, 44–45
socioeconomic welfare: civic agriculture and, 25–27; community food projects and, 335–40; farmers' markets and, 77–81; food deserts and, 208, 208, 209, 209–13; sustainable model of agriculture and, 318–20
Southeast Asian Farm Development, 76
space and the food system, 10–13
staffing, food policy council, 134–35, 136
state food policy councils, 123–24
Stefani-Ruff, Dianne, 95
Stevenson, Garry, 6, 10, 179, 236, 349
Stockton Farmers' Market, 81
Stone, K., 203
St. Paul Food and Nutrition Commission, 125, 126, 129–36
Strange, Marty, 23, 319
study circles, 192–93
subjective relevance, 36
subsidies, government, 351–52
supermarkets. *See* retailers, food
supplies, regional food, 171–74
surveys: close-ended questions on, 88; community supported agriculture, 109–14; cross-tabulation analysis of, 89; dot, 87–92; early vs. late responders to, 88; rapid market assessments and, 85, 92–95; usefulness of, 90; Washington state agriculture, 240–42
sustainability: community food projects, 336–40; environmental, 36
sustainable model of agriculture, 318–20; collaboration between producers in, 325; community-based, 320–21; community-supported agriculture and, 326–27; food distribution and, 327; regulation and, 327–28; research on, 321–29; support groups, 323–25

Target, 203
third-world farmers, 49–50
Thomson, Joan, 39, 137, 185, 188, 346
Tohono O'odham Reservation, 339–40, 341
Tolbert, C. M., 27
Toronto Food Policy Council, 129–36
Torres, R., 27
trade agreements, 48
Tuckermanty, Elizabeth, 13, 217

Ulmer, Melville, 26, 27
United Farm Workers Union, 44–45
U.S. Conference of Mayors, 122–23
USDA. *See* U.S. Department of Agriculture (USDA)
U.S. Department of Agriculture (USDA), 163, 335, 341, 343

Vermont Land Trust, 323–24
visibility: food policy councils, 133; local food, 70–71

Wal-Mart, 68, 84, 203–4, 260; organic foods and, 300; in Puerto Rico, 264–66, 274
warrior work, 42, 44–45, 51–52, 55–56
Washington (state): direct marketing by farmers in, 242–50; globalization and, 235–37, 254–55; relocalization and, 238–40; research on, 240–42, 254–57; transition in agrifood system of, 236–38
Washington Apple Commission, 235
Washington State Potato Commission, 235
Washington Wheat Commission, 235
weaver work, 43, 46–47, 56, 349
Weinberg, A., 39
Welch, R. M., 174
Welsh, R., 21, 27
Whatmore, S., 236
White Rose, 265–66
Wholesome Harvest, 155–56, 158–60, 161n18
WIC. *See* Women, Infants, and Children (WIC)
Wilkins, Jennifer, 195, 217, 260

Williamsburg Farmers' Market, 78–79
Wilson, Bob, 122
Wilson, N., 292, 293
Women, Infants, and Children (WIC), 54, 81, 262; FMNP Itinerant Markets, 270–71, 274
women farmers, 104–5
World Trade Organization, 44, 48–50, 149, 351, 352

Young, C. E., 173

Zald, M., 35

In the Our Sustainable Future series

Ogallala: Water for a Dry Land
John Opie

Building Soils for Better Crops: Organic Matter Management
Fred Magdoff

Agricultural Research Alternatives
William Lockeretz and Molly D. Anderson

Crop Improvement for Sustainable Agriculture
Edited by M. Brett Callaway and Charles A. Francis

Future Harvest: Pesticide-Free Farming
Jim Bender

A Conspiracy of Optimism: Management of the National Forests since World War Two
Paul W. Hirt

Green Plans: Greenprint for Sustainability
Huey D. Johnson

Making Nature, Shaping Culture: Plant Biodiversity in Global Context
Lawrence Busch, William B. Lacy, Jeffrey Burkhardt, Douglas Hemken, Jubel Moraga-Rojel, Timothy Koponen, and José de Souza Silva

Economic Thresholds for Integrated Pest Management
Edited by Leon G. Higley and Larry P. Pedigo

Ecology and Economics of the Great Plains
Daniel S. Licht

Uphill against Water: The Great Dakota Water War
Peter Carrels

Changing the Way America Farms: Knowledge and Community in the Sustainable Agriculture Movement
Neva Hassanein

Ogallala: Water for a Dry Land, second edition
John Opie

Willard Cochrane and the American Family Farm
Richard A. Levins

Down and Out on the Family Farm: Rural Rehabilitation in the Great Plains, 1929–1945
Michael Johnston Grant

Raising a Stink: The Struggle over Factory Hog Farms in Nebraska
Carolyn Johnsen

The Curse of American Agricultural Abundance: A Sustainable Solution
Willard W. Cochrane

Good Growing: Why Organic Farming Works
Leslie A. Duram

Roots of Change: Nebraska's New Agriculture
Mary Ridder

Remaking American Communities: A Reference Guide to Urban Sprawl
Edited by David C. Soule
Foreword by Neal Peirce

Remaking the North American Food System: Strategies for Sustainability
Edited by C. Clare Hinrichs and Thomas A. Lyson

Crisis and Opportunity: Sustainability in American Agriculture
John E. Ikerd

Printed in the United States
116159LV00005B/64-72/P

9 780803 224384